What people are saying about

Woody's Last Laugh

This delightful book reviews the enduring fascination with alleged sightings of the "extinct" American ivory-billed woodpecker to illuminate the role of well-understood cognitive biases and reasoning errors in science and conservation. Who would have thought that an "extinct" bird can tell us so much about the quirks of the human mind?
Dr. Norbert Schwarz, University of Southern California

An excellent book, revealing how scientists and environmentalists, like all of us, suffer from the same dangerous judgment errors called cognitive biases. These mental blind spots played a devastating role in the narrative around the ivory-billed woodpecker, even around the basic Schrödinger's cat-like mystery of whether it's dead or alive. Dr. Haney uses an excellent probabilistic thinking approach to have a much more nuanced and thoughtful approach to the question rather than the simple binary approach promoted by cognitive biases. Along the way, *Woody's Last Laugh* illustrates the numerous dangerous judgment errors that we all need to avoid if we want to have a truthful view of reality.
Dr. Gleb Tsipursky, behavioral scientist, trainer in addressing cognitive biases, and best-selling author, including of *Pro Truth: A Practical Plan for Putting Truth Back Into Politics* (Changemakers Books, 2020)

Woody's Last Laugh

How the "Extinct" Ivory-billed Woodpecker Fools Us into Making 53 Thinking Errors

Woody's Last Laugh

How the "Extinct" Ivory-billed Woodpecker Fools Us into Making 53 Thinking Errors

J. Christopher Haney

CHANGEMAKERS BOOKS

Winchester, UK
Washington, USA

JOHN HUNT PUBLISHING

First published by Changemakers Books, 2021
Changemakers Books is an imprint of John Hunt Publishing Ltd., No. 3 East Street,
Alresford, Hampshire SO24 9EE, UK
office@jhpbooks.com
www.johnhuntpublishing.com
www.changemakers-books.com

For distributor details and how to order please visit the 'Ordering' section on our website.

ISBN: 978 1 80341 004 3
978 1 80341 005 0 (ebook)
Library of Congress Control Number: 2021946931

A CIP catalogue record for this book is available from the British Library.

Design: Stuart Davies

UK: Printed and bound by CPI Group (UK) Ltd, Croydon, CR0 4YY
Printed in North America by CPI GPS partners

Contents

If a man will begin with certainties, he shall end in doubts; but if he will be content to begin with doubts, he shall end in certainties.
—Francis Bacon, *The Advancement of Learning*

Preface Siren of Reason

Man is a rational animal who always loses his temper when he is called upon to act in accordance with the dictates of reason.
—Oscar Wilde

On a crisp fall afternoon, I stare vacantly out my office window, straining to placate this stubborn unease. In my hands I grasp the farfetched – a new, draft endangered species recovery plan from the U.S. Fish & Wildlife Service, one aimed to conserve a bird deemed by most as already extinct. Yet even this absurdity does not conceal a more particular disquiet. Recalling a destitute bargain I struck with a fellow graduate student decades ago, I glance over to my bookshelf to see if I still possess Arthur Cleveland Bent's *Life Histories of North American Birds*. Yes… there it is… I seize a paperback volume on woodpeckers, flip pages to reach the germane account, and confirm what I thought I remembered. But after comparing passages in each document, I shake my head in disbelief at such glaring disparities in depictions of this bird's diet. It makes no sense… how does barefaced contradiction this obvious get flouted?

Epiphany need not reveal itself as striking illumination. A summons to scholarly exploration may consist of ephemeral glimmers, seemingly trivial, followed by long years of slow gestation. Certain things, apparently, do not season otherwise. Once the pull could be denied no longer, I then had to overcome my own stiff internal resistance, due mainly to the extensive coverage on a notoriously disputed topic. No way could I ignore the social perils that have ensnared those who might express ardent interest in the big woodpecker, or discount the harsh penalties levied on those who dared question conventional wisdom about the bird's providence. Despite my best efforts to strangle the reckless inclination, I came to realize that there endured a tightly encrypted story about the American ivory-billed woodpecker *Campephilus principalis*.[1] Perhaps not a story our own species wishes to hear, exactly, but one that deserved a fair hearing nonetheless.

Aside from my window-staring episode, other vignettes proved crucial for taking on this venture. After a putative rediscovery of the ivory-billed woodpecker in 2004 was reported in *Science* and received such wide media coverage, I tried to keep tabs on challenges made to the identity of a large woodpecker recorded in the infamous David Luneau video. Shot in the swampy bottomlands of Arkansas's Big Woods, this blurry amateur film went on to become a penetrating gauge for the woodpecker's uncanny knack for inciting us. During that epoch, however, my attention was engrossed with a life-and-death matter at home. All the uproar about this woodpecker certainly piqued my interest, indeed to the point of total astonishment, but I was not free at the time to delve into intricacies of that controversy.

My curiosity in the destiny of ivory-bill was reignited later after a colleague insisted that I scrutinize more closely Arthur A. Allen and P. Paul Kellogg's "Recent observations of the ivory-billed woodpecker" reported in the April 1937 issue of *The Auk*. While doing so, I was struck by a black-and-white photograph of the giant woodpeckers busily occupied on their Florida feeding grounds – some open and rather scrawny-looking pine flat woods, not a majestic stand of tall southern hardwoods or ancient cypress trees. Once again, I registered startling incongruity in various accounts for this bird, my reaction prompting marvel, if not a demand for the instant explanation.

Revisiting chronicles about ivory-billed woodpecker reached the decisive tipping point after I circled back to the bird's final recovery plan, made available to the public in 2010. Scanning 168 pages of dense text, I lingered especially at all of the fascinating material the plan presented in its Appendix E. The more I read, the more disconcerted I got. That same bewilderment returned, a puzzling sense of too many things that seemed to never add up. What made least sense to me were our own screwball reactions to a bird.

But I also felt that I had been… well… deceived, placed at the butt end of a bad joke. Not for the last time would I be struck with a sensation of being pranked, much like Wally Walrus who long served as clueless foil to the comic mischief perpetrated by Woody the Woodpecker. Two questions really mystified me. Why did every "appearance" by this bird after the 1940s get attacked with such viciousnous?[2] And how even could there be so many "incidents" of finding a presumably extinct bird, incidents that numbered from at least the scores to a hundred or more?

To no avail I tried to resist that treacherous call from the siren of reason. I tried to sit with my discomposure, to see what might emerge just with waiting. As much for probing whatever repudiation there might be for the thesis I was pondering as anything else, I then poured over various accounts of the woodpecker, focusing attention on books and articles that appeared during the last few decades. From these I verified that although an author might edge tentatively toward themes that I wished to exhume, they would back away from any deep treatment of the cognitive and sociological facets to the ivory-bill's story. Narratives charted for the giant woodpecker seemed to never shed much light on the murky origins behind so much human conflict centered around this bird.

So it was that a broad outline was birthed for this book. *Woody's Last Laugh* seeks to fathom the many tricks inflicted on us by a perplexing species. It strives to answer questions that have long exasperated us. Why did a very material thing, a large, clawed-and-feathered woodpecker, become transformed into the realm of make-believe? How did reality convert to fable? Why is folklore encasing this species so potent? What is it about *this* bird that sparks so much obsession, indecorous quarrelling, and faulty thinking among us all? How does a mere bird drive us batty?

Historical outlooks must have swayed our perceptions about the American ivory-billed woodpecker, but my exploration is arranged chronologically largely out of convenience. In trying to puzzle out the strange bonds we forged with the ivory-bill, I relied instead on three other means of orientation: embracing contradiction, both its origin and potential resolution; consenting frankly to large uncertainties; and, seeking whatever clarification might be available from the body of best available science. With respect to the last, I purposely sought aid from the social disciplines, as these are at best overlooked and often snubbed during attempts to decipher biology. In this sense, my book aspires to integrate rather than to fragment understanding about one of nature's abiding mysteries.[3]

Readers at the very outset may ask (even demand) to know on which side of that grand divide I belong, that is, whether I personally deem the ivory-billed woodpecker to be still living or instead forever gone. But you tell me: how, exactly, ought we even to pose this question to ourselves? What standard for a logical belief in either stance would be judged convincing? By whom, exactly? And why does absolute certainty matter so much? At the outset of my voyage, I had far too many questions to just wave away with an uninformed retort. Moreover, my journey revealed that a fundamental mystery, whether this bird has been long dead, only deepened once some of the lesser secrets were exposed.

On the other hand, I will concede that both firm believers and rigid deniers in any living persistence for this stately American icon might bristle at content presented in this book. But that is all well and good, as there are some matters that really ought to trouble us each and every one. Just like Woody, the cartoon woodpecker, the ivory-bill has long provoked the infirmities of our fallible minds, tormenting its antagonists and protagonists alike.

J. Christopher Haney
July 25, 2021
Washington, DC

Endnotes

1 Based on monophyly and the approximately equal genetic distances among the three North American taxa of *Campephilus, C. principalis* here distinguishes American ivory-billed from Cuban ivory-billed woodpecker *C. bairdii* throughout the text. See Fleischer, R.C., J.J. Kirchman, J.P. Dumbacher, L. Bevier, C. Dove, N.C. Rotzel, S.V. Edwards, M. Lammertink, K.J. Miglia, and W.S. Moore. 2006. Mid-Pleistocene divergence of Cuban and North American ivory-billed woodpeckers. *Biology Letters* 11: 466–469.

2 "The ivory-bill debate has been waged with a depth of feeling, and a shallowness of civility, usually reserved for arguments about such vital institutions as sex, politics, and religion." Wright, R. 2007. Taking it personal: where the ivory-bill survives. *Birding* March/April 2007: 48–52, p. 50.

3 My treatment is not content to just grant the rampant "obsession" about all things ivory-bill. While conceding due respect to the personal, psychological, cultural, and historical dimensions to the woodpecker's story, *Woody's Last Laugh* probes the primordial origins behind the faulty thinking that is so prevalent in all of human nature.

Acknowledgments

To Margaret and Nat Halverson, my deepest appreciation for the discipline instilled by two superb educators who ignited my life-long curiosity about the natural world, and who so early on introduced me to the fascinating realm of birdlife. Chuck Robertson, thanks for keeping the explorer instinct alive during tough years, and for urging all of us to put to field testing our book learning about wild places using a decent set of wits and a good edible plant guide. Edgar Grundset and Carl Swafford were decisive in encouraging my first long forays into international travel as an alternative means to secure a genuine education.

I am indebted greatly to Joshua Laerm and David S. Lee, both of whom provided early mentoring on how scholarly life could be intellectually rewarding and hugely amusing all at once, especially given the harmonizing libations. In particular, the late Dave Lee pressed me to take on this very subject as a sort of check on the extravagant claims we sometimes make in our practice of conservation. Early versions of all or some portion of the manuscript benefited from keen insights and helpful suggestions by Kurt Fristrup, Tim Ward, Will Mackin, Greg Aplet, Jon Andrew, Dan Thornhill, Chuck Hunter, and William Hayes. From the late Bill Pulliam, a fellow graduate from the University of Georgia, I derived much inspiration to boldly confront the sociological aspects of the ivory-bill's story. I also thank others whose efforts and thinking encouraged or motivated me to persist in collaring such a prickly topic: Mark Michaels, Mike Collins, Mary Kay Clark, and my wife, Nashwa Beach. Ryan Covington assisted graciously with some of the illustrations.

A work like this benefits from a great many visions and inspirations. Diverse opportunities to quench my manifold curiosities benefited from contrasting perspectives afforded over the years with governmental, private, and non-profit sectors of wildlife conservation. This book was an outgrowth of discussions with many others, some of it casual, some of it quite to the contrary. Despite, or maybe because of, all of these disparate influences, the views expressed in this book are entirely my own. This topic would never form such rich ore to be mined however absent the colorful and opinionated personalities long linked to our deep fascination with the ivory-billed woodpecker. To those who looked for it, those who didn't, those who were successful, and those not, thanks to you all: believers, deniers, and the still unsure alike.

Introduction

Few animals provoke such heated disputes among us as the American ivory-billed woodpecker. The bird's puzzling habits have since the 1800s inflamed speculation and rushed us into a bogus certitude, ingredients that set us up for thinking errors. But why are rumors of its persistence so prevalent, unlike other recently extinct animals? How is it that we cannot agree even on whether a bird is alive or dead? Why are we so bad mannered with each other about a bird? *Woody's Last Laugh* uncovers how the wiring inside our heads explains a lingering mystery, and how uncertainty gets woven into myth as we struggle to deal with large unknowns. By exploring uncharted borders between conservation and mental perception, new ways of evaluating truth and accuracy are unmasked.

Besides a cautionary tale, *Woody's Last Laugh* illustrates at least 50 common, every-day mental shortcuts that don't work, and how to recognize their traits and consequences. What I discovered is that just a single point of uncertainty, whether a woodpecker was still living or not, could and did go on to trigger a host of serious cognitive errors and thinking fallacies in us. This was a sobering realization, in no small part because I recognized myself in so many of these same mental mistakes. And it shook the foundations of what I thought I knew, including an assumption that our practice of wildlife conservation always rested on reliably accurate footings.

During my 35 years as a wildlife researcher, I ran into my share of wrong conclusions, unsubstantiated claims, and faulty decisions. But it was not so much this blundering itself that surprised me (to err is human, after all). Rather, it was utter breakdowns in detectable logic that so floored me, the obvious clues betrayed by barefaced contradictions. Nothing ever confounded me more, however, than the strange deeds that we act out with each other concerning a big woodpecker that may or may not be extinct. It just didn't make sense to me to get this worked up over a bird, so I struggled to figure out why it was the locus of so much inflexible disagreement. This journey did not take me in the direction that I expected. Instead of uncovering some arcane biological mystery as a potential explanation, the journey invited me instead to peer much deeper into the inner workings of our minds.

My book is among the first to expose cognitive derailment inside the environmental and conservation sciences, disciplines for framing knowledge that we normally consider less prone to such blunder. Yet across all human pursuits, mental mistakes routinely upset our need to find and apply reliable information. Human reason is fallible in science and conservation, too. Because we put up stiff resistance against ambiguity, we commit errors of thinking and action in order to achieve closure, a process sometimes referred to as "belief relief." None of us are immune from this cognitive bias, either. Indeed, several of our social identities and conventions tend to just magnify these errors in us.

While tracing the environmental history of a bird that over time came to represent a lost version of America, you will learn about dozens of prevalent mental gaffes that each of us make. You will also discover that although we think we are freer of bias than our peers, this in itself is just another form of bias! To assist with new concepts about cognitive error and fallacy, I highlight the key terms throughout the entire book using bold-faced italics. You can find more detail about these terms in four locations: the narrative itself, accompanying text boxes, within the endnotes where original research provides the experimental findings for these biases, and finally, in a comprehensive glossary toward the end of the book.

We all wish we could make fewer mistakes, especially those that are costly to us financially and socially. Whether we are aware of it or not, our world today is directing vast amounts of disinformation at us, knowing full well that our thinking processes are vulnerable to getting hijacked by these covert falsehoods. To further assist you with your own particular journey into cognitive diligence, and to build some offensive protection as well, I provide as a supplement to this book a list of other reading resources that can help improve our thinking skills.

How our mental shortcuts lead to costly mistakes. Imagine these scenarios. Your normally friendly neighbour snaps when you greet her, leading you to avoid her from then on. The salesperson at a car dealership refuses to lower a lease price from well above what the manufacturer advertises. A nephew insists you invest in a stock on Fridays because the price rose 18% the following Monday each of the last three weeks. An executive at your workplace doubles down on a business plan that reduces company profits. Your teenager resists all efforts to get eyeglasses for a vision problem. In vain you try to dissuade an uncle from buying airplane crash insurance. A colleague worries constantly that abdominal aches portend impending cancer. You watch a television show where the media pits an economist against an epidemiologist for a debate on how to handle a disease outbreak. You can't remember the last four digits of your social security number, unless you repeat the entire sequence from the beginning.

Have you struggled over how to react to someone's unseen motives, or even to understand what was truly going on, when something in you or others just made no sense? Do you want to make more accurate (or at least better informed) decisions about important life matters?

Would you benefit from recognizing some of the most common psychological traps that ensnare us all? If you have ever floundered with the more illogical side of our selves, then this book is for you.

Every day all of us routinely process complex information without having all the relevant facts. We also find ourselves in a global, information-driven war for our attention, wondering how to possibly sort out reliable data from all the disinformation spread by social media. We generally handle this great uncertainty by taking mental shortcuts. Some of these shortcuts work, but others don't. Indeed, some shortcuts may result in wrenching tragedy if our cognitive biases run counter to reality.

In 2010, workers on the *Deepwater Horizon* drilling platform conducted a well integrity test after assuming the cementing process for the well bore had been successful. Despite getting test results that clearly indicated potential for dangerous problems, workers believed the test itself was faulty, then went ahead and opened the well, with disastrous outcomes. Overconfidence in pre-existing views, a tendency to find only what we're looking for, is called ***confirmation bias***. This error is one of the more prevalent types of mental gaffes that we make.

You are more likely to learn something by finding surprises in your own behavior than by hearing surprising facts about people in general.
—Daniel Kahneman, *Thinking, Fast and Slow*

A Time before Myth

Chapter 1

Parallax View

We are actually living in a million parallel realities every single minute.
—Marina Abramović

Time after time, the ivory-billed woodpecker has confounded our grasp on what constitutes truth. Despite oft repeated and very strident declarations of its certain extinction, some ruse keeps thwarting an orderly storyline that could wrap up once and for all some consensus fate for this bird. Either we are prone to a peculiar, excessive guesswork about this species' very existence, or we have allowed our minds to swindle us in ways that no other forest creature has ever provoked. If we hope to escape this relentless trap, shifting our vantage point seems like a good place to start.

Viewed along different lines of sight, an object can change in appearance, illustrating the physical principle known as parallax displacement. A figurative parallax denotes our "seeing" an object from some new angle, gaining a fresh perspective. So what does ivory-billed woodpecker look like if we fine-tune this viewpoint? How might the bird appear if we reach beyond readings that were fastened selectively and extrapolated so wildly? What habits would the woodpecker be apt to display in order to have survived *and* dodged our casual detection,[1] yet still hold fast only to principles that regulate avian life and govern the rest of the natural world?

A parallax woodpecker

Overview

Ivory-billed woodpecker exploited large forest mosaics that yielded transitory foods supplied by disturbance, crown dieback,[2] and synchronized masting.[3] A chisel-shaped bill broadened its feeding niche without compromise to foraging capabilities that overlapped and duplicated those of smaller woodpeckers. Narrower wings and rapid flight granted the big woodpecker consummate facility at long-distance dispersal. Nomadic, at times irruptive, it moved efficiently across extensive landscape mosaics. Ivory-bills did not defend a conventional breeding territory nor occupy a fixed range.[4] A year-long social system based on small family units fortified it from deleterious Allee effects linked to the smallest populations. After protracted human persecution, behavioral plasticity in ivory-bill included a disposition for contextual wariness attained through habituation, social facilitation, cultural learning, and/or natural selection.

Distinctive morphology

In addition to other unique anatomy,[5] the bill and wing shape distinguished the ivory-bill from all other North American woodpeckers. A more narrow wing profile separates it from at least some other *Campephilus* woodpeckers, too.[6] Previous ecological portrayals of the ivory-bill tended to stress its bill shape (leading to misguided belief in overly-narrow, year-round specialization on habitat and/or diet),[7] but ignore or downplay the adaptations for a nomadic lifestyle enabled by the novel wings and rapid flight.

Narrow, stiff wings empowered the ivory-bill to transit vast distances with economy. Whereas other woodpeckers own rounded or paddle-shaped wings for more leisurely flap-bounding flight,[8] the ivory-bill flew direct and fast, like a loon or a pintail,[9] granting it a proficiency to search quickly and forage efficiently over extensive areas. Flight speeds were never sampled in ivory-bill, but lower bounds on velocities of loons and ducks[10] hint that flight speeds in this woodpecker[11] might comfortably reach 60 km hr^{-1}.

Itinerant movement and versatile habitat use

Ivory-bills did not invariably reside in the same locality[12] for more than 2–4 years.[13] "They disappeared from those areas when the amount of woodpecker food diminished."[14] Searchers emphasized how "ivory-bills appeared to be constantly on the move"[15] and "they appear to roam the country like gypsies."[16] Tanner attested to non-resident birds present even in the Singer Tract: "Mr. Kuhn happened to see a trio of wandering birds."[17] Nomadism was in keeping with an irruptive nature[18] of their largest, most profitable food items,[19] giving the bird unrivalled versatility for exploiting the shifting forage patches characteristic of large, ever-changing forest landscapes.[20]

Ivory-bill was neither strictly an old-growth nor a "disaster" obligate,[21] but rather could feed in a variety of stand ages. Foraging patches for insect prey might be as small as single limbs on recently dying trees, expanding inward and downward to the trunk as tree death progressed. Ivory-bill could exploit small stands that suffered concurrent die-back, or massive forest blow downs and very large burns characterized by full tree mortality. Wherever and whenever suitable patch dynamics[22] and food supplies recurred long enough to sustain its needs within a larger landscape matrix, a small population[23] of the large woodpecker might become "facultatively sedentary."[24]

Tree mortality was ceaseless in the original forests of the southeast. Forage patches were created by natural die back[25] as trees reached their ages of maximum longevity.[26] Other mortality was caused by windstorms (downbursts, tornadoes, hurricanes),[27] outbreaks of fire,[28] and physiological stress aggravated by drought, flooding,[29] beaver damming,[30] and pulsed irruptions of forest insects (e.g., tree-damaging beetles and canopy-defoliating caterpillars).[31] Mature forests undoubtedly benefited the ivory-bill, since older trees are more vulnerable to die-back and more likely to produce mast in significant quantity.[32] Masting typically cycles within a tree species every 2–4 years,[33] over spatial scales of 0.1 to 10 km,[34] and occurs unevenly across stands of different composition,[35] all in keeping with this woodpecker's sporadic habits and localized distribution.

Ivory-bill need not have restricted its living requirements to old-growth forest in part because unusually large tree trunks were not required for nest excavation. Tanner listed diameters at cavity height ranging from only 33 to 56 cm.[36] Given inside diameters for nesting cavities of 18 to 27 cm, the ivory-bill was unlikely to have ever run out of cavity prospects at large spatial scales. Few regions lost all their large trees at once. Other large bottomland cavity-dependent birds (e.g., wood duck *Aix sponsa*; barred owl *Strix varia*) managed to navigate the historical bottlenecks that truncated availability of this resource.

Incomplete evidence hints that ivory-bill never occurred steadily even within the same larger-scale regions (***Image 1.1***).

Indeed, during an era in which the woodpecker was already regarded as in decline, two accounts divulge how the bird's numbers changed markedly in response to local food availability. Tanner noted that ivory-bills occupied Wadmacaun Island, South Carolina, during only the "good mast and berry years" of 1935–1937.[37] But the birds then left this locale shortly afterwards because "the following years were poor mast years."[38] Similarly, ivory-bills appeared two years after a cyclone had "ripped a swath through the forest" in a section of the Singer Tract.[39]

Apparent ivory-bill detections increased by order of magnitude in an area of northern Florida during just two years. In 1890 Brewster and his traveling companions found but two ivory-bills in a weeks-long search down a 120-mile stretch of the Suwannee River.[40] Merely two years later, however, again during

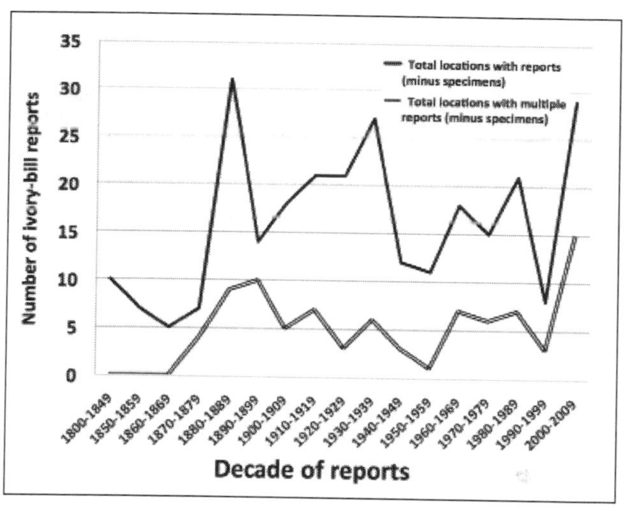

Image 1.1. Historical trends (minus collected specimens) in the total number of geographic locations from which the ivory-billed woodpecker was reported during the nineteenth, twentieth, and early twenty-first centuries. Peaks correspond generally to times when public interest in the woodpecker was greatest: the rush for specimen collections in the late nineteenth century (Chapter 3), the 1930s Tanner study and fight to save the Singer Tract (Chapter 10), and the first decade of the twenty-first century as prompted by search efforts in Arkansas and Florida (Chapter 12). Adapted from Hunter, W.C. 2010. Interpreting historical status of the ivory-billed woodpecker with recent evidence for the species' persistence in the southeastern United States. Appendix E, Recovery Plan for the Ivory-billed Woodpecker (*Campephilus principalis*), U.S. Fish & Wildlife Service, Atlanta, GA

early spring, Arthur T. Wayne tallied 23 ivory-bills taking the same river route.[41] Wayne noted that those birds used a large, recent burn-out of pines. Neither he nor later scholars seemed to take notice of that episode's bearing on the bird's potential for irruptive movement (***Table 1.1***). Such a rapid increase cannot be attributed readily to population growth, but nomadic shifts could account easily for any locally strong numerical response of ivory-bills over such a short time.[42]

Table 1.1. Five separate hypotheses, any of which could account for either an apparent or a real change in the numbers of ivory-billed woodpecker reported from a given locale. Human fixation on just one of these explanations, such as habitat loss or direct persecution, leads us into cognitive blunders caused by ***congruence bias***[43]

"Retires from the advance of civilization to … where it is not so liable to be molested."[44]
Local population is shot out or depleted by lethal human persecution.[45]
Once-suitable forest habitat is destroyed or too altered for the woodpecker to continue using.[46]
Nomadism leads to shifting patch use for optimal feeding and reproductive efficiency.[47]
Changes in search effort, or random sampling error, give only a facade of population change.

Dietary breadth, energetics, and spatial requirements

Ivory-bill had a catholic, opportunistic diet[48] of plant and animal matter (Chapter 7). Some of its foraging differed little from other North American Picidae.[49] "They chisel into the sap and heartwood for borers like other woodpeckers."[50] Ivory-bills fed on the ground like flickers *Colaptes*, or drilled deep into heartwood for wood-boring insects "exactly like" hairy *Leuconotopicus villosus* and pileated woodpeckers *Dryocopus pileatus*.[51] Smaller items would have been excavated without difficulty by the formidable ivory-bill, and more easily than by smaller-bodied woodpeckers with which it shared habitat.

Ivory-bills even klepto-parasitized and cached acorns from squirrel nests, as do acorn woodpeckers *Melanerpes formicivorus*.[52] Tanner over-stressed breeding season diet of a single pair at one site (Chapter 8). Hence, any individual, temporal, and regional flexibility in ivory-bill diet has gone widely unheeded.

A mighty, wedge-tipped bill[53] may have enabled the ivory-bill to strip tight bark off recently stressed trees to extract the very largest of wood borers. But this unique proficiency would *expand,* not contract, its foraging and dietary niche relative to the other eastern woodpeckers.[54] With a robust bill that doubled as an imposing weapon, and a body mass almost twice that of pileated, ivory-bill had nothing to fear from its closest competitors,[55] and it was little bothered either by most natural predators.[56]

Ivory-bill was never demonstrated to be a food-limited, narrow foraging specialist. Blaming its imperilment on a conjectured reliance on wood boring larvae[57] arose from ***illusory correlation***.[58] The big woodpecker showed behaviors contrary to those expected from a species hard-pressed to find a single, scarce food type. Tanner chided it for being "the last bird of those woods to arise in the morning."[59] Then, as early as 10 a.m., "they would become quiet, almost cease feeding, and do little but sit during the middle of the day."[60] They might even return to roosting cavities before resuming a bout of late-day foraging. Such behavior countermands a belief in food limitation.

Body mass in the ivory-bill was approximately 450–570 g,[61] almost double that of the smaller-bodied pileated woodpeckers that are characteristic of the southeastern United States (230–300 g).[62] Based on formulations that scale the energy requirements of birds to their body mass,[63] the ivory-bill would have required around 677–788 kilojoules per day compared to about 431–511 kilojoules in the smaller pileated.[64] This energetic difference represents an extra demand on the ivory-bill of just 32–82%.

Box 1.1. ***Illusory correlation*** arises when our mind leaps to form an association when in truth there is none. In Louisiana during the mid-1930s, Tanner watched ivory-bills provision young with large grubs, readily visible to the naked eye as they were brought to the nest cavity in bills of adult birds. From that entirely valid observation, beliefs were launched in which the woodpecker was seen as having such a specialized diet that a supposed reliance on large grubs could be met only from large, recently dying trees of aged forests in wilderness settings. Later authors would extend that ***illusory correlation*** even more broadly to claim that ivory-bills were incapable of delivering smaller food items like ants and termites to their young.

Since ivory-bill could extract prey *not* accessible to the smaller pileated, one can infer that the bigger woodpecker would have needed no more than twice (and as little as a third) more forest area to sustain its daily energetic needs, all else equal. Tanner's claim that ivory-bill had spatial demands 36 times larger than the pileated cannot be justified using fundamental principles of vertebrate physiology.[65] "Tanner may have … greatly overestimated the space needs of populations."[66] Instead, real disparity between local densities of the two big woodpeckers[67] can be explained by other reasons, including but not limited to the larger species' rarity and locally depleted numbers – there were not enough ivory-bills still remaining to occupy fully whatever suitable habitat did remain.[68]

Behavioral wariness

Acute guardedness in the ivory-bill was stressed by writers who accentuated this very trait for *two centuries* (some 30 citations;

see Chapter 7). How do we reconcile such indisputable evidence of the bird's supreme caution with Tanner's testimony that "the ivory-bill is not unusually wary of man nor seriously affected by man's presence"?[69] Setting aside witness that the Singer Tract birds habituated to benign human presence,[70] and that Madison Parish had received substantial protection just prior to Tanner's field study (Chapter 8), behavioral plasticity in birds can lead to the expression of both wariness *and* acclimation in the same species at the same time.[71]

"Flighty" species tend to remain cautious.[72] Several factors interposed a strong anti-predator caution in the ivory-bill. For at least a thousand years, and likely far longer, the big woodpecker had been relentlessly pursued and killed by native Americans for cultural use as ceremonial and trading artifacts.[73] A single millennium equates to roughly 70 generations in the ivory-bill,[74] a sufficiently long duration for population-level selection to have occurred. Had the big woodpecker never adjusted to this sustained predation from native peoples, we can have little conviction that European colonists or historical scholars would have had the good fortune to detect much of a living bird at all.[75]

Individual birds of the same species display a spectrum of personality traits, ranging from fast, bold, aggressive, and routine-forming, or instead shy, non-aggressive, but also more innovative.[76] Heritable even absent behavioral copying,[77] traits for avian wariness are selected genetically at the polymorphisms associated with fear in animals: the dopamine receptor and serotonin transporter genes. Even in mobile species, a genetically distinct proclivity for wariness can differentiate bird populations that are separated by just 30 km.[78]

Ivory-bill owned life history traits that predispose birds to exhibit cautious behavior around humans. Two if not three of four nests found in the Singer Tract in 1935 were abandoned by adult ivory-bills soon after the Cornell team approached and

Box 1.2. Fear reactions are greater (flight initiation distances longer) in adult birds, if more predators are around, when past experience has already shown risk, and in rural instead of urban habitats. Because year-one ivory-billed woodpeckers stayed with their parents for at least a year after fledging, an extra level of "watchfulness" provided by the adults may have provided greater security against predation over the near term, as well as facilitated acquisition of critical learning skills for identifying potential future threats.

inspected them.[79] In southern Georgia a pair deserted a nest cavity site immediately after disturbance by an egg collector.[80] Avian fearfulness and flightiness co-evolve with other life history traits.[81] Flight initiation distances of large species are greater and escape flights begin sooner than in smaller birds.[82] Species that capture live prey and show greater sociality are also more wary. Birds also show strong, spatially explicit responses to persecution, moving away from human threats quickly to areas that serve as refuge.[83]

Evolutionary proclivities for wariness can be veiled in acclimated birds, but express and become amplified once any discouraging or threatening actions by humans trigger this latent tendency. Whereas bird populations can habituate to people if left undisturbed, other populations of non-habituated birds from that same species will still increase their flight initiation reactions around people.[84] Once human dangers have been identified, aversions to the particular threat and recognition of the same person(s) exhibiting that threat will persevere in the individual birds that have been exposed.[85]

Predator avoidance by birds is socially facilitated, or learned, across individuals.[86] Predator recognition can be transmitted along a social chain of at least six individual birds.[87] Wariness in the ivory-bill, like other birds, may also have been culturally transmitted.[88] Facilitated by social learning outside kinship lines and across generations, these behavioral traits are relatively long-lived in birds.[89] Wild birds also eavesdrop on and learn the novel alarm sounds of other species. In as little as a few days, this kind of anti-predator behavior can spread rapidly through wild populations.[90]

Resistance to Allee effects

Strong flight and unconventional sociality (for woodpeckers) granted the ivory-bill a measure of protection from deleterious consequences of Allee effects[91] that so worsen the extinction risk in tiny populations. Structurally capable of flying hundreds of kilometers per day,[92] long-distance dispersal by young-of-the-year or unmated ivory-bills into new regions would increase possibilities of founding new pairs, promote outbreeding, and curtail the harmful consequences of inbreeding, all expected in a more sedentary species.

A social system of small family units gave the woodpecker other protections against harmful Allee effects. Because ivory-bill pairs foraged, fed, and traveled "closely together at all times except ... incubation,"[93] the large woodpecker was less exposed to predation as it spent little expense revealing itself for mate selection and territory defense. Ivory-bill was even more wary during breeding than at other seasons.[94] Ivory-bill pairs also consisted of mobile, self-contained reproductive units free to establish across landscape mosaics wherever suitable conditions might be found. Skewed sex ratios found in declining, small, but sedentary populations would have been markedly abated.[95]

Traveling together, in steady but discrete contact,[96] pairs and attendant young had more eyes to seek out food and warn against predators, too. Because it endured until the start of the next breeding season,[97] this social system hastened adaptive learning in the year-one birds. Able to witness their parents respond to novel feeding opportunities and react to unfamiliar threats over longer periods,[98] this heightened social learning could have elevated the species' prospects for survival.

Unlike passenger pigeon *Ectopistes migratorius,* its giant flocks delimited into fixed colonies easily shot out, and Carolina parakeet *Conuropsis carolinensis,* vulnerable for its hapless tendency to return again and again to fellow members gunned down earlier from the same flock, the ivory-bill caught on to our exceedingly lethal natures. For good reason early naturalists called them "the shyest and most cunning of anything that wore feathers."[99] So great was the ivory-bill's wariness that avid collectors often failed to secure the second member of a mated pair that they so eagerly coveted for their cabinet.[100] Countermeasures to human predation granted the woodpecker an edge at survival never afforded the ill-fated pigeon and parakeet.

Demographic potential and population regulation

The large-bodied ivory-bill probably had a lifespan longer than the 15 years normally attributed to it.[101] Investment by parents into prolonged care of their year-one offspring increased survival rates in what is typically the most vulnerable age cohort in avian populations. Speed, powerful morphology, and other habits of the adult birds tended to keep annual survival high as well. Because cautious birds tend to pick breeding sites that are less susceptible to human disturbance,[102] ivory-bill chose nest site locations that were hard to find and/or difficult to reach by people.[103]

Reproductive output in ivory-bill was variable,[104] but potentially high. In years with fewer resources ("bust" years), reproductive output of a pair might be nil or invested in rearing but a single offspring. Ivory-bill clutch size reached 4 to 6 eggs, however, indicating that with a resource surplus ("boom" years), a pair might fledge higher than average numbers of offspring. A rather protracted breeding season (January through as late as July)[105] enabled the woodpecker to re-nest if the first clutch failed, and possibly to time breeding so as to better match any seasonality in food quantity or quality.[106] As a result, the lifetime fecundity of this long-lived bird could be relatively high.

Absent large forest disturbances, ivory-bill breeding pairs were highly dispersed and occurred at very low densities. In such conditions, and because the bird was able to move across such large areas, the ivory-bill was difficult to find. If foraging conditions were markedly enhanced, as in years immediately following very large disturbances like forest fires, local densities of the nomadic ivory-bill might increase, and the species might appear then to be *locally* common. In this sense, ivory-bill demography may have resembled that of another enigmatic, disturbance-reliant picid in North America, the black-backed woodpecker.[107]

Disappearing acts

Overview

For some indefinite time after the 1930s, the ivory-billed woodpecker persisted at a very low population size, perhaps no more than lowest double to lowest triple digits. This small population was maintained by source-sink dynamics that rescued the species from extinction only at a meta-population level.[108] Small population size combined with lack of territoriality, high mobility, strong sociality, weak and sparse vocalizations, plus other cryptic behaviors, all coalesced to generate exceptionally low probabilities of detection. Disappearing acts[109] by the ivory-bill can be illustrated in a simple albeit incomplete fashion using multiplication and addition rules of probability, along with numerical estimates applied to simulate some of the bird's biological traits.

Small population size

Barely more than guesses, we still have at least three estimates for how small the population of ivory-bill might have been in reality or could have been theoretically. In 1939 Tanner listed only 24 individuals in five areas of Louisiana, Florida, and South Carolina.[110] But Tanner's projections overlooked 12 ivory-bills in 5,000 ha of virgin timber known as the Allan Grey Estate south of Rosedale, MS, a site from which these birds left immediately after that tract was logged during the 1940s.[111] Acknowledging the profound limitations to Tanner's survey protocol outside the Singer Tract,[112] these two sets of numbers give a minimum population of 36 ivory-bills.

Bertram Murray applied a deterministic model for a proposed minimum population[113] that could survive for as long as 50 years, up until the time of the Arkansas episode (Chapter 12). He used life history tables to emulate populations of the ivory-bill based on fixed estimates for its annual survival rate (0.73 age-specific survival in adults; 0.23 in one-year-old birds), mean clutch size (3 eggs), age at first breeding (2 years), and longevity (up to 20 years).[114] His reasoning led to a theoretical minimum population of 30 post-juvenile ivory-bills, consisting of as few as 10 breeding pairs.

A third study conducted a more realistic appraisal of population viability in which population size was allowed to interact with an extinction probability. Mattsson and colleagues framed this exercise around a very practical question: "under what conditions could the species have persisted today?"[115] Using a stage-based single-population model, they varied randomly several key demographic variables[116] to estimate the ivory-bill's initial population size, its survival and fecundity, while also allowing for Allee effects, and then ran the model under a wide variety of scenarios.

Astonishingly, an initial population size of as few as five females gave statistical assurance for likely persistence of the ivory-bill through modern times. This persistence would occur if on average each female produced just 1.1 female per year[117] and adult survival was moderate at ≥80%. If the population consisted of 30 or more females, then the ivory-bill could have persisted despite a strong Allee effect, if either survival (90%) *or* fecundity (1.65 recruited females per adult female) averaged on the high side of the model inputs that were used.

Under the very worst cases, the median time to extinction in the ivory-bill was estimated to be 7 years, but also as little as one to as long as 32 years. And when Allee effects were modeled as weak, as justified in the narrative given here in this chapter, if the bird was given a moderate survival or a low fecundity the likelihood of extinction in a population of 17 or more adult female ivory-bills fell below 5%. Consequently, if ivory-bills "maintain relatively high demographic rates, they can likely persist as small, and difficult-to-detect, populations."[118]

Bookmaking with ivory-bill

Profound struggles to find ivory-bills even *if* present are illustrated using three scenarios. In the first, odds of site occupancy are estimated across what is typically perceived as the most suitable habitat. Second, chances for re-finding a single ivory-bill are estimated once one is seen from a precise location. Finally, the likelihood of re-finding a bird is estimated under the best set of conditions imaginable, one in which a possible roosting or nesting cavity seems to have been located.

Conventional depictions of the ivory-bill's historical range[119] reach about 760,000 km^2. Remaining suitable habitats (bottomland hardwoods or forested wetlands), however, are now put at just 26,000 to some 120,000 km^2.[120] If all ivory-bills occurred in an area delimited by the smaller estimate, a naïve probability[121] of our selecting a 1-km^2 block with any ivory-bills runs from just less than 0.05% to no more than 2% for hypothetical population sizes that range from 12 up to 480 individuals. After we account for ivory-bill sociality (adjustment of 0.5),[122] for areas missed by standard survey methodology that cannot cover an entire 1-km^2 block all at once (0.385),[123] and for failure to find large woodpeckers present but not seen or heard (detection probabilities ≤0.18–0.36),[124] the odds of our finding one or more ivory-bills in any of these individual blocks plummet to a mere 0.00016%–0.00032%.

To improve chances of finding ivory-bills, we might sample a larger number of 1-km^2 blocks all at once. But if the total population size is 40 birds (10 three-bird family units plus 10 singletons), it would require us to survey 4,740 1-km^2 blocks to achieve just over a 50% likelihood of finding at least one bird! And even if the total population size were as high as 480, it would still take 395 1-km^2 blocks to reach odds as good as a coin toss. Finding an equal number of qualified observers, training and supervising them, then running all surveys near simultaneously,[125] would entail a logistical program having enormous and almost certainly prohibitive costs.

Now suppose we attempt to re-find an ivory-bill seen once. Without knowing the location of any central place (roost or nest),[126] we must allow for it having been detected anywhere inside and toward the very outer edge of a circle having a diameter at least twice its daily foraging range. Tanner offered some guidance on these central place foraging distances: ivory-bills wandered daily at least 2.4 and 5.6 km away from roosting sites during breeding and non-breeding seasons, respectively.[127] In just minutes of straight-line flight, though, an ivory bill could easily transit 5–10 km from a cavity.[128]

Total areas in which a single ivory-bill might recur, then, verge on circles with radii of 4.8 to 20 km when accounting for the unknown position of a sighting relative to the bird's central place. Since we are not likely to know the bird's reproductive status, we allow for the largest distance, giving us sampling areas that run from 72 km^2 to just over 1,256 km^2. Assuming we deploy one observer per day in any 1-km^2 block inside the smaller circle, we obtain only a 0.096%–0.2% chance of re-finding the bird. If we try the same survey tactic in the larger circle, our odds fall proportionately, to 0.022–0.044%. Our chances improve with complete survey coverage, but again logistical costs become extravagant for deploying scores of observers. Even then our probability improves to no better than 6.9%. We still end up with more than a 90% chance of *not* finding the bird.

Our best chances for re-finding an ivory-bill arise upon seeing one actually exit a tree cavity.[129] Assuming that the bird is incubating eggs or feeding young, we need survey only 18 km^2 of the surrounding area with the assumptions described above, but with the result that the odds of re-finding the bird *away* from the cavity are still only 0.39%–0.77% in any single 1-km^2 block. If the original sighting was made in June through December, then a larger non-breeding foraging radius gives us more than 98 km^2 to survey, so the probability of finding this bird in any 1-km^2 block falls to just 0.071%–0.141%. Using many more observers to achieve complete coverage of all the 1-km^2 blocks for each circle again raises our odds to only 6.9%.

So, apart from Tanner's singular instance of that one pair linked to roosting or nesting cavities, our *not* finding or easily re-finding the ivory-billed woodpecker has long been the overwhelming statistical prospect. Conversely, if we might ever wish to conclude that our not finding the bird was actually strong evidence of its truly not being there, we would need to sample the entirety of even an 18-km^2 area around a putative cavity at least 14 times.[130] Only then would we achieve enough statistical power (≥90%) to make a responsible inference that the bird's absence was in fact real (i.e., not due to our fallible detection abilities).

"Hunting for localities where ivory-bills were, and in these localities trying to find the birds, was like searching for an animated needle in a haystack."[131] Even after being pinned down in the Singer Tract by its full-time wardens, Tanner protested that the ivory-bills "traveled so fast that I was unable to keep up with them."[132] No real evidence exists that ivory-bills were ever particularly sedentary, or that they were anything other than highly-mobile and constantly on the move. Indeed, Tim Spahr used an algorithm for rare events based on Tanner's summary of the species' daily movements to estimate that a single ivory-bill could avoid indefinitely its detection by up to 20 observers.[133]

Dictates of probability hence erect a nearly insurmountable barricade to our gaining any *numerical* assurances for confirming the ongoing presence (never mind density) of a highly-vagile species from a very small population perched at the edge of extinction. Rules for probability make even more difficult any trustworthy decisions about the bird's absence. And "if one wants to protect against the error of declaring an extant species as extinct, the threshold for 'reasonable doubt' should be set high, perhaps higher than the commonly used 95% level."[134]

Paradoxical no more

When it comes to the ivory-bill, a little parallax can go a long way in settling much paradox. Although the parallax woodpecker portrayed here obeys the laws of biological plausibility, that feature does not make it the ivory-bill of inevitability. And indeed, other ivory-bills may well be discerned through parallax shift. Although we must not fall to wishful thinking in a thought experiment such as this, we can enlarge legitimate portrayals of this woodpecker using knowledge gleaned from interdisciplinary biology, insights that reveal how birds adapt to their natural world in complex ways. That world is one shaped invariably by biological and mathematical principles alike.

Obscured by uncertainty, this "other" ivory-bill makes itself known to us primarily through the ***improbability principle***[135] working in opposite directions on two sets of controls. On the one hand, certain overlooked or discounted life history traits act as "probability levers"[136] that *raise* odds for the ivory-bill's persistence. On the other hand, a very small population size coalesced with some of these same traits to operate other probability levers that *lower* the bird's detectability. Counter-prevailing probability levers offer a logical means to dissolve many paradoxes of perception that have so besieged and bedeviled our warring beliefs about this enigmatic bird.

Yet not even a parallax view can shelve the bitter disputes that divide us over this bird's fate. To understand the origins and consequences of so many rancorous controversies appended to the ivory-billed woodpecker, we must first voyage into, then pass through, and finally take leave of mythology, staying attentive all the while to the manifold deceits inflicted by the inner workings of our very own minds.

Endnotes

1 Reference to the ivory-bill surviving or avoiding detection refers to any time span past the 1930s, including the 1944 date often used as a convention for the species' putative extinction based on Eckleberry's widely promulgated ***anchoring*** portrait of a supposed "last ivory-bill" (see Chapter 9). All the same, I reach no fixed conclusion whatsoever that the ivory-bill in fact survived undetected up through the present day, or that it survived to any other date certain. Moreover, my thought-experiment for this parallax ivory-bill is strictly an hypothesis, one of several possible ones, not a verified fact.

2 Tree dieback is measured as the percent of branch tips in the upper portion and outer edges of the live crown that are dead from causes other than shading or competition, that is, branch tips expected to be alive if the tree were not subject to other stressors. Dieback varies across years and among species, being more common in hardwoods. http://fhm.fs.fed.us/fhncs/chapter2/crown_dieback.htm (accessed 16 January 2016).

3 Masting refers to pulsed production of hard seed (e.g., from

oaks, beech, hickories, chestnuts) and soft fruit (e.g., from gums, dogwoods, cherries) by forest plants. Masting varies across tree, shrub, and vine species, with some but almost never all plant species masting during the same year.

4 Lack of territoriality carried notable consequences. First, more than one family group of the woodpecker could be localized in a small area. Eleven ivory-bills in a Florida cypress swamp were seen together in trees "dead and sufficiently decayed to offer fine feeding ground for the larvae of a kind of large boring beetle ... a busy and noisy scene never to be forgotten." Scott, W.E.D. 1898. *Bird Studies: An Account of the Land Birds of Eastern North America*. G.P. Putnam's Sons, New York, p. 309. Twelve ivory-bills were found "in a small patch of dead pine trees." Ellis, J.B. 1917. Forty years ago and more. *Oologist* 34: 2–4. Secondly, with no need to defend territories, little broadcast vocalization was needed just prior to and during breeding. Indeed, "these woodpeckers are very silent at all times so far as their voices are concerned. Especially is this true during the breeding season." Higley, W.K. 1906. *Birds and Nature*, Volume III, A.W. Mumford and Company, Chicago, p. 122.

5 Even the foot structure of the ivory-bill may have been unique. It was claimed that all four toes could point forward, in one plane, so that the feet were held to the side of the body and directed at an angle to vertical, thereby giving it a "very firm support against both the downward and outward components of gravity." Bock, W.J., and W.D. Miller. 1959. The scansorial foot of the woodpeckers, with comments on the evolution of perching and climbing feet in birds. *American Museum Novitates* 1,931: 1–45. However, others witnessed "three toes turned forward and one ... turned back" in the similar Cuban ivory-billed woodpecker. Dennis, J.V. 1948. A last remnant of ivory-billed woodpeckers in Cuba. *The Auk* 65: 497–507.

6 The ivory-bill's living relative, the Magellanic woodpecker *Campephilus magellanicus,* does not seem to have this same distinctively-narrow wing profile. Paddle-shaped wings in that species of *Campephilus* found in the far southern hemisphere are

evident in several of the photographs that are posted here, http://travellingbirder.blogspot.com/2011/12/tierra-del-fuego.html (accessed 14 January 2016), and here, http://www.pbase.com/picidpics/magellanic_woodpecker (accessed 14 January 2016). Wing shapes in ivory-bill and imperial woodpeckers were quite alike, however, inferring that both of these closely-related species had comparably strong flight ability. See Lammertink, M., T.W. Gallagher, K.V. Rosenberg, J.W. Fitzpatrick, E. Liner, J. Rojas-Tomé, and P. Escalante. 2011. Film documentation of the probably extinct imperial woodpecker (*Campephilus imperialis*). *The Auk* 128: 671–677.

7 And thereby greatly increasing a likelihood of error from ***illusory correlation***.

8 Tobalske, B.W. 1996. Scaling of muscle composition, wing morphology, and intermittent flight behavior in woodpeckers. *The Auk* 113: 151–177.

9 Tanner, J.T. 1942. *The Ivory-billed Woodpecker*. Dover Publications, Inc., Mineola, NY, p. 58.

10 Waterfowl flight speeds range from 65 to 115 km hr^{-1}. http://www.ducks.org/conservation/waterfowl-biology/amazing-waterfowl-facts, and https://www.allaboutbirds.org/guide/common_loon/lifehistory (both accessed 14 January 2016).

11 An approximate speed of 15.2 m s^{-1}, close to 60 km hr^{-1}, was estimated over a short distance (66.5 m) from videography of a woodpecker thought to be an ivory-bill in 2008. See: Collins, M.D. 2011. Putative audio recordings of the ivory-billed woodpecker (*Campephilus principalis*). *The Journal of the Acoustical Society of America* 129: 1,626–1,630; Collins, M.D. 2017. Periodic and transient motions of large woodpeckers. *Scientific Reports* 7: 12551.

12 Locality here refers to a relatively small area matching the approximate size of a territory or home range in other birds. Ivory-bills might remain within areas defined at the scales of much larger regions, especially if the environmental conditions remained suitable within an overall landscape mosaic.

13 "Numbers of wood-boring insects are most abundant in wood two

to three years dead. Twice in the Singer Tract, an unusual amount of timber was killed, and about two years afterward ivory-bills moved into the vicinity of each deadening and remained there for two or more years." Tanner, p. 46. Not even the John's Bayou pair in the Singer Tract was documented using the exact same locality for more than 3–4 years. Indeed, "pairs of birds and individuals, especially young birds, have disappeared from the Singer Tract, and the simplest explanation is that they have wandered away, for there was no evidence of their having been killed either by predators or by shooting." *Ibid.*, p. 33. Ivory-bill's nested temporarily in Royal Palm Hammock, Florida, as late as 1917, apparently also within a 2- to 4-year window, after which "no indications of breeding were discovered." Howell, A.H. 1921. A list of the birds of Royal Palm Hammock, Florida. *The Auk* 38: 250–263.

14 Tanner, p. 47.

15 Eastman, W. 1958. Ten-year search for the ivory-billed woodpecker. *Atlantic Naturalist* 13: 216–228.

16 *Ibid.*, p. 218.

17 Tanner, p. 34.

18 Forest disturbance from fire furnishes more than twice the foraging success with wood-boring beetles in certain woodpeckers whose populations are regulated by such patchy, episodic food resources. Rota, C.T., M.A. Rumble, C.P. Lehman, D.C. Kesler, and J.J. Millspaugh. 2015. Apparent foraging success reflects habitat quality in an irruptive species, the black-backed woodpecker. *The Condor* 117: 178–191.

19 The ivory-bill moves "from one area to another in search of a variable food supply that has always been more or less eruptive and undependable." Tanner, p. 53. "Considering all the evidence, I believe that ivory-bills were not sedentary birds, but sometimes wandered considerable distances." *Ibid.*, p. 35. (Wood-boring beetle larvae, soft and hard mast, all large items, would tend to be sporadic but energetically-profitable compared to smaller dietary items).

20 Tanner's depiction of ivory-bill and old-growth were swayed

conceptually by "balance of nature," an ecological view in which forests were believed to return to a previous equilibrium after disturbance. Newer conceptions in ecology instead emphasize patch dynamics as embedded within hierarchy theory, and place greater weight on spatial and temporal patterns, non-equilibrium, scale, and heterogeneity. Wu, J., and O.L. Loucks. 1995. From balance of nature to hierarchical patch dynamics: a paradigm shift in ecology. *The Quarterly Review of Biology* 70: 439–461. A fundamental tenet to this more modern view is that ecological "stability" is strictly relative, and more likely to be approached only at larger spatial and longer temporal scales. Riitters, K.H., J.D. Wickham, and T.G. Wade. 2009. An indicator of forest dynamics using a shifting landscape mosaic. *Ecological Indicators* 9: 107–117. The prominence of disturbance regimes in shaping the multi-scale patterns of forest ecosystems was not well known (if recognized much at all) during Tanner's era. Cf. Jentsch, A., C. Beierkuhnlein, and P.S. White. 2002. Scale, the dynamic stability of forest ecosystems, and the persistence of biodiversity. *Silva Fennica* 36: 393–400.

21 Compound sightings of American and Cuban ivory-bill in severely cut-over forest, some of which constituted little more than scrub, ought to be persuasive that the big woodpecker was not restricted to old-growth. "Such habitat would have been immediately ruled out as potential ivory-bill habitat by believers in dependency of the species on virgin forests," and "how are we to rationalize the surprisingly long persistence of the last known Cuban population in degraded habitat?" Snyder, N., D.E. Brown, and K.B. Clark. 2009. *The Travails of Two Woodpeckers*. University of New Mexico Press, Albuquerque, pp. 56–57. Even in the Singer Tract many "of the trees in the area were second-growth ash and rock and red elm, plus oaks, gums, and maples. There were a few very big oaks and gums." Steinberg, M.K. 2008. *Stalking the Ghost Bird: The Elusive Ivory-billed Woodpecker in Louisiana.* Louisiana State University Press, Baton Rouge, p. 41. A common denominator to the ivory-bill's use of highly diverse forests, however, was similar feeding

prospects offered by very recently dead or still dying trees, the same conditions used by other woodpeckers.

22 Parts of northern Florida were once frequented by good numbers of ivory-bills, possibly due to juxtaposition of inland swamp and pine flat wood forests interlaced in a matrix that created especially favorable landscapes for the woodpecker. This "forest mosaic [was] ... substantially different," and "peninsular Florida presented more of a habitat mosaic" than elsewhere. Jackson, J. A. 2006. *In Search of the Ivory-billed Woodpecker*, 2nd ed. HarperCollins, New York, pp. 51, 53.

23 But not necessarily individual pair(s). Small population refers here to only a few pairs. "It is our belief that most individuals spend their entire lives within a few miles of the place where they are hatched and develop little ivorybill communities." Allen, A.A., and P.P. Kellogg. 1937. Recent observations of the ivory-billed woodpecker. *The Auk* 54: 164–184. Belief, however, cannot be equated with accuracy.

24 In Tanner's words (p. 99): "When the food supply is sufficient, the woodpecker is probably resident or sedentary." Gauged by decades of occupancy, ivory-bills seemed to have been facultatively sedentary at the *regional* scale near Avery Island and in Madison Parish, Louisiana (McIlhenny, E.A. 1941. The passing of the ivory-billed woodpecker. *The Auk* 58: 582–584), and possibly parts of Florida (see Allen and Kellogg, p. 166). Each region then still contained extensive forest tracts, making feasible any ecological adaptations that birds may have had for exploiting larger landscape mosaics of shifting patches of food-producing trees.

25 Tree selection for cavity building by the related Magellanic woodpecker depends on crown dieback and tree growth decline, the latter triggered in part by droughts and massive defoliations by caterpillars. Ojeda, V.S., M.L. Suarez, and T. Kitzberger. 2007. Crown dieback events as key processes creating cavity habitat for Magellanic woodpeckers. *Austral Ecology* 32: 436–445.

26 Typical ages of longevity in trees vary greatly. Sweet gum *Liquidamber*

styraciflua have typical and maximum ages of mortality at 200 and 300 years, respectively, whereas for bald cypress *Taxodium distichum* these ages reach 600 and 1,200 years, respectively. See: Appendix 1 in Loehle, C. 1988. Tree life history strategies: the role of defenses. *Canadian Journal of Forest Research* 18: 209–222.

27 Putz, F.E., and R.R. Sharitz. 1991. Hurricane damage to old-growth forest in Congaree Swamp National Monument, South Carolina, USA. *Canadian Journal of Forest Research* 21: 1,765–1,770; Doyle, T.W., B.D. Keeland, L.E. Gorham, and D.J. Johnson. 1995. Structural impact of Hurricane Andrew on the forested wetlands of the Atchafalaya Basin in south Louisiana. *Journal of Coastal Research* SI 21: 354–364; King, S.L., and T.J. Antrobus. 2001. Canopy disturbance patterns in a bottomland hardwood forest in northeast Arkansas, USA. *Wetlands* 21: 543–553.

28 Gagnon, P.R. 2009. Fire in floodplain forests in the southeastern USA: insights from disturbance ecology of native bamboo. *Wetlands* 29: 520–526.

29 Massive changes altered the Mississippi River's natural hydrology and flooding regime. By the 1970s, almost 6,000 km of levees had been constructed along the river's main stem and tributaries. Saikku, M. 2005. *This Delta, This Land: An Environmental History of the Yazoo-Mississippi Floodplain*. University of Georgia Press, Athens, p. 238.

30 Townsend, P.A., and D.R. Butler. 1996. Patterns of landscape use by beaver on the lower Roanoke River floodplain, North Carolina. *Physical Geography* 17: 253–269. Sweet gum, one tree shown by Tanner as targeted by foraging ivory-bills, is particularly susceptible to girdling mortality from beavers. See: Bullock, J.F., and D.H. Arner. 1985. Beaver damage to nonimpounded timber in Mississippi. *Southern Journal of Applied Forestry* 9: 137–140. Introduced American beavers (*Castor canadensis*) even helped facilitate habitat mosaics favored by Magellanic woodpeckers in a Chilean nature reserve. Soto, G.E., P.M. Vergara, M.E. Lizama, C. Celis, R. Rozzi, Q. Duron, I.J. Hahn, and J.E. Jiménez. 2012. Do beavers improve the habitat

quality for Magellanic woodpeckers? *Bosque* 33: 271–274.

31 Leininger, T.D., and J.D. Solomon, J. D. 1995. A forest tent caterpillar outbreak in the Mississippi delta: host preference and growth effects. Pp. 126–131 in Eighth Biennial Southern Silvicultural Research Conference, Auburn, AL, Nov. 1–3, 1994.

32 Most tree species in southeastern forests require a few decades before they seed at all, i.e., produce mast in substantial quantity. American beech *Fagus grandifolia* first produces nuts around 40 years, but its average age at first reproduction is later still, at 60 years. Appendix 1 in Loehle. Older trees are typically larger, so they also produce greater mast volumes than would be expected from younger trees.

33 Sork, V.L., J. Bramble, and O. Sexton. 1993. Ecology of mast-fruiting in three species of North American deciduous oaks. *Ecology* 74: 528–541.

34 Liebold, A., V. Sork, M. Peltonen, W. Koenig, O.N. Bjørnstad, R. Westfall, J. Elkinton, and J.M.H. Knops. 2004. Within-population spatial synchrony in mast seeding of North American oaks. *Oikos* 104: 156–164.

35 Greenberg, C.H., D.J. Levey, C. Kwit, J.P. Mccarty, S.F. Pearson, S. Sargent, and J. Kilgo. 2012. Long-term patterns of fruit production in five forest types of the South Carolina upper coastal plain. *The Journal of Wildlife Management* 76: 1,036–1,046.

36 Tanner, p. 70.

37 *Ibid*, p. 26.

38 *Ibid*.

39 *Ibid*., p. 46. Also, "about two years afterward ivory-bills moved into the vicinity of each deadening and remained there for two or more years." *Ibid*.

40 Brewster, W., and F.M. Chapman. 1891. Notes of the birds of the lower Suwanee River. *The Auk* 8: 125–138. In line with nomadic and or irruptive habits, Brewster's field notes also indicated that natives "in the woods daily" recognized ivory-bill as once present but then "not seen … for three months or more." *Journals of William Brewster,*

1871–1919 Inclusive, untitled p. 61. http://transcribebhl.mobot.org/display/read_work?ol=w_tab_read&page=13&work_id=9 (accessed 22 August 2016).

41 "The locality where this bird is to be found at all times is what the people call 'burn-outs.' These are large tracts of heavy timber which the forest fires have destroyed; and the dead trees harbor beetles, etc." Wayne, A.T. 1893. Additional notes on the birds of the Suwanee River. *The Auk* 10: 336–338. Although differences in search effort between the two expeditions might also account for a 2-year rise in the apparent numbers of ivory-bills, the fact that relative abundance could vary so much for *any* reason calls into question the certainty we can place on much of what we think we know about the species' natural history.

42 Ivory-bill might still differ importantly from those North American woodpeckers having habits that are superficially alike. Population regulation in the irruptive black-backed woodpecker *Picoides arcticus* is largely a demographic process of colonization driven by natal dispersal into recently disturbed forest, with populations declining 6–10 years post-fire, in line with lifespans of early colonizers and the offspring they produce. Siegel, R.B., Tingley, M.W., R.L. Wilkerson, C.A. Howell, M. Johnson, and P. Pyle. 2016. Age structure of black-backed woodpecker populations in burned forest. *The Auk* 133: 69–78. Given life spans rather longer than some forest disturbance cycles, any numerical responses in the ivory-bill were likely to have reflected irruptive movements by experienced breeding adults as well.

43 ***Congruence bias*** is a thinking blunder caused by our failure to consider alternatives, or to evaluate too few hypotheses, when attempting to account for some phenomenon. Cognitive biases, logical errors, and reasoning fallacies such as the ***congruence bias*** that are relevant to a fuller understanding of our relationship with the ivory-billed woodpecker are highlighted in bold-faced italics whenever they appear in the main body of this text. Additional definitions, examples, and citations for these biases and fallacies

are provided also as a Glossary.

44 In this hypothesis, the ivory-bill's *behavioral* propensity to seek out seclusion from human disturbance is the underlying mechanism. One reason suggested for presence of ivory-bill in Arkansas was an extensive system of private hunting lands that served as secluded refuge from disturbance. Unlike public lands that allow unrestricted hunter access over wide areas, private hunting clubs tend to grant permits only to confined, blind-based hunting, they are posted against trespass using enforcement year-round, and they serve consequently as relatively quiet, secluded havens for most of the year. Bivings, A.E. 2006. Rediscovery and recovery of the ivory-billed woodpecker. *Journal of Wildlife Managemen*t 70: 1,495–1,496.

45 A more thorough scrutiny of whether ivory-billed woodpecker succumbed to human persecution via market forces that were brought on by the anthropogenic Allee effect is provided in Appendix A.

46 Plausibility of the ivory-bill as a forest habitat specialist is examined in Appendix B.

47 Although actually depicting black-backed woodpecker, the ensuing illustrates a form of population regulation that is proposed for the ivory-bill in my parallax rendering. "A highly irruptive species; numbers may be low in unburned or undisturbed forest, but depending upon demographic bursts or recruitment related to temporarily superabundant foods (mostly beetles that attack dead trees in new stand-replacement burns), these numbers can dramatically increase and thus are clearly regulated by extent of fires [disturbances]." Dixon, R.D., and V.A. Saab. 2010. Black-backed woodpecker (*Picoides arcticus*). *The Birds of North America Online* (A. Poole, ed.). Cornell Lab of Ornithology, Ithaca, NY. http://bna.birds.cornell.edu/bna/species/509 (accessed 2 June 2016).

48 Other *Campephilus* are not dietary specialists, instead taking a broad spectrum of prey. Magellanic woodpeckers consume and provision young with wood-boring larvae, adult insects, caterpillars, pupae, arachnids, and even small vertebrates, including bats, lizards, and

eggs and nestlings of other birds. Ojeda, V.S., and M.L. Chazarreta. 2006. Provisioning of Magellanic woodpecker (*Campephilus magellanicus*) nestlings with vertebrate prey. *The Wilson Journal of Ornithology* 118: 251–254.

49 "They are fully as adaptable as other woodpeckers in their food and feeding habits." Allen and Kellogg, p. 167.

50 Tanner, p. 41.

51 Allen and Kellogg, p. 166.

52 "They would sit on top of the nest and with a few strokes of their bill scatter it in all directions." Bendire, C. 1895. Life histories of North American birds, from the parrots to the grackles. *Special Bulletin of the U.S. National Museum*, p. 44.

53 This bill was highly versatile, "a Swiss Army Knife for the primeval forest," able to function as a "chisel … probe, dagger, and pincers all wrapped into one tool." Jackson, J.A. 2002. Ivory-billed woodpecker (*Campephilus principalis*), *The Birds of North America* (A. Poole, Ed.). Ithaca: Cornell Lab of Ornithology. http://bna.birds.cornell.edu/bna/species/711 (accessed 16 January 2016).

54 "Thus, the ivory-bill is able to reach a supply of food that is not easily accessible to other woodpeckers." Tanner, p. 44.

55 "The ivory-billed woodpecker has no real competitors for its food supply" and it was "not at all affected by the numbers of other woodpeckers." *Ibid.*, p. 54. As Tanner was a direct observer, Jackson and Lester Short's speculation that ivory-bill competed with pileated for cavities and food lacks support. Jackson, 2006, p. 41. Tanner watched four ivory-bills and two pileateds together in the same tree for a half hour without any interaction; he also once saw a single larger woodpecker drive off two pileateds. Tanner, p. 54.

56 We have no substantial "evidence of natural predators of any kind, nor evidence of that happening." *Ibid*. Tanner did not regard the widely sympatric red-shouldered hawk *Buteo lineatus* as large and strong enough to take down an ivory-bill. *Ibid*. The big woodpecker's keeping to dense forest canopies likely made its eggs and young the most vulnerable to predation. However, no nest predation was

detected in 22 breeding events followed in the related Magellanic woodpecker. Ojeda, V.S. 2004. Breeding biology and social behavior of Magellanic woodpeckers (*Campephilus magellanicus*) in Argentine Patagonia. *European Journal of Wildlife Research* 50: 18–24.

57 Averring that food limitation functioned in ivory-bill primarily (or only) during reproduction, because adults then provisioned young with beetle larvae, including both large and small grubs, is still unsupported. Virtually all forest-dwelling birds that breed in the northern hemisphere seek out protein-rich, high-energy foods during this part of their seasonal cycle, i.e., maximizing protein intake is hardly a unique behavior.

Over time the notion of food specialization was extended without evidentiary backing to encompass corollaries, including assertions that ivory-bills (and other *Campephilus*) cannot regurgitate smaller prey brought back to their young. "The downside for *Campephilus* woodpeckers is the fact that they don't regurgitate." http://www.ibwo.net/forum/showthread.php?p=6622 (accessed 1 May 2016). And: "ivory-billed woodpeckers don't eat ants or termites and don't regurgitate." Michaels, M. 2016. Digging deeper into Tanner, part 3 of 3 – prey species. https://projectcoyoteibwo.com/2016/03/24/digging-deeper-into-tanner-part-3-of-3-prey-species/ (accessed 1 May 2016). Yet Tanner's mentors noted: "Certainly the ivory-bills did not do enough digging while we were watching them to uncover any additional borers, so they may have been picking up such termites as appeared in the gash." Allen, A.A., and P.P. Kellogg. 1937. Recent observations of the ivory-billed woodpecker. *The Auk* 54: 164–184.

Tanner contradicted speculation ivory-bills were unable to deliver small prey via regurgitation (or its functional equivalent) from the larger food loads made possible by the ivory-bill's huge bill. Tanner (p. 76) noted adult ivory-bills "had trouble feeding large grubs to the [small] nestling." Tanner also described behavior (p. 75) by a male ivory-bill in which it "seemed to be jerking as though working food from the back of its mouth," the sort of motion expected in regurgitation or its functional equivalent. Tanner also noted behavior expected prior to regurgitation: he watched a "female secure a grub from the tree, put in the back of her mouth, and continue scaling away bark after more food." *Ibid.*

No morphological limitations were ever demonstrated that would have prevented ivory-bill from delivering multiple smaller prey items. Their economic flight, too, would offset costs associated with frequent foraging trips. Indeed, proficiency at finding prey gave adults ample foraging budgets. Tanner, p. 57. He watched a pair feed a young nestling 30 times in one day. *Ibid.*, p. 76. And he emphasized pairs still rested 3–4 hours from feeding young, again reinforcing that ample time reserves were available for provisioning.

58 ***Illusory correlation*** occurs when we mistakenly over-emphasize one outcome due to its purely incidental association with some other factor, ignoring all other possible links or associations that could also explain that outcome. Hamilton, D.L., and R.K. Gifford. 1976. Illusory correlation in interpersonal perception: a cognitive bias of stereotypical judgments. *Journal of Experimental Social Psychology* 12: 392–407. This stereotypical error in thinking is even more likely to happen when what seems to be some unusual action (e.g., eating very large grubs) takes place in a less familiar group or category (e.g., the rare ivory-billed woodpecker). See especially: Risen, J.L., T. Gilovich, and D. Dunning. 2007. One-shot illusory correlations and stereotype formation. *Personality and Social Psychology Bulletin* 33: 1,492–1,502.

59 *Ibid*, p. 57.

60 *Ibid.*

61 Jackson, 2002.

62 Bull, E.L., and J.A. Jackson. 2011. Pileated woodpecker (*Dryocopus pileatus*), *The Birds of North America* (A. Poole, Ed.). Ithaca: Cornell Lab of Ornithology. http://bna.birds.cornell.edu/bna/species/148 (accessed 18 January 2016).

63 Lasiewski, R.C., and W.R. Dawson. 1967. A re-examination of the relation between standard metabolic rate and body weight in birds.

The Condor 69: 13–23.

64 The relationship of daily energy requirements to body mass in birds is given by FMR = $10.5M_b0.681$, where FMR = field metabolic rate (in kilojoules) and M_b = body mass in grams. Body mass in this equation describes almost 94% of the variation in field metabolic rate measured across bird species. See Nagy, K.A. 2005. Field metabolic rate and body size. *The Journal of Experimental Biology* 208: 1,621–1,625.

65 Magellanic woodpeckers (265 to 365 g) have territory sizes as little as 100 ha, or only about 6% of the forest area Tanner claimed as required by ivory-bill. Ojeda, V.S., and L. Chazarreta. 2014. Home range and habitat use by Magellanic woodpeckers in an old-growth forest of Patagonia. *Canadian Journal of Forest Research* 44: 1,265–1,273. Feeding territory size shows a strong positive relationship to body mass in birds, yet the density of acceptable and accessible food items as gauged by biomass per unit area actually decrease in larger bird species. On the whole, then, larger birds offset some of their greater energetic needs with higher (not lower) energetic efficiency. Schoener, T.W. 1968. Sizes of feeding territories among birds. *Ecology* 49: 123–141.

66 Snyder et al., p. 31.

67 Residents of the Fort Drum region in southeast Florida during the 1920s reported "often seeing many ivory-bills during a day's activities." They also considered the big woodpecker to be "once even more abundant in this location than the pileated woodpecker." *Ibid.*, p. 20. But that observation is consistent with at least two hypotheses: 1) the ivory-bill was once more abundant, or 2) ivory-bill aggregated at higher densities when environmental conditions were suitable (possibly due to nomadic irruption).

68 Attributing relatively low density of ivory-bill exclusively to unmet food needs embodies ***congruence bias*** because it fails to consider or disprove alternative hypotheses that can explain the same observation. A third hypothesis that might explain differences in density between the two large woodpeckers could be because

ivory-bills were more aggregated, whereas pileated woodpeckers were more even, in their distribution across some forest mosaics. What appeared to be low density by human standards also could be merely an artifact of the woodpecker's very large ambits (total spatial extent used during the species' life history).

69 Tanner, p. 100.

70 *Ibid.*, p. 63. See also Allen and Kellogg, p. 170–171. Moreover: "gradually the birds became accustomed to our presence." Steinberg, p. 44.

71 Cooke, A.S. 1980. Observations on how close certain passerine species will tolerate an approaching human in rural and suburban areas. *Biological Conservation* 18: 85–88.

72 Blumstein, D.T., L.L. Anthony, R. Harcourt, and G. Ross. 2003. Testing a key assumption of wildlife buffer zones: is flight initiation distance a species-specific trait? *Biological Conservation* 110: 97–100.

73 Ivory-bill artifacts, widely sought for ceremonial purposes and traded across many North American native tribes, were recovered from archaeological sites dating from 1,000 CE and outside the bird's presumed original range: north Georgia, Illinois, Missouri, and Colorado. Jackson, 2006, pp. 77–89.

74 A maximum longevity of about 15 years is generally attributed to the ivory-bill, although 20 or more years seems plausible, e.g., see: Fitzpatrick, J.W., M. Lammertink, M.D. Luneau, Jr., T.W. Gallagher, B.R. Harrison, G.M. Sparling, K.V. Rosenberg, et al. 2005. Ivory-billed woodpecker (*Campephilus principalis*) persists in continental North America. *Science* 308: 1,460–1,462.

75 Arrival of the Clovis culture is believed to have eradicated completely many entire lineages of North America's larger fauna. See for example: Stuart, A.J. 1991. Mammalian extinctions in the Late Pleistocene of northern Eurasia and North America. *Biological Reviews* 66: 453–562; Gill, J.L., J.W. Williams, S.T. Jackson, K.B. Lininger, and G.S. Robinson. 2009. Pleistocene megafaunal collapse, novel plant communities, and enhanced fire regimes in North America. *Science* 326: 1,100–1,103.

76 van Oers, K., P.J. Drent, P. de Goede, and A.J. van Noordwijk. 2004. Realized heritability and repeatability of risk-taking behavior in relation to avian personalities. *Proceedings of the Royal Society B* 271: 65–73; Exnerová, A., K.H. Svádová, E. Fučíková, P. Drent, and P. Štys. 2009. Personality matters: individual variation in reactions of naïve bird predators to aposematic prey. *Proceedings of the Royal Society B* 277: 723–728.

77 Bize, P., C. Diaz, and J. Lindström. 2012. Experimental evidence that adult antipredator behaviour is heritable and not influenced by behavioural copying in a wild bird. *Proceedings of the Royal Society B* 279: 1,380–1,288.

78 van Dongen, W.F.D., R.W. Robinson, M.A. Weston, R.A. Mulder, and P.-J. Guay. 2015. Variation at the DRD4 locus is associated with wariness and local site selection in urban black swans. *BMC Evolutionary Biology* 15: 253.

79 Bales, S.L. 2010. *Ghost Birds: Jim Tanner and the Quest for the Ivory-billed Woodpecker, 1935–1941*. University of Tennessee Press, Knoxville, p. 64.

80 Thompson, M. 1885. A red-headed family. *The Elzevir Library* 4: 5–21.

81 Blumstein, D.T. 2006. Developing an evolutionary ecology of fear: how life history and natural history traits affect disturbance tolerance in birds. *Animal Behaviour* 71: 389–399.

82 Díaz, M., A.P. Møller, E. Flensted-Jensen, T. Grim, J.D. Ibáñez-Álamo, J. Jokimäki, G. Markó, and P. Tryjanowski. 2013. The geography of fear: a latitudinal gradient in anti-predator escape distances of birds across Europe. *PloS One* 8(5): e64634.

83 Casas, F., F. Mougeot, J. Vinuela, and V. Bretagnolle. 2009. Effects of hunting on the behaviour and spatial distribution of farmland birds: importance of hunting-free refuges in agricultural areas. *Animal Conservation* 12: 346–354.

84 Clucas, B., and J.M. Marzluff. 2012. Attitudes and actions toward birds in urban areas: human cultural differences influence behavior. *The Auk* 129: 8–16.

85 Marzluff, J.M., J. Walls, H.N. Cornell, J.C. Withey, and D.P. Craig. 2010. Lasting recognition of threatening people by wild American crows. *Animal Behaviour* 79: 699–707; Griffin, A.S., and K. Haythorpe. 2011. Learning from watching alarmed demonstrators: does the cause of alarm matter? *Animal Behaviour* 81: 1,163-1,169.

86 Conover, M.R. 1987. Acquisition of predator information by active and passive mobbers in ring-billed gull colonies. *Behaviour* 102: 41–57; Griffin, A.S. 2004. Social learning about predators: a review and prospectus. *Learning & Behavior* 32: 131–140.

87 Curio, E., U. Ernst, and W. Vieth. 1978. Cultural transmission of enemy recognition: one function of mobbing. *Science* 202: 899–901.

88 Freeburg, T.M. 1998. The cultural transmission of courtship patterns in cowbirds *Molothrus ater*. *Animal Behaviour* 56: 1,063–1,073.

89 Payne, R.B., W.L. Thompson, K.L. Fiala, and L.L. Sweany. 1981. Local song traditions in indigo buntings: cultural transmission of behavior patterns across generations. *Behaviour* 77: 199–221.

90 Magrath, R.D., T.M. Haff, J.R. McLachlan, and B. Igic. 2015. Wild birds learn to eavesdrop on heterospecific alarm calls. *Current Biology* 25: 2,047–2,050.

91 The Allee effect refers to positive effects conferred on individual fitness from presence of conspecifics, e.g., sociality. For example, an individual bird is more likely to find a mate when population density is high. In the context of conservation, the Allee effect suggests that population growth can be negative when individuals become ever more rare, potentially leading to an extinction vortex from which the population cannot recover. See, e.g., Stephens, P.A., and W.J. Sutherland. 1999. Consequences of the Allee effect for behaviour, ecology and conservation. *Trends in Ecology & Evolution* 14: 401–405.

92 Even the smaller, less powerful black-backed woodpecker is capable of irruptive flight distances of at least 600 km. Yunick, R.P. 1985. A review of recent irruptions of the black-backed woodpecker and three-toed woodpecker in eastern North America. *Journal of Field Ornithology* 56: 138–152.

93 Wayne, p. 60.

94 Higley, p. 123.

95 Dale, S. 2001. Female-biased dispersal, low female recruitment, unpaired males, and the extinction of small and isolated bird populations. *Oikos* 92: 344–356.

96 "Four birds would travel loosely, not in a close flock; they called little while feeding, but managed to keep together." Tanner, p. 61. Feeding sounds, too, in the large *Campephilus* are not invariably loud or distinguishable from smaller bird species. See Short, L.L. 1970. The habits and relationships of the Magellanic woodpecker. *The Wilson Bulletin* 82: 115–129.

97 Young Magellanic woodpeckers remain with their family group for up to 2 years or more, all the while being fed by adults. Ojeda, V.S. 2004. Breeding biology and social behaviour of Magellanic woodpeckers (*Campephilus magellanicus*) in Argentine Patagonia. *European Journal of Wildlife Research* 50: 18–24.

98 Younger birds are more naïve and vulnerable to predation, but learn rapidly to adopt more cautious behavior of adults towards the end of the first six months beyond fledging. Dhindsa, M.S., and D.A. Boag. 1989. Influence of age on the flushing distance of marked and unmarked black-billed magpies. *Ornis Scandinavica* 20: 76–79.

99 Hoyt, R.D. 1905. Nesting of the ivory-billed woodpecker in Florida. *The Warbler* 1: 53–55.

100 "After securing the male … [the female] was not to be deceived in that way." Ridgway, R. 1898. The home of the ivory-bill. *Osprey* 3: 35–36.

101 Some accounts surmise that the ivory-bill's lifespan could even reach as long as 20–30 years. http://animals.nationalgeographic.com/animals/birds/ivory-billed-woodpecker/ (accessed 1 February 2016).

102 Carrete, M., and J.L. Tella. 2010. Individual consistency in flight initiation distances in burrowing owls: a new hypothesis on disturbance-induced habitat selection. *Biology Letters* 6: 167–170.

104 Hoyt, p. 53.

104 Only in large, narrow-winged *Campephilus* (ivory-billed and imperial woodpeckers) did the number of fledglings produced per nest typically exceed one. See Chazarreta, M.L., V.S. Ojeda, and A. Trejo. 2011. Division of labour in parental care in the Magellanic woodpecker *Campephilus magellanicus*. *Journal of Ornithology* 152: 231–242.

105 Within the Singer Tract, J. J. Kuhn encountered four young just out of the nest in late May 1936 and in July 1931. Tanner, p. 68.

106 Puzzled by what seemed like a long breeding season in the ivory-bill, Tanner (pp. 68–69) discussed this possibility, though conceding there were "no facts to support or contradict this supposition." On the other hand, later nests were found consistently to have more eggs and young than earlier ones, in keeping with an expectation of greater food resources later in the season. *Ibid.*, p. 72.

107 Dixon and Saab.

108 This view concedes that ivory-bills were extirpated locally or regionally through unsustainable killing or by habitat-driven bottlenecks (population "sinks"). However, for some period of time after the 1930s the meta-population as a whole avoided extinction due to successful localized reproduction elsewhere (population "sources"). See, e.g., Naranjo, E.J., and R.E. Bodmer. 2007. Source–sink systems and conservation of hunted ungulates in the Lacandon Forest, Mexico. *Biological Conservation* 138: 412–420.

109 Unless our cognition stays focused relentlessly on probabilistic aspects of ivory-bill persistence and detection, the human mind strains to grasp what seem to be (but actually are not) stark contradictions. "It requires a remarkably fine balancing act for there to be enough ivory-bills continuously reproducing successfully over 7 decades, yet so few as to be undetectable or little encountered. The bird gets seen, but then rarely re-seen; it is heard, but then rarely found; its sign is observed, but it doesn't return to it; it is spotted by a single individual, but virtually never by a group, nor remote camera – this species either does NOT exist in the places we are looking for it, or, if present, it is essentially invisible to human

eyes." Cyberthrush. 2016. Explaining the inexplicable… http://ivorybills.blogspot.com/ (accessed 1 May 2016).

The exertion described by this writer arises from our cognitive vulnerability to ***neglect of probability***. This same cognitive toil manifests with other rare birds (Chapter 13). But for other vertebrates, premature extinction claims seem to be avoided despite the fact that no human has ever seen the species alive. Spade-toothed beaked whale *Mesoplodon traversii* is represented only by two fully intact dead specimens, plus two partial sets of head bones, reported since its original discovery in 1872. Thompson, K., C.S. Baker, A. Van Helden, S. Patel, C. Millar, and R. Constantine. 2012. The world's rarest whale. *Current Biology* 22: R905–R906.

110 Tanner, p. 99. Tanner's searches for ivory-bill beyond the Singer Tract were exasperatingly brief and highly arbitrary, out of season for achieving best results (not always in winter or early spring), and largely governed by highly stereotyped perceptions of ideal habitat as "virgin" forest of sweet gum-oak hardwoods. That arbitrariness seriously biased downward any sort of robust estimate for any remaining ivory-bills and their habitat across the southeastern United States.

111 Jackson, 2006, pp. 61–63.

112 Ivory-bill was never counted reliably anywhere, and indeed might not have been countable given its usually cryptic lifestyle. Regardless, "Tanner … was somewhat cavalier in his approach to evaluating potential habitat … he's more likely to have missed populations in areas that he rejected for being suboptimal and not expansive enough." Michaels, M. 2015. Tanner and population density. http://projectcoyoteibwo.com/2015/12/14/tanner-and-population-density/ (accessed 20 January 2016); "Stoddard's perspective [that potential habitat was huge and difficult to survey] leaves room for the possibility that Tanner grossly underestimated the ivorybill population in his monograph, something that might lead to very different projections about the likelihood of survival today." Michaels, M. 2014. The two faces of James T. Tanner – a study

in selective memory and blind spots. http://projectcoyoteibwo.com/2014/10/24/the-two-faces-of-james-t-tanner-a-study-in-selective-memory-and-blindspots/ (accessed 20 January 2016).

113 Murray, B.G. 2011. Demography and population dynamics of the ivory-billed woodpecker. Pp. 111–149 in *What Were They Thinking? Is Population Ecology a Science? Papers, Critiques, Rebuttals, and Philosophy*. Infinity Publishing, Conshohocken, PA.

114 "It is conceivable that ivory-billed woodpeckers could have a natural potential longevity of 20-plus years." Jackson, 2006, p. 240.

115 Mattsson, B.J., R.S. Mordecai, M.J. Conroy, J.T. Peterson, R.J. Cooper, and H. Christensen. 2008. Evaluating the small population paradigm for rare large-bodied woodpeckers, with implications for the ivory-billed woodpecker. *Avian Conservation and Ecology* 3: 5. http://www.ace-eco.org/vol3/iss2/art5/ (accessed 19 January 2016).

116 Key demographic variables like annual adult survival and mean clutch size were interpolated using measurements from 20 populations or species of the world's woodpeckers. *Ibid.*, Figure 3.

117 Eleven female offspring surviving to reproductive age each decade is hardly an extravagant fecundity. It could arise a number of ways, and include years of complete reproductive failure. To whit, successful broods of 3, 2, 4, and 2 females would produce enough offspring to meet this threshold while also conceding total reproductive failure in 6 out of 10 years. Clutch size in ivory-bill could reach 5 or 6 eggs. Tanner, p. 71. Broods of up to at least four young fledged successfully. Jackson, 2006, p. 37. Potential for rather high life-long fecundity aligns with life history traits expected in an irruptive species for which (maximally) successful reproduction might occur only during boom years of surplus food.

118 Mattsson et al. Apparently the only attempt to figure a *maximum* population size, one based loosely on approximate field encounter rates with this species since 1920, was estimated at ~100 ivory-bills (order of magnitude). Collins, M.D. 2019. Statistics, probability, and a failed conservation Policy. *Statistics and Public Policy* 6: 67–79.

119 E.g., Jackson, 2002, Figure 1. The already long odds for finding ivory-

bill plummet even further if we use the area of the species' original range for estimating occupancy. Likelihoods of the woodpecker occupying any 1-km^2 block in this case would range from less than 2 to only 63 chances in a hundred thousand for a total population size of 12 up to 480 birds, respectively.

120 12 million acres of this forest type are given for the Mississippi Alluvial Valley: Guilfoyle, M.P. 2001. Management of bottomland hardwood forests for nongame bird communities. EMRRP Technical Notes Collection, ERDC TN-EMRRP-SI-21, U.S. Army Engineer Research and Development Center, Vicksburg, MS; 30 million acres across the entire southeastern United States: Meadows, J.S., and J.D. Hodges. 1997. Silviculture of southern bottomland hardwoods: 25 years of change. *Proceedings of the Twenty-fifth Annual Hardwood Symposium* 1997: 1–16.

121 A naïve probability ensues if one or more key factors are not incorporated into the estimate, in this instance from a flawed premise that occupancy for a population of ivory-bills can be assessed using the total numbers of birds instead of the real social units, i.e., pairs or family groups. Reliance on naïve probability would thus lead to an estimate of detection biased upward (too high).

122 This adjustment assumes that one-half of all birds are paired, with a single year-one bird still accompanying these pairs, and all remaining individuals distributed as singletons. Adjusting the probability of detection follows because "clumping (e.g. pairs or family groups) actually makes detections happen less often … the uniformly spread randomly moving bird is the EASIEST to find if you don't already know where to look. Clumping and repeated movement patterns make it harder to find initially, but then easier to relocate." Pulliam, B. 2010. Shroedinger's woodpecker. *Notes from Soggy Bottom,* p 6. http://bbill.blogspot.com/2010/04/schroedingers-woodpecker.html (accessed 5 February 2016).

123 Using standard point counts of radius 50 m with centers spaced 150 m apart, we could place a total of 49 such counts inside a 1 km × 1 km block, covering a total of 384,845 m2. Thus, ≥60% of

this 1-km^2 box would still remain un-surveyed. Allowing an observer 5 minutes to watch/listen at the count center, 3 minutes to manage other logistics, then 5 minutes to transit between points over sloughs, dense vegetation, and other obstacles expected in swamp forest, it would take more than 10 hours, around a full day, to cover just 38.5% of one block. For more details on point-count methodology, see: Hamel, P.B., W.P. Smith, D.J. Twedt, J.R. Woehr, E. Morris, R.B. Hamilton, and R.J. Cooper. 1996. *A Land Manager's Guide to Point Counts of Birds in the Southeast*. U.S. Department of Agriculture, Forest Service, Southern Research Station, General Technical Report SO-1 20, Asheville, NC.

124 Detection probabilities for the relatively loud pileated woodpecker are just 18–36%. See: Farnsworth, G.L., K.H. Pollock, J.D. Nichols, T.R. Simons, J.E. Hines, and J.R. Sauer. 2002. A removal model for estimating detection probabilities from point-count surveys. *The Auk* 119: 414–425; Watson, K. 2013. A comparison of single-day versus multiple-day sampling designs using occupancy modeling. Bachelor of Science thesis, College of William and Mary. Ivory-bills would not be expected to have detection probabilities this high. Hoyt, p. 55, indicated pileated vocalizations carried 15 times further than the ivory-bill's softer calls. "The woodpeckers are very silent at all times so far as their voices are concerned." Higley, W.K. 1906. *Birds and Nature,* Volume III, A.W. Mumford and Company, Chicago, p. 122. Even when the ivory-bill was actively foraging, "the ensuing noise scarcely equaled in volume the work of the downy woodpecker [*Dryobates pubescens*]." Dennis, p. 503.

Wary habits (Chapter 7) would further lower the ivory-bill's detectability. "A wary bird may move away from the search path before it can be detected." Collins, M.D. 2021. The role of bioacoustics in the conservation of the ivory-billed woodpecker (*Campephilus principalis*). *Journal of Theoretical and Computational Acoustics.* Consequently, use of a much lower probability value for its detection would be entirely justified, e.g., based on "… a set of factors related to behavior and habitat, it must be millions of times harder to obtain a photo of an ivory-billed woodpecker than... a hypothetical... species of comparable rarity." *Ibid*. The inevitable result of evasion around humans is that the odds of finding or re-finding a bird are reduced even further in examples provided here.

125 Surveys would need to be run concurrently (same days) because otherwise ivory-bill movement(s) by this vagile species across different survey blocks confounds more the ability to detect any birds present.

126 "can always find them near the spot where they have their nest or winter home, from which place they are hard to drive away." E.A. McIllhenny, quoted in Bendire, p. 43. However, Singer Tract warden J. J. Kuhn "could only remember seeing young ones twice in his lifetime." Bales, p. 64.

127 Tanner, p. 33. Also, the ivory-bills usually "traveled in a straight line." *Ibid.*, p. 22.

128 These ranging distances might still be too low. "Both parents feed them, often going quite a distance into the open country in search of food." E.A. McIllhenny, quoted in Bendire, p. 43.

129 Even this technique is/was not fool proof. Ivory-bills could be "erratic ... in using the same hole." Their use of the same roosting holes or even the same roosting grounds was only a reliable clue "most of the time." Tanner, p. 59. The big woodpecker also might use deception when it perceived threats because it would "go in and out of every old nest in the swamp, but steer wide of the new [active] one." Hoyt, p. 52.

130 This level of effort demands 252 observer-days and walking at least 1,890 km to cover the total area (252 km^2). The sampling power for justifying an inference for a true absence in a rare species is explained by Reed, J.M. 1996. Using statistical probability to increase confidence of inferring species extinction. *Conservation Biology* 10: 1,283–1,285.

131 Tanner, as quoted in Bales, p. 123.

132 Tanner, p. 33.

133 Rosenberg, K.V., R.W. Rohrbaugh, and M. Lammertink. 2005. An

overview of ivory-billed woodpecker (*Campephilus principalis*) sightings in eastern Arkansas in 2004–2005. *North American Birds* 59: 198–207.

134 Scott, J.M., F.L. Ramsey, M. Lammertink, K.V. Rosenberg, R. Rohrbaugh, J.A. Wiens, and J.M. Reed. 2008. When is an "extinct" species really extinct? Gauging the search efforts for Hawaiian forest birds and the ivory-billed woodpecker. *Avian Conservation and Ecology* 3: 3. http://www.ace-eco.org/vol3/iss2/art3/ (accessed 5 February 2016).

135 The ***improbability principle*** refers to everyday situations in which what at first seem to be highly unlikely incidents are really to be expected, even viewed as commonplace. This principle arises from five influences, any of which can operate singly or together: the Laws of Inevitably, of Truly Large Numbers, of Selection, of the Probability Lever, and of Near Enough. Hand, D.J. 2014. *The Improbability Principle: Why Coincidences, Miracles, and Rare Events Happen Every Day*. Scientific American/Farrar, Straus and Giroux.

136 The Law of the Probability Lever basically apprises us "that a slight change in circumstances can have a huge impact on probabilities." *Ibid.*, p. 142.

Advent of Myth: 1731–1932

Chapter 2

Leaving Utopia

Such treaties may be alright for men who are too old to hunt or fight. As for me, I have my young warriors about me. We will hold our land.
—Tsi'yu-gunsini (Dragging Canoe), eighteenth-century Chickamauga-Cherokee war chief

Days had passed since his last meal. Hunger's craving gave so much audible protest that he expected gastric rumbling to give away his location. In such treeless terrain, only the most patient and quiet stealth could ever place him close enough to strike his quarry. The Tsalagi hunter-warrior was entirely prone, snaking bit by bit through dense sedges and grasses. A wary gobbler's bronze wings reflected bright sunlight as it turned to face in the direction of the young hunter. The bowman froze, holding his gaze steady, until the huge bird resumed feeding around the edges of a small seep, a long, deep-pink wattle quivering as it pecked at the damp ground. Only then did the hunter inch forward again.

They called him Little White Owl, and he belonged to the Ani'-Wa'Ya or wolf clan of his Tsalagi people. Outsiders would label them Overhill Cherokee for living on the west side of the Appalachian range, but they referred to themselves as *Aniyun-Wiya,* the Principal People. Today he was hunting game, alone, deep inside tribal lands, a few miles north from Amaye'li-gunah'ta, a meandering creek with tulip poplars columned high along the tributary's narrow flood plain. He was stalking game now in a slightly rolling meadow well east of the big river at

Tsatanu'gi.

Just within range of the bird now, he reached slowly over his shoulder into a deerskin quiver, placed the cane-shafted arrow across his bow carved out of cured black locust, drew back, and fired. His first projectile narrowly missed the big bird, striking just to the near right. On full alert now, the wild turkey (*Meleagris gallopavo*) had not yet fled; it was still within range. Very slowly reaching for another missile, he drew back on a bowline made of deer sinew, and...

"James Christopher Haney, come here right this instant!!!"

Thoroughly startled, my six-year-old self drew up slowly in the un-mowed back yard, slung the bow over my shoulder, and glanced back in the direction of our small home on Zeigler Road in East Brainerd. Meanwhile, the house sparrow *Passer domesticus* I was stalking flew off chirping into the shelter of a nearby privet hedge. My mother stood just outside our kitchen screen door, hands on her hips, patently agitated. Dragging back in her direction as slowly as possible, I stopped a safe distance from the bottom of the steps. Hearing my entire name spoken at full volume by any adult was never a good sign.

"What in heaven's name are you doing?" she demanded.

"I'm just hunting," I replied, guardedly.

"Dressed like that?" she snapped, jabbing her finger in the direction of my attire.

"Well, Indians wore clothes like this," my young logic responded.

"You are not an Indian. And you're naked!" she replied with consternation.

Well, not quite. To be in character, I had borrowed a belt from my father that morning, cinched it tightly around my waist, and tucked two small washcloths, front and back, between belt and torso to mimic a buckskin loincloth. In order for these accessories to not slip out, I had doubled the washcloths over, such that virtually no material remained for any secondary purpose of social modesty. There was also a crow feather stuck in a headband, which I hopefully pointed out to her, but that did not help.

"My heavens, whatever will our neighbors think," she admonished, casting furtive glances over at the house across the driveway. She then ordered me inside to change right off into something respectable. I was crestfallen, but knew better than to defy her openly. Better to just wait another opportunity to costume up.

To say I possessed a vibrant imagination as a young boy would be like calling Mount Everest a slight incline. I lived inside a constant realm of brilliant fantasy, yet one that was still bound by its own logic for space and time. It was a realm not without objective heritage, but I did not know that then. Invariably my dreaminess fastened on American[1] frontier life; it mattered not whether it was the Native or the European side. Learning to read just after the loincloth caper, I begged each week to go to our public library where I devoured books on Lewis and Clark, Daniel Boone, Kit Carson, David Crockett, the French and Indian Wars, Tecumseh, and anything else related to Native Americans, especially if they flaunted an open defiance of the white man. American frontier history from the 1700s[2] and 1800s became my solid anchor for personal grounding.

In first grade I learned that Ms. Halverson, my only teacher for the next four levels, was an ardent bird-bander. Under her and her husband's watchful tutelage, we were soon retrieving birds out of traps and mist nets set out behind the classroom. At our tiny two-room cinder-block schoolhouse set in rural Standifer Gap, Tennessee, this activity became recess, science, and math all rolled into one. Natural history did require some slight modification to my role-playing, as sketching creatures now vied with hunting the local wildlife.[3]

Rummaging through the modest school library I added some new explorer-heroes: Alexander Wilson, John James Audubon,

Louis Agassiz Fuertes, Spencer Fullerton Baird, and a variety of military naturalists like Elliott Coues and Charles Bendire,[4] all of whom roamed and reported on America's vast frontier. I could not fathom more interesting occupation than describing new animals and plants while trekking around on horseback.[5] When our local Audubon society hosted Roger Tory Peterson, I waited patiently after his presentation so that my *Field Guide to the Birds*, so threadbare already that the binding was in total ruins, could be autographed by our stellar author-visitor from New England. By age nine I had taken up binoculars for good, and already marked off carefully all species that I considered to have been satisfactorily detected.

Blending fascination in frontier exploration with my growing fixation on birds, I was not above the occasional prank on the unwary. One of these involved elaborate fictions of the ivory-billed woodpecker hanging on in the more remote acres of vacant second-growth to be found behind our family home. Regaling my gullible buddy Dale with tall tales of what an incredible discovery this would make, I succeeded in luring him out for days into what was, of course, nothing more than a taxonomically-twisted snipe hunt. He took it surprisingly well, when at last I came around to fessing up to the ruse.

Growing up so near Chattanooga, we played out more realistic sagas on the region's Civil War battlefields. Our Zeigler Road home was also just a few bike miles away from Audubon Acres and its Elise Chapin Wildlife Sanctuary. When I reached college, I became for a while the sanctuary's weekend naturalist. The sanctuary featured a log cabin built by Cherokee naturalist Spring Frog, whose homestead of the same name later became residence of poet Robert Sparks Walker. We searched for arrowheads on the sprawling property, a late eighteenth-century home of Little Owl and his renegade brother Dragging Canoe.[6] Archaeologists credit the locale, too, as a Mississippian-era village between 1540 and 1560, a point of first contact between indigenous peoples and Spanish explorers Hernando de Soto and Tristan de Luna.[7]

My parents introduced my sisters and I early to primitive camping in more remote areas of the Appalachian Mountains. From that beginning I learned to enjoy backpacking, wilderness trekking, white water canoeing, caving, and mountain climbing. I took courses in wild edible plants and survival skills. By age 17, I convinced my parents to allow me and three other lads all my age or younger to go on an adult-free camping drive across the American West for most of a summer. Highlights included homage to the frontier boomtown of Tombstone, although I recall a much greater awe at seeing Cochise's natural stronghold high in the Chiricahuas. Climbing the Chisos Mountains of west Texas to see Colima warbler *Leiothlypis crissalis* was a distant runner-up. Outdoor exposures like these merely reinforced the childhood idealizations that were founded in historical memory.

In college, George Miksch Sutton was added to my roster of esteem. Borrowing freely from his book *At a Bend in a Mexican River*, I began years of travel to that country, usually for birds, but mostly for simple adventure. I was rewarded, often, including a restless night in the Sierra Madre Oriental, listening to a jaguar grunting somewhere outside the thin fabric of my tent. Howler monkeys, toucans, fer-de-lance. Once I roamed for days utterly lost high in the Sierra de Tuxtla of Veracruz, looking for its endemic quail-dove, rescued only by dogged persistence and a spring-fed cattle trough that refurbished my expired water supply. My family often exclaimed that they could not fathom from where I got this restless curiosity.

This eclectic personal history led me ultimately to graduate school, applying to Louisiana State, Tennessee, and Georgia. I was quite fortunate to wind up in Athens. There my wide-ranging interests meshed well with ecosystem theory, although only after a feral detour to study the cryptic white-throated jay *Cyanolyca mirabilis*, found in narco-trafficked mountains of Guerrero, Mexico, proved far too dangerous even for my romantic adventurism. Despite steady encouragement from the University of Florida's John William Hardy and others, I just could not secure enough data with that project to fulfill the quantitative expectations for a graduate degree. After ultimately coming to my pragmatic senses with a doctorate in marine sciences instead (still focused on birds), I carved out a decent living in environmental fields, a career that usually included extensive travel and occasional field assignments.

Among other identities and affiliations held, these days I am both an environmental scientist and a birder. I keep an American Birding Association life list, as well as some state lists, regularly take part in Christmas Bird Counts, chase some rarities, and have been known even to plan a modest big day on the rare occasion. I am proud of a northern Virginia suburban yard bird list that peaked at 126 species, including a Mississippi kite *Ictinia mississippiensis* added during a rare cicada outbreak.

My first vivid recollection of the ivory-billed woodpecker was from illustrated books gifted me at birthdays and Christmas by my maternal grandfather. From these I learned of extinction, about its sobering, utter finality. As a young birder I felt cheated by some mysterious runaway process. It didn't seem fair to no longer have opportunity to go find great auks *Pinguinus impennis*, Labrador ducks *Camptorhynchus labradorius*, passenger pigeons, and Carolina parakeets. I pondered idly if the ivory-billed woodpecker might have passed into its own oblivion early during my lifetime, perhaps a few still hanging on somewhere while I was playing Cherokee along South Chickamauga Creek during the 1960s.

Extinction clashed with my idealized values of unfettered possibility that is said to make for a peculiarly American outlook. For one class assignment, I sought an unorthodox reconciliation for this internal disquiet. Supported with color drawings, I labored to explain away permanent loss of our native birds, including the ivory-bill, as a sort of trade off. My immature thesis proposed extinction was softened because our continent had gained about as many bird species via alien introduction as those that had been lost forever. Numerical substitution by human hands did little to bring comfort, though. Not even ring-necked pheasant *Phasianus colchicus* rivals the elegance of the ivory-bill.

Teenage introversion eroded away the idle imagination. But birds and outdoor adventuring still served as outlet for my friends and I, giving us a shared identity to offset those exclusive cliques found in high school. Our birding identification, status, and distribution skills were closely mentored and diligently tutored by more experienced adults. In this circle of fellowship, we were given as much responsibility as we could handle – Christmas Bird Counts, Breeding Bird Surveys, tending mist nets, and other semi-professional tasks. We learned quickly that rarities were subject to close scrutiny, expert deliberations, and uncertain judgments, with decisions that could go either for or against a particular sighting that we might bring forward for an evaluation.

In 1972, Dianne Tennant, an administrator at an elementary school where I worked in the afternoon, stopped me in the hall. She was a close friend of my mother, so she well knew of my keen interest in birds. At first burdened by that pathological timidity of adolescence, I could not concentrate fully on what she was saying.

"You might really enjoy a visit down to our homestead along the Tombigbee River," she invited cordially.

"Uh huh…," I muttered, without the least bit of conviction.

"We see lots of water birds there, the white egrets are everywhere, and ospreys," she went on.

At this point, it sunk in that there were birds in southern Alabama that would make for nice additions to my life list, so I began to pay closer attention.

"We inherited tracts of bottomland forest from my great-grandfather, and we try to keep it in a natural state," Miss Tennant added helpfully.

The next exchange came as total astonishment, no less due to its matter-of-fact tone and transparent context. Thinking by now only of fish crow *Corvus ossifragus* and anhinga *Anhinga anhinga*, I finally asked the question I was really dying to know:

"Do you have crows there that sound like they have a cold?"

She answered instantly: "Oh sure, we have fish crows."

"What else?" I perked up.

"Well, we see the big woodpeckers. They are our favorites."

A bit deflated, because I wanted so badly to see the anhinga, and thinking only of the obvious, I replied dejectedly: "pileated woodpecker."

And then, her shocking answer drifted back guilelessly: "Yes, oh yes, we see the pileated woodpecker all the time! But we see the ivory-bill too once in a while, in the swamp woods. They seem to be rarer, shy even, so they don't often approach close to the house."

For several minutes we chatted amicably about the ivory-bill, although I did have to process my first cognitive friction on this topic – the woodpecker was supposed to be extinct. Yet here was somebody with no discernable agenda insisting without the least bit of defensiveness that they still lived. My mother was not rare-bird savvy enough to tip off a hopeful meaning of ivory-bill to Miss Tenant, who also seemed oblivious to much ornithological value in her revelations. Knowing her devotion to moral standards in other arenas, it was inconceivable to me that she was fibbing. So, instinctively, I filed away for memory's retrieval this small consolation: "At least I talked with a living person who has seen the ivory-billed woodpecker…"

After diligent reflection, I cannot recall ever having nurtured an unbending conviction one way or the other about ivory-billed woodpecker in the present tense. Perhaps this was because there was never a real need. I certainly possessed a wistful longing for wanting to have lived in a place or during a time during which I *could* have seen the grand woodpecker.[8] Yet this desire was not detached from the same historical nostalgia[9] that encompassed my coveting a chance to have lived like a Native American and hunt bison at a salt lick. It was just one seamless fantasy all fixed on vanished opportunity, a natural heritage that was now lost to each and every one of us.

Those dreams were perhaps augmented by a distant hope that a few ivory-bills had managed to survive miraculously somewhere. During the 1970s and 1980s, the most likely place thought to harbor such a marvel would be Cuba, an island where I had very tangential links through my uncle's wife and her family, refugees from revolution. My Aunt Lillian was an effusive advocate for any adventure: from her I learned to love Spanish and its diverse culture. Thanks to that influence, I would eventually visit Cuba twice in the 1990s searching for another imperiled bird, the black-capped petrel *Pterodroma hasitata*.

A long time passed before the great woodpecker and I crossed paths again. In 2007, The Wildlife Society asked me to formally review the draft recovery plan for the ivory-bill. I had known about the Arkansas searches well before their public announcement. Recovery team leader Jon Andrew was a close personal friend since we had worked together for the U.S. Fish & Wildlife Service in Alaska, that still extant Last Frontier.[10] Jon kept me abreast of various tantalizing rumors, all the while asking me to abide by secrecy. I was hopeful, though not especially vested in the bird's existence. The organization where I worked was intrigued, but not zealous to engage publicly on a putative rediscovery. I found myself most captivated with certain biological assertions that I read in the draft recovery plan, especially those that seemed so blatantly contradictory.

During this same era, I received a surprise windfall about the saga and origins of my father's family. My Aunt Phyllis first had traced us back to Scots-Irish immigrants from Northern Ireland, a group whose stubborn traits are glibly stereotyped by a supposed fierce individualism, persistent egalitarianism, mistrust of authority, and a tendency to take up arms and put them to use.[11] A distant cousin in Michigan, however, furnished the most revealing background.

Around 1727, landing first in Delaware, my great-great-great-great-great grandfather immigrated to frontier Pennsylvania from rural Ulster, north of Londonderry. His 31-year-old son, Charles Haney, fought in the colonial army between September 1776 and January 1777 with General Washington at the Battle of Fort Lee. After a furlough, Charles resumed service with the colonials in Virginia under Colonels Rawleigh P. Downman and John Lucas. Eventually mustering on to Hillsborough, North Carolina, he and the rest of the Virginia militia joined fighting led by General Horatio Gates at Camden, South Carolina, where they were routed spectacularly on August 16, 1780 by British General Cornwallis. Charles escaped capture, finally settling near Marion, North Carolina in McDowell County. He applied for and received in 1832 a $31.00 annual pension for his war services,[12] passing away in 1836 at the respectable age of 91.

One of his sons, great-great-great grandfather Samuel, moved west to settle along the Nolichucky River on the far side of the Unaka and Bald Mountains in frontier Greene and Washington counties, Tennessee. His son Samuel Calvin later served from 1861 to 1863 in the Civil War with his brother James in Company E of the 4th Regiment Tennessee Volunteer Infantry, Union Army. After their service in Kentucky was over, and they resumed civilian life back with their families, Confederate sympathizers murdered both brothers in October 1864 at their homesteads. Samuel Calvin's grandson, my grandfather James Franklin Haney, would marry Carmen Fortner, one-quarter Cherokee by her maternal grandfather. All of which goes to say that my frontier appetites as a juvenile were perhaps not quite so entirely baffling as my parents had supposed.[13]

Whatever do personal revelations like these have to with a celebrated woodpecker or its presumed extinction? Each of us, including this writer, is the creation of idiosyncratic experience.[14] None of us escape a pull of our genes or the push of our environments when assimilating the perceptions that we come to hold. Our memories are selectively processed, too, varying on how negative or positive we perceived them inside the diverse social circumstances in which they were experienced.[15] Even our core values point toward a need to find cognitive resolution for the uncertainties that we bump into during everyday life.[16] We never react nor judge as strictly rational beings wholly capable of delineating some irreducible, objective reality. To believe we are immune from such ambiguity is to fall victim to the thinking delusion of ***naïve realism***.[17]

Box 2.1. ***Naïve realism*** is a strictly pragmatic view of the world as being exactly how *we* sense it; subjective experience does or should not influence that fundamental reality. ***Naïve realism*** holds that only the objective ivory-bill is real. Whether or not we were birders or instead media reporters would (or ought) not have any influence on the impartial depiction of this species.

Bias blind spot is a widespread breakdown in the human condition that leads us to think we are less likely to be prejudiced in our beliefs or decisions than are other people. This meta-bias often leads us to conclude that whenever conflicts arise across social identities or from individual opinions, only other people, and not us, are under a delusion of some cognitive defect or philosophical fallacy that has unduly influenced a mental error.

My own daughter can be remarkably shrewd at nailing some of my ample contradictions. In late autumn, when I return from pheasant hunting somewhere out in the prairie states, she challenges me: "Dad, how can you be a scientist, count birds, work for a conservation organization, and then shoot them?" Up until now I have resisted any facile defense using that expedient passage from *Song of Myself*.[18]

Failing to see the biases present in one's own judgment can manifest incoherently as a result of this ***naïve realism***. One particular impediment leads us to develop the condition of ***bias blind spot***.[19] Surveys show that we will consistently rate ourselves as less prone to various biases than the "average American." We seem to view our peer's limitations quite clearly, but then we consider our own awareness to be utterly free of such partiality. Not only are we unable to loosen ourselves from the grip of this meta-bias through our introspective efforts alone, the higher is our cognitive intelligence, the larger is our ***bias blind spot***.[20]

Consequently, nothing is ever exactly quite what we think it is. Our cognition, our memory, and especially our social relationships inform all that we know about the ivory-billed woodpecker. ***Naïve realism*** holds that only an ivory-bill perceived directly with our physical senses is one to be taken seriously. If to measure is to know, only a corporeal bird really matters. But I beg to differ, and instead urge us to confront head on the starkly opposing *perceptions* about the ivory-bill that were cultivated over and over again, especially throughout the twentieth century.

As fallible human beings, we resort to metaphor and allegory to construct yarns that help us grasp the "powers that set in motion the forces of history and rule the world of nature."[21] This tendency for story-telling is as true for our interest in the lives of animals as it is for understanding the course of human affairs. To be sure, such myth can help us organize the many conundrums that inevitably pierce our awareness. Enough contradictions will arise eventually, however, so that our "cognitive dissonance or 'discontent'"[22] will force us into confronting the very tiredness entrenched in these makeshift representations. "Myths are not history. Myths are what we wish history had been."[23]

Over time we came to embellish the ivory-billed woodpecker with so much inflated folklore that the mythology framing our depictions of this striking bird became extravagant. Too much illusion clouds the efficacy of conservation practice when our myth and its associated rhetoric become this extreme.[24] Today, the vast mythology about the ivory-bill no longer serves an original, useful purpose in offering us a cohesive narrative about this bird. Too many paradoxes exist within this species' conventional storylines for us to just keep ignoring all of the many puzzles.

Why did the ivory-bill become for some an object of near-sacred quest, a sort of avian Holy Grail? How did the living bird for so long attain pariah status as an untouchable third rail of professional ornithological science? Why are rumors of ivory-bill persistence so prevalent, unlike other near or recently extinct species? What does this species activate that encourages such bitter contention and polarized division among us? From where come the irrational impulses whenever this species becomes the topic? Why are we so bad-mannered with each other about a bird? Most vitally, what is it about this particular species that makes achieving any social consensus[25] about the bird's ultimate fate near to impossible?

Although we may need to abandon the tidy comfort of our little utopias during a voyage for answers, we will discover how it came to be that this object of blood and feathers ultimately manifested so contrarily among us. After all, "highly charged emotions and sometimes irrational behavior have figured into the search for the ivory-billed woodpecker."[26] If we render proper respect to the thinking errors that arise from our inborn ***naïve realism***, a tendency that is unavoidable because we are humans and not androids, we might become more receptive to alternative narratives[27] for this riveting species. Perhaps we can even illuminate with greater clarity the woodpecker's original habits and its ecological fate. Most of all, we will learn much about ourselves.

Endnotes

1 Unless indicated otherwise, and purely for convenience, America throughout this work refers to the United States, with full awareness that our neighbors to the north and south are entitled to the same usage.

2 Travels through South Carolina and Georgia by English naturalist Mark Catesby, 1722–1726, are generally credited as the first European description of ivory-billed woodpecker. Catesby's painting of the species appeared in *The Natural History of Carolina, Florida and the Bahama Islands,* published in 1731.

3 My play-acting embodied unintentional blasphemy in the devout community populated by our nuclear family. The frontier hunter rejected utterly the Eden myth about man's fall from grace and his exile from the original earth garden. Instead of viewing hunting as a lustful, worldly expression of moral failings, as rendered in the Christian faith, frontier Native Americans and Europeans alike viewed pursuit of the wild beast as a quest for deeper union with the land, and even a preferred initiation rite into a higher state of being. "It is striking that the American myth of the hunter so closely resembles the creation myths of the Indians." Slotkin, R. 1973. *Regeneration Through Violence: The Mythology of the American Frontier*. University of Oklahoma Press, Norman, p. 307.

4 One of my favorite stories was Bendire's obsession to secure an egg of a zone-tailed hawk *Buteo albonotatus* while on cavalry patrol with the U.S. Army in central Arizona in 1872. After sending his troops ahead to set up camp, he climbed a promising nest tree alone. After

an Apache scout fired a round over his head, he shoved the egg quickly into his mouth (where it had to be retrieved later through such force that it broke a tooth), shimmied hurriedly down the tree, and rode hard back to camp with the Apaches behind him in full pursuit. See: http://www.arlingtoncemetery.net/charles-bendire.htm (accessed 20 May 2015).

5 My grasp over the astonishing treks of these frontier naturalists still remained quite rudimentary. Growing up in the East, I was entirely unaware of the long, rich history of naturalist-explorers who worked for centuries in California. Beidleman, R.G. 2006. *California's Frontier Naturalists*. University of California Press, Berkeley.

6 Conley, R.J. 2007. *A Cherokee Encyclopedia*. University of New Mexico Press, Albuquerque, pp. 87–88.

7 Armstrong, Z. 1931. *The History of Hamilton County and Chattanooga, Tennessee*, Vol. 1. Lookout Publishing Co., Chattanooga, pp. 4–8.

8 Instead, I just succumbed to resignation, e.g., "the ivory-billed woodpecker and many other species that were alive when I was born are now ghosts." Barlow, C.C. 1997. *Green Space, Green Time: The Way of Science*. Springer-Verlag, New York, p. 90.

9 Ladino, J. 2004. Longing for wonderland: nostalgia for nature in post-frontier America. *Iowa Journal of Cultural Studies* 5: 88–109.

10 Kollin, S. 2001. *Nature's State: Imagining Alaska as the Last Frontier*. University of North Carolina Press Books, Chapel Hill.

11 Griffin, P. 2001. *The People with No Name: Ireland's Ulster Scots, America's Scots Irish and the Creation of a British Atlantic World, 1689–1764*. Princeton, New Jersey; Montgomery, M. 2004. Scotch-Irish or Scots-Irish: what's in a name? *Tennessee Ancestors* 20: 143–150; Vann, B.A. 2008. *In Search of Ulster Scots Land: The Birth and Geotheological Imagings of a Transatlantic People*. University of South Carolina Press, Columbia.

12 Pension application of Charles Haney, S8661, October 22, 1832, Burke County, North Carolina.

13 My role-playing as an Indian and the later nomadic restlessness

to explore geographies of open space can be traced to a prevalent cultural reaction to the sweeping changes brought on by a post-frontier America, i.e., the "widespread ambivalence about modernity." Huhndorf, S.M. 2001. *Going Native: Indians in the American Cultural Imagination*. Cornell University Press, Ithaca, NY, p. 14.

14 As admonished by Edward R. Murrow: "Everyone is a prisoner of his own experiences. No one can eliminate prejudices – just recognize them."

15 Skowronski, J.J., and D.E. Carlston. 1987. Social judgment and social memory: the role of cue diagnosticity in negativity, positivity, and extremity biases. *Journal of Personality and Social Psychology* 52: 689–699.

16 Persons who place greater value on security, conformity, and tradition tend to have the strongest need to escape ambiguity and thus strive urgently to reach cognitive closure. See: Calogero, R.M., A. Bardi, and R.M. Sutton. 2009. A need basis for values: associations between the need for cognitive closure and value priorities. *Personality and Individual Differences* 46: 154–159.

17 Ross, L., and A. Ward. 1996. Naïve realism: implications for social conflict and misunderstanding. Pp. 103–135 *in* T. Brown, E. Reed, and E. Turiel, eds. *Values and Knowledge*. Lawrence Erlbaum Associates, Hillsdale, NJ. Or, and more poetically, "without philosophy to save us from naïve realism, science will be stuck defending the indefensible." Chopra, D. 2014. Naïve realism, or the strange case of physics and fake philosophers (Part 2). http://www.huffingtonpost.com/deepak-chopra/naieve-realism-or-the-str_1_b_5687816.html (accessed 4 May 2016).

18 "Do I contradict myself? Very well, then I contradict myself, I am large, I contain multitudes." – Walt Whitman.

19 Pronin, E., D. Lin, and L. Ross. 2002. The bias blind spot: perceptions of bias in self versus others. *Personality and Social Psychology Bulletin* 28: 369–381.

20 West, R.F., and K.E. Stanovich. 2012. Cognitive sophistication does

not attenuate the bias blind spot. *Journal of Personality and Psychology* 103: 506–519.

21 Slotkin, 1973, p. 269.

22 Slotkin, R. 1998. *Gunfighter Nation: The Myth of the Frontier in Twentieth-century America*. University of Oklahoma Press, Norman, p. 6.

23 Spurgeon, S.L. 2005. *Exploding the Western: Myths of Empire on the Postmodern Frontier*. Texas A&M University Press, College Station, p. 3.

24 Adams, J.S., and T.O. McShane. 1992. *The Myth of Wild Africa: Conservation without Illusion*. University of California Press, Berkeley; Peterson, M.N., J.L. Birckhead, K. Leong, M.J. Peterson, and T.R. Peterson. 2010. Rearticulating the myth of human-wildlife conflict. *Conservation Letters* 3: 74–82.

25 Lack of consensus is hardly a prerequisite to successful implementation of conservation. For example, see: Peterson, M.N., M.J. Peterson, and T.R. Peterson. 2005. Conservation and the myth of consensus. *Conservation Biology* 19: 762–767. However, in very few varieties of wildlife other than ivory-billed woodpecker are the differences in social opinion so bitterly divided that even the species' very existence is openly (and so rigidly) contested.

26 Steinberg, M.K. 2008. *Stalking the Ghost Bird: The Elusive Ivory-billed Woodpecker in Louisiana*. Louisiana State University Press, Baton Rouge, p. 19. "Informants for this book, for example, have described physical threats, lawsuits, and accusations of fraud against them after they have claimed to have seen the bird." *Ibid*.

27 Nasie, M., D. Bar-Tal, R. Pliskin, E. Nahhas, and E. Halperin. 2014. Overcoming the barrier of narrative adherence in conflicts through awareness of the psychological bias of naïve realism. *Personality and Social Psychology Bulletin* 40: 1,543–1,556.

Chapter 3

Frontier Lament

Americans ... are forever searching for love in forms it never takes, in places it can never be. It must have something to do with the vanished frontier.

—Kurt Vonnegut, *Cat's Cradle*

Vanishing birds were hardly the foremost insecurity garnering domestic U.S. attention as the nineteenth century drew to an end. Nevertheless, the ivory-billed woodpecker ran into trouble at a momentous turning point in American history. The bird did not disappear in a cultural vacuum. It vanished from much of its range during a rambunctious epoch, one marked by extraordinary changes in every sector of American society. Indeed, our narratives for charting the large woodpecker's demise cannot be extricated from how the nation came to terms with the literal and symbolic closing of the frontier,[1] a cultural juncture that spawned considerable national anxiety,[2] and not a little grieving as well.

Had there never been a physical frontier as (mostly) European immigrants lived it out in all of its extractive intemperance, some of our lost birds might still be here. Passenger pigeon, heath hen *Tympanuchus cupido cupido*, and Carolina parakeet might have survived for a bit longer. To be sure, indigenous peoples arriving tens of thousands of years ago also transformed their environment.[3] Advent of these immigrants had its own severe, adverse impact on native plants, animals, and ecosystems.[4] That influence is especially germane for loss of North America's large mega-fauna. South of Alaska, 34 Pleistocene genera of mammals[5] did not survive into the Holocene, and two entire orders of mammals were eliminated totally from this continent.[6]

Any human culture is capable of killing to extinction, especially upon arrival at a new ecological frontier that contains numerous large animals naïve to our deadly ways.[7]

Threats from European-driven settlement were identified consistently as the culprits behind the ivory-billed woodpecker's decline. Two theories gained traction, but one has been far more wide reaching, indeed near immovable in its supremacy. The un-replicated study by James T. Tanner from 1937 to 1939 in the Singer Tract of northeast Louisiana took a resolute view that the ivory-billed woodpecker succumbed from near-complete logging of old-growth bottomland hardwood forest across the southeastern U.S. According to this hypothesis, lumbering practices that removed primeval woods all but eliminated the bird's specialized food supply, large beetle larvae extracted from just below the bark of huge, newly dead trees.[8]

More recently Noel Snyder and colleagues stressed that the woodpecker's demise ensued from direct killing by humans, including exploitation for purposes of scientific collection, commercial taxidermy, subsistence hunting, and just idle curiosity.[9] This hypothesis better matches some (but still not all) aspects of the timing for the bird's decline. Under this thesis, we can add supplements. Little attention has been granted to how the grinding, life-threatening poverty that plagued southern residents after the Civil War and during the Great Depression may have contributed to very heavy, widespread subsistence hunting throughout much of the bird's range.[10] Real dangers of human starvation could have placed intense pressure on wild game between 1865 and 1890, and again during the early 1930s. These eras are when prospects for the ivory-bill apparently turned for the worst, but also when such matters received little study from science of their day.

Not everyone who exploited America's now-lost frontier and its once super-abundant resources wished to subjugate nature entirely, never mind to misuse species to the point of extinction.

Historians note that fur trappers, missionaries, herdsmen, and others depended on preserving natural areas for sustaining their fundamental business pursuits. But when the Subduers at last replaced the Users, demands for an ever-growing economy usually doomed some part of the natural world.[11] Regardless of motives and intentions behind our predecessors' actions, my exploration takes no real issue with an anthropogenic blame for the ivory-bill's decline. We are likely at fault.

Looking back now, what strikes us as so compelling about the ivory-bill's fate is the remarkably tight correspondence between the timing of its vanishing and so many momentous cultural, social, and historical changes that occurred in near perfect synchrony. To contextualize these parallels, we first take note that numbers of museum-archived specimens of the woodpecker topped out between the years 1890 and 1900.[12] Sadly, this is almost the only hard data that we have for monitoring any historical trends in this bird.

Despite a cultural voracity for ivory-bill skins and taxidermy mounts in that time, the ornithological thinking of the day still usually fell short of predicting an imminent extinction for this bird.[13] In 1891, Edwin M. Hasbrouck surmised with what today sounds like astonishing optimism:

> *[W]hile by no means as abundant as* Conurus *[Carolina parakeet], [ivory-billed woodpecker] is still found in greater or less numbers in many parts of the southern United States, the Mississippi Valley, and in Texas.... There are thousands of square miles of swamp throughout the Mississippi Valley and Gulf States that never will or can be reclaimed or sett'led, country that is admirably suited to this bird, and in which, as I have shown, it is much more common today than elsewhere; and here, it is safe to say, it will be found indefinitely; for, into those swampy fastnesses in which it most delights, few care to penetrate, at certain seasons none dare....*[14]

A map prepared by Hasbrouck and depicting a then-construed range for the woodpecker still showed extensive areas of occupation along the Gulf and south Atlantic coasts. The main region of apparent disappearance, then, was the far interior Mississippi River system and its northern tributaries, regions already well-settled by then. Hasbrouck used a dividing line of 1880 to distinguish the bird's former from its then-present distribution. Nowhere in his review did Hasbrouck speculate that either logging or direct take was then posing an irreversible threat to the bird's continued survival.

Even ten years later, the ivory-bill was considered by some writers to be common in places. There is anecdotal suggestion, too, that certain populations within portions of the bird's range still could be relatively large. In a letter sent by Arthur T. Wayne to Frank M. Chapman on August 23, 1905, Wayne declared:

> *The Ivory-bill is* still *an abundant bird in certain localities in N.W. Florida as I well know ... in different portions of N.W. Florida I have seen upwards of 200 of these magnificent birds.*[15]

Once again, there is no mention in either of Wayne's 1905 letters to Chapman about threats to the woodpecker. Indeed, Wayne remarked "it would require years to explore (even a part of) the great inland swamps in Levy, Jefferson, Lafayette and Taylor Counties, Florida." Moreover, Wayne directly challenged claims that ivory-bills had been exterminated locally on the Wacissa River by his shooting competitors. And as late as 1914, ivory-bills in Oklahoma were "not considered by the local hunters as any great rarity."[16] In the Fort Drum region of south-central Florida as late as the 1920s, Minor McGlaughlin recalled seeing "many ivory-bills" on a daily basis, with the species "even more abundant in this location than the pileated...."[17]

Regardless of its actual abundance, the number of collected specimens accelerated in the 1880s, but then decelerated very rapidly, so that by about 1915 the mania (or the ability) to find, collect, and catalogue remnants of this species seemed to have substantially dissipated. This pattern is often assumed to represent a rush by avid collectors to get the last of their kind from an ever-dwindling supply,[18] followed by a rapid drop in ability to find birds later due to the species' ever increasing rarity.

That hypothesis presumes that the acceleration in collection was *motivated* by sufficient awareness that the bird was rapidly getting so scarce that waiting longer for a chance at future acquisition might jeopardize its availability. A deceleration in collections was assumed to arise from the birds having been depleted so rapidly from over-collection during earlier decades. This interpretation is hardly the only one, however. Progressive social movements taking root in the American experience (Chapter 4) also explain the very abrupt trend observed in the historical record, especially a steep decline after 1900.

Specimen collection of the ivory-bill declined steadily after 1895, an era during which the nation's commercial taxidermy markets were devastated by the Financial Panic of 1893. The United States was hit by its worst-ever economic crisis, one caused mostly by overbuilt and speculatively financed railroads, although other businesses were also over-extended. After popular runs on banks, suspension of cash payments, and multiple banking failures, the nation's first harsh economic depression spread even wider to commerce, industry, and manufacturing. This depression lasted until 1897, accompanied by a national unemployment rate that reached a staggering 25%.

Bird protection efforts strengthened notably after the 1893 panic.[19] Drops in market demand and changing social mores ruined many of the commercial taxidermy businesses that once thrived to feed our strange Victorian taste in interior decorating. Fewer buyers were now interested in acquiring mounted birds. In the words of one disgruntled taxidermist commenting on the prospects for his once-thriving livelihood: "It is simply dead.... Those ... Audubon Societies and bird books and new-fangled laws are just crowding us out. The men are afraid to shoot or handle them in any shape."[20]

Peak collection of ivory-billed woodpecker also brackets with near exactitude the legendary speech by Frederick Jackson Turner in 1893 announcing the ending to the once vast American frontier. Delivered at the Columbian Exposition World's Fair in Chicago, Turner's conclusion about the sunset of this iconic American experience could hardly have been any more succinct: "The West is now closed," he proclaimed solemnly. And just what was the trigger for his sweeping conclusion?

Three years prior to Turner's speech, the Superintendent of the United States Census for 1890 had summarized thusly the nation's latest count and distribution of its widely scattered citizens:

> *Up to and including 1880 the country had a frontier of settlement, but at present the unsettled area has been so broken into by isolated bodies of settlement that there can hardly be said to be a frontier line. In the discussion of its extent, its westward movement, etc., it cannot, therefore, any longer have a place in the census reports.*[21]

Compiling data with a modicum of scientific procedure, a government bureaucrat had detected a key transition in the life history strategy of a continent-wide predator – us. Yet without analogous population information for the woodpecker, one consequence was that huge amounts of uncertainty were generated for this bird at the dawning of the twentieth century. That uncertainty was then *incessantly propagated,* masking the species' genuine ecological needs, its fundamental adaptations, and its actual vulnerabilities. Large uncertainties usually demand to be filled with something, anything. Too often, these voids lead us to adopt blinding speculation and then fall to its sidekick, ***cognitive bias***.[22]

> **Box 3.1.** ***Cognitive bias*** refers to systematic departures in our thinking, judgment, or decision-making that are due to inherent limitations in how the human mind can process (often complex) information. These departures are flawed in fundamental ways, e.g., they fail to conform to standards of rationality or accuracy. They arise because we typically have only limited time, knowledge, or mental capacity available to make decisions, so we invariably fall to using faulty shortcuts. Such mental shortcuts are called ***heuristics***.

Turner's speech denoted more than the permanent passing of a geographic frontier demarked somewhere on a map. His elegy also centered on the symbolic roles of this physical space in shaping what was then widely believed to form a distinctly American identity.[23] Both Turner and later historians would represent that identity (including our supposed exceptionalism)[24] as something intrinsic to the notion of pristine wilderness, those wide open landscapes having one or more of the following cultural functions: 1) a space of unlimited, unexploited resources; 2) a societal safety-valve, and; 3) sites of closeness and conflict.[25] It matters less whether these beliefs and corollaries like Manifest Destiny were universal, because many Americans in certain eras thought them true. Our perceptions of (and reactions to) the ivory-billed woodpecker could not help but fall under a spell of magical thinking about frontier qualities.

Turner's somber declaration implied that we no longer possessed boundless stocks of natural resources to exploit. Economic potential in frontier spaces had long motivated

European exploration away from the Atlantic seaboard. New spaces served both to supply population centers with raw materials (passenger pigeons) while acting as markets for urban merchandise (guns). "City and country formed a single commercial system, a single process of rural settlement and metropolitan economic growth."[26] Watching land fill up, only to be followed by a roster of vanishing wildlife, was unsettling to anyone who might be paying attention. How could such abrupt change happen so fast? And how would the country's social institutions function in this new era of diminished potential?

Indeed, the country's social institutions were not up to the task. Strains were plainly evident. Surely one of the most chaotic episodes of civic and spatial planning in U.S. history occurred on April 22, 1889, when along the Arkansas and Texas borders some 50,000 homesteaders raced virtually unchecked to stake out claims inside the Oklahoma territories, lands once again appropriated from Native Americans.[27] Further west, land conflicts escalated between smaller-scale ranchers and the Wyoming Stock Growers Association, tensions ultimately leading to seething resentment, partisan division, and murder in the 1892 Johnson County War.[28] (Today the same environmental tensions simmer with antipathy over federally subsidized grazing rights allowed to private interests on western public lands.)[29]

A national sense was pervasive that resources were running out. Best to get whatever one could while the getting was still possible. In a Gilded Age marked by greed, not even a thin layer of gold frill from political corruption could mask the social and economic unfairness satirized so memorably by Twain.[30] During an era known for its lusty land grabs, range wars, and other resource stampedes, individuals so inclined might surely seek to acquire any other "last of their kind" items, including such gaudy renderings of a vanishing natural world as the splendid ivory-billed woodpecker. Indeed, for a while hundreds of commercial specimen dealers fed a curious natural history craze for these private taxidermy mounts.[31]

Additionally, the nation was being starkly confronted with loss of its frontier as a societal safety valve.[32] A purifying frontier had been envisioned to serve as an agricultural garden to domesticate the masses ever since the nation's inception. In a 1787 letter to James Madison, agrarian Thomas Jefferson sagely contended that:

> *Our people will remain virtuous so long as agriculture is our principal object, which will be the case while there remain vacant lands in America. When we get piled on one another in large cities, as in Europe, we shall go to eating each other as they do there.*[33]

As if to validate Jefferson's prediction, and as the physical frontier collapsed upon itself, certain forms of legendary lawlessness reached a zenith in the American West. Many Old West gangs achieved a level of notoriety for their daring exploits in the late nineteenth century.[34] Several of these operated between 1880 and the early 1900s, including the infamous Wild Bunch, Dalton, James Younger, and Hole-in-the-Wall gangs.

Social disorder of this magnitude came surprisingly late in the century if one supposed that a denser settlement would lead to a more orderly civic life and law-abiding citizenry. Nearly all of the famous gangs were active during periods in which ivory-billed woodpeckers were so sought after. From this societal impotence over the rule of law, one can surmise that even had suitable conventions existed at the time, any regulations aimed at protecting a diminishing bird like the ivory-bill would be ineffective during the unruly decades of the late nineteenth century. Until then there had been little need to place any checks on our unlimited takes of wild bounty from the land.

Just how much loss of the frontier meant to an average citizen at the time can be better appreciated by granting sweeping impacts of frontier themes on our future cultural expression. A principal manifestation of this loss would be a deep reverence for and strong romanticizing of frontier and pioneer themes. This myth-making[35] became fundamental to our ongoing bond with the frontier.[36] Even as the frontier was still in the process of disappearing, Buffalo Bill Cody's showmanship[37] already portended our inventive adoration of that lost era. Myth-making through performance art bridged our reluctant transition out of a frontier America.[38] Once frontier realities had disappeared in fact, Western-themed genres would go on to color all corners of American life, including its politics, art, music, literature and, perhaps most of all, cinema.[39]

Cowboys and other frontier characters,[40] both quiet (*Shane*) and not (*True Grit*), outlaws (*Butch Cassidy and the Sundance Kid*), range wars (*Heaven's Gate*), and gunslingers (*The Magnificent Seven*) entertained us at movie theaters through the end of the next century. Our native mythologies were so compelling that they were borrowed thematically (and sometimes literally) by art crafted in other lands (e.g., Australia's *Ned Kelly*, Italian Sergio Leone's *The Good, the Bad and the Ugly*). So it was that "the American Western supplies a good test of cinema's capacity to sometimes mirror, contest, or even shape, our perception of late-nineteenth century history of the American nation as one of coming to terms with a rapidly closing frontier."[41]

Elsewhere in art, Aaron Copland's (1900–1990) compositions drew inspiration from the American West in ballet works *Billy the Kid* and *Rodeo*. Pearl Zane Grey (1872–1939) mythologized the Old West by integrating characters with landscapes, then making the land essential to mythical storylines.[42] Louis L'Amour (1908–1988) extended similar historical fiction deeper into the new century through more than 100 novels and short stories. John Steinbeck's work also showcased fascination with our pioneer past.[43] In his 1960 *Travels with Charley*, a farmer laments to the wayfaring Steinbeck: "This used to be a nation of giants. Where have they gone?" Could not the same be said candidly for our gigantic woodpecker? Is the enigmatic vanishing of ivory-bill any less gripping than the alleged death, rumored survival, and disputed fate of Butch Cassidy?[44]

Later historians would contest[45] Frederick Jackson Turner's scholarly assumptions and his overly broad (and minority exclusive) generalities about the American character. Even that reliable staple, the western film genre, eventually morphed from soaring, operatic dirge for the frontier's last days (*Once Upon a Time in the West*) to revisionist anti-western that portrayed morally-vacuous protagonists (*Unforgiven*), the latter surely anathema to the America so idealized by Turner. Nevertheless, much of the world today recognizes cowboy presidents and cowboy diplomacy, not to mention cowboy boots. And almost every U.S. citizen understands that "cowboy-up" is an action verb.

Ending of the frontier dealt this country enormous yearning to adapt quickly to a brand new and still ill-defined era. Powerful anxiety evoked by so much cultural change meant that the nation needed prompt remedies to unforeseen dilemmas. One historian noted that Turner's 1893 speech on the frontier's passing did not receive an enthusiastic reception from his audience.[46] Another historian surmised listeners found it outright depressing.[47] Frontier anxiety appeared in embryonic form in the 1870s and crystalized national temperament by the 1880s. Virtually no preparation had been made anywhere by anyone for the frontier's ultimate end. Widespread social uncertainty of this magnitude routinely demands its closure.[48]

As national attention turned away from a physical frontier, the United States looked elsewhere in both inward and outward directions for surrogates toward the end of the 1890s. The country fought an imperial war with Spain for more real estate.[49] Internally, the country faced daunting trials, many of which were blamed on a now-subjugated continent. A closed frontier challenged us to rethink public land policy, immigration, labor and capital, international relationships, government reach, racism, industrialization, irrigation, tenant farming, class struggle, and the philosophy for shoring up a nascent environmental movement. Our national turning point meant we needed to find ways to better manage and conserve what nature still remained to us.

Closing the American frontier did not invent the mythology that would in due time shroud the ivory-billed woodpecker – rather, the frontier's end would *fuel* much of the ensuing myth-making that occurred. When feelings of nostalgia, loss, foreboding, and mourning for a bygone era provoke deep cultural questioning, our individual and collective responses need not be complete or accurate, merely useful and unifying. Even though it was hardly uppermost in priority, America would eventually need sturdy answers for our precipitous wildlife extinctions. For the ivory-bill, those answers would fasten deeply on a distorted notion of vanished wilderness.

We are not a patient people; but we are pragmatic. What better way to oblige our fixed beliefs and reconcile our notions of pioneer exceptionalism than to compose whatever legends

Box 3.2. Our minds are incredibly resistant to open-ended uncertainty. Resolving this hesitation motivates much of human behavior. Our aversion to ambiguity and the desire for firm answers, then, drive us to seek ***cognitive closure***. This need for closure varies among individuals. When closure happens it generally consists of "seizing" and "freezing" on an answer that seems plausible, or one that is simply at hand. Our rigidity with these answers then goes on to predispose us to a wide array of cognitive biases and decision errors.

might be required to alleviate any doubts still harbored in the national psyche. Whenever the need might arise, why not fall back on old-fashioned folklore, and expound with no more introspection than necessary a good rationalization for the strange disappearance of the mighty ivory-bill?

Endnotes

1 Turner, F.J. 1935. *The Frontier in American History*. Henry Holt & Sons, New York.

2 Wrobel, D.M. 1993. *The End of American Exceptionalism: Frontier Anxiety from the Old West to the New Deal*. University Press of Kansas, Lawrence.

3 Depictions of native peoples in North America as non-polluting, conservationist, and environmentalist are problematic for a variety of reasons. A much fuller exposition of why native peoples' relationships with the continent's flora and fauna were often not so benign is given by Krech, S. 1999. *The Ecological Indian: Myth and History*. W.W. Norton & Company, New York.

4 Gill, J.L., J.W. Williams, S.T. Jackson, K.B. Lininger, and G.S. Robinson. 2009. Pleistocene megafaunal collapse, novel plant communities, and enhanced fire regimes in North America. *Science* 326: 1,100–1,103.

5 Most extinctions were of mammals ≥44 kg (about 100 pounds) such as saber-toothed cats, mammoths, and mastodons, as well as the North American horses, camels, and tapirs. Smaller mammals like short-faced skunk and the giant beaver also went extinct, however.

6 Koch, P.L., and A.D. Barnosky. 2006. Late Quaternary extinctions: state of the debate. *Annual Review of Ecology, Evolution, and Systematics* 37: 215–250.

7 Flannery, T. 2002. *The Eternal Frontier: An Ecological History of North America and its Peoples*. Grove Press, New York.

8 Tanner, J.T. 1942. *The Ivory-billed Woodpecker*. Dover Publications, Inc., Mineola, NY.

9 Snyder, N.F. 2007. *An Alternative Hypothesis for the Cause of the Ivory-billed Woodpecker's Decline*. Western Foundation of Vertebrate Zoology; Snyder, N., D.E. Brown, and K.B. Clark. 2009. *The Travails of Two Woodpeckers*. University of New Mexico Press, Albuquerque.

10 In his review of Snyder's 2007 monograph, Hill also points to a likely reliance of rural residents in the Deep South on taking any wild game when pressed by hard times. Hill, G.E. 2008. Book review: An alternative hypothesis for the cause of the ivory-billed woodpecker's decline. *The Condor* 110: 808–810.

11 Billington, R.A., and M. Ridge. 2001. *Westward Expansion: A History of the American Frontier*. University of New Mexico Press, Albuquerque, pp. 7–8.

12 Jackson, J. A. 2006. *In Search of the Ivory-billed Woodpecker*, 2nd ed. HarperCollins, New York, Figure 3.4, p. 75. Slightly more than 400 ivory-bill specimens exist, Jackson, p. 74.

13 Among the earlier and more succinct predictions, however, was one given by Thompson in 1885: "the species will probably be extinct within a few years." Thompson, M. 1885. A red-headed family. *The Elzevir Library* 4: 21.

14 Hasbrouck, E.M. 1891. The present status of the ivory-billed woodpecker (*Campephilus principalis*). *The Auk* 8: 174–186.

15 Letter of Arthur T. Wayne to Frank M. Chapman sent from Mount Pleasant, South Carolina, August 23, 1905, as given in Snyder et al., p. 135.

16 Cooke, W.W. 1914. Some winter birds of Oklahoma. *The Auk* 31: 473–493.

17 Snyder et al., p. 20.

18 If less abundant species suffer disproportionately from exploitation as their rarity makes them more valuable, this dangerous spiral is known as the anthropogenic Allee effect. The tendency of humans to push species into density ranges where the Allee effect can become problematic is a well-documented setback in conservation biology. See, for example: Gault, A., Y Meinard, and F. Courchamp. 2008. Consumers' taste for rarity drives sturgeons

to extinction. *Conservation Letters* 1: 199–207; Angulo, E., and F. Courchamp. 2009. Rare species are valued big time. *PLoS ONE* 4: e5215. Potential ramifications of this anthropogenic Allee effect on ivory bill are explored in greater detail in Appendix A.

19 Barrow, M.V., Jr. 1998. *A Passion for Birds: American Ornithology after Audubon*. Princeton University Press, Princeton, p. 139.

20 Witmer Stone furnished this quote in an 1898 bird protection committee report for the American Ornithologists' Union. Stone used it to emphasize the substantial decrease in birds then being brought in to taxidermy shops. Barrow, pp. 136–137.

21 Porter, R., H. Gannett, and W. Hunt. 1895. *Report on Population of the United States at the Eleventh Census: 1890, Part 1*. Department of the Interior, Census Bureau.

22 Tversky, A., and D. Kahneman. 1974. Judgment under uncertainty: heuristics and biases. *Science* 185: 1,124–1,131.

23 "No one force did more to Americanize the nation's people and institutions than the repeated reconstruction of society on the western edge of settlement during the three centuries required to occupy the continent." Billington and Ridge, p. 1.

24 American exceptionalism may manifest in any of five domains: distinctive rights, different labels, a "flying buttress" mentality, double standards, and global leadership and activism. Koh, H.H. 2003. On American Exceptionalism. *Stanford Law Review* 55: 1,479–1,527.

25 Slotkin, R. 1973. *Regeneration Through Violence: The Mythology of the American Frontier, 1600–1860*. University of Oklahoma Press, Norman; Gouge, C. 2007. The American Frontier: history, rhetoric, concept. *Americana: The Journal of American Popular Culture (1900 to present)* 6: 16.

26 Cronon, W. 1992. *Nature's Metropolis: Chicago and the Great West*. W.W. Norton and Co., New York.

27 Russell, S. 2012. *Dreams to Dust: A Tale of the Oklahoma Land Rush*. University of Oklahoma Press, Norman.

28 Davis, J.W. 2012. *Wyoming Range War: The Infamous Invasion of*

Johnson County. University of Oklahoma Press, Norman.

29 Lancaster, J. 1991. Public land, private profit – a modern version of an old-fashioned range war. *Journal of Forestry* 89: 20–22. See also: Sheridan, T.E. 2007. Embattled ranchers, endangered species, and urban sprawl: the political ecology of the new American West. *Annual Review of Anthropology* 36: 121–138.

30 Twain, M., and C.D. Warner. 1973. *The Gilded Age: A Tale of Today*. American Publishing Company.

31 Barrow, M.V. 2000. The specimen dealer: entrepreneurial natural history in America's Gilded Age. *Journal of the History of Biology* 33: 493–534.

32 Von Nardroff, E. 1962. The American frontier as a safety valve: the life, death, reincarnation, and justification of a theory. *Agricultural History* 36: 123–142.

33 Dunlap, J. R. 1903. *Jeffersonian Democracy: Which Means the Democracy of Thomas Jefferson, Andrew Jackson and Abraham Lincoln*. The Jeffersonian Society.

34 Patterson, R.M. 1985. *Historical Atlas of the Outlaw West*. Big Earth Publishing, Boulder, Colorado.

35 Throughout my exploration, myth is not equated inevitably to erroneous belief, but rather seen "as the body of tales, fables, and fantasies that help a people make sense of its history. Like history, myth finds meaning in the events of the past." Faragher, J.M. 2016. The myth of the frontier: progress or lost freedom. *History Now: The Journal of the Gilder Lehrman Institute*. https://www.gilderlehrman.org/history-by-era/art-music-and-film/essays/myth-frontier-progress-or-lost-freedom (accessed 15 February 2016).

36 Slatta, R.W. 2010. Making and unmaking myths of the American frontier. *European Journal of American Culture* 29: 81–92.

37 Cody's depictions of frontier mythology were more oriented towards American domestic life than usually assumed. Warren, L.S. 2003. Cody's last stand: masculine anxiety, the Custer myth, and the frontier of domesticity in Buffalo Bill's Wild West. *Western Historical Quarterly* 34: 49–69.

38 Hall, R.A. 2001. *Performing the American Frontier, 1870–1906*. Cambridge University Press, New York.

39 Rollins, P.C., and J.E. O'Conner, Eds. 2005. *Hollywood's West: The American Frontier in Film, Television, and History*. University Press of Kentucky, Lexington.

40 Davis, W.C. 1992. *The American Frontier: Pioneers, Settlers, & Cowboys, 1800–1899*. University of Oklahoma Press, Norman.

41 Springhall, J. 2011. Have gun, will travel: the myth of the frontier in the Hollywood Western. *The Historian* Winter 2011: 20–24. Between 1910 and the 1960s, Western-themed movies made up at least 20% of all cinematic titles in the U.S.

42 Gruber, F. 1969. *Zane Grey: A Biography*. Amereon Ltd, Mattituck, New York.

43 Busch, C.S. 1993. Longing for the lost frontier: Steinbeck's vision of cultural decline in the "White Quail" and "The Chrysanthemums." *Steinbeck Quarterly* 23: 81–89.

44 Pointer, L. 2013. *In Search of Butch Cassidy*. University of Oklahoma Press, Norman.

45 "No less familiar than the Turner thesis itself, of course, are the complaints against it made by Turner's critics. In the half-century since Turner's death, his reputation has been subjected to a devastating series of attacks which have left little of his argument intact." Cronon, W. 1987. Revisiting the vanishing frontier: the legacy of Frederick Jackson Turner. *The Western Historical Quarterly* 18: 157–176. Nevertheless, even Cronon salutes Turner's lasting influence on us. "[Turner] remains one of the pathfinders whose well-blazed trail we continue to follow. And whether or not we ultimately abandon the frontier thesis, we are unlikely ever to escape its narrative implications." *Ibid.*, p. 176.

46 Kyff, R. 1993. Frederick Jackson Turner and the vanishing frontier. *American History Illustrated* 28: 52.

47 Wrobel, D.M. 1991. The closing gates of democracy: frontier anxiety before the official end of the frontier. *American Studies* 32: 49–66.

48 Our need to find closure in various social settings affects both the *content* and the *process* whereby we seek and find this cognitive resolution. See De Grada, E., A.W. Kruglanski, L. Mannetti, and A. Pierro. 1999. Motivated cognition and group interaction: need for closure affects the contents and processes of collective negotiations. *Journal of Experimental Social Psychology* 35: 346–365.

49 Ironically, one outcome of that war was a right exercised by the United States to intervene politically in the internal affairs of Cuba, a demand that included resource extraction policies that did nothing to improve prospects for the Cuban ivory-billed woodpecker.

Chapter 4

Swept Along

Time is a sort of river of passing events, and strong is its current; no sooner is a thing brought to sight than it is swept by and another takes its place, and this too will be swept away.
—Marcus Aurelius

Frontier closure fashioned just an opening for myth making about the ivory-billed woodpecker. Specific contours for the mythology that would eventually cloak this species were molded from the outset by a rise to eminence of four social movements during the Progressive Era. These progressive revolts in part sought to wrest away exclusive control of natural resources from private corporations, promoting instead a stewardship of the country's natural heritage for the wider public welfare.[1] Each progressive movement fledged quickly around the end of the nineteenth century, prompted by brand new expectations of public life and participatory democracy that arose in a post-frontier America.

Since each social movement had strong links to the natural world, cooperation and rivalry among these progressive ***frames*** clamored for the professional, cultural, and social spaces that were carved out to mold and hold our ensuing perceptions about the ivory-bill. Understanding this rivalry is important – all social identities succumb to unverified rumor for handling ambiguity as they formalize an internally coherent (but not necessarily accurate) narrative about the natural world.[2] Consequently, the various ***frames*** for processing our knowledge about the ivory-bill came to diverge substantially.[3]

Four movements in particular outlined our grasp of the woodpecker during the early twentieth century. The first one

> **Box 4.1.** ***Frames*** are cognitive shortcuts in language, habits, or personal attributes that influence our perceptions and help us organize complicated knowledge. ***Framing effects*** describe a systematic bias in how we perceive information or data. A positive frame can lead us to avoid risk whereas a negative frame might prompt us to seek risk. Because ***framing*** varies so much across different social identities, the ***framing effect*** may exacerbate social conflict due to incompatible interpretations placed on the same events.

certified ornithology as an adjudicating body for professional avian science (American Ornithologists' Union; founded 1883). The second gave credence to wild lands preservation (Sierra Club; founded 1892). A third launched bird protection as an amateur, citizen-driven venture (National Audubon Society; founded 1905). And the last arranged natural sciences into a new, inter-disciplinary order called ecology (Ecological Society of America; founded 1915). To a great extent, all four of these movements could be seen as reactions of the increasingly settled, educated, and urban country to a natural world that had become suddenly much diminished.

American Ornithologists' Union

Professional ornithology's seal of approval would validate most of what could be said as constituting a legitimate body of biological knowledge about the ivory-bill. In addition to sponsoring publications on avian distribution and arbitrating the rules, precedents, and practices for naming America's birds,[4] professional ornithology would tussle with amateur bird enthusiasts for control over some of the primary narratives that depicted the woodpecker's fate.

Despite a very few ornithologists decrying the "gap between

the amateur naturalists and the pure field-workers on the one side, and the trained biologists on the other,"[5] reliance upon sight records as a means to document bird distribution and abundance was routinely derided as an unreliable method by early professional ornithologists.[6] These differing data standards were a major provocation for a scrimmage between American professional ornithology and amateur bird watching, one that never truly resolved itself. Indeed, evidentiary standards played a sizeable role in fueling the passionate disputes over an Arkansas "rediscovery" of ivory-bill in 2004 (Chapter 12).

More technically oriented ornithologists in the late nineteenth century sought foremost to create a profession and discipline distinct from popular bird watching.[7] Although not the earliest ornithological body formed in the United States (the Nuttall Ornithological Club was the first, in 1873), the American Ornithologists' Union (AOU) would go on to become the nation's largest and most prominent. On September 26, 1883, a three-day meeting attended by 23 ornithologists organized the AOU at the American Museum of Natural History in New York. Bylaws forged at this invitation-only gathering would create a periodical publication, *The Auk*, classes of membership (the largest being associate, which had no voting rights), and several select committees, all attributes that instilled an organizational structure that was both socially hierarchical and exclusive.[8]

Although the AOU tried to accommodate needs of its non-professional aspirants, relying on this much larger membership class for its financial solvency, tensions between professional ornithology and amateur bird devotees started off high. Ironically, only five of the AOU's founding members even made a living in science, and just two of these (J. A. Allen and Robert Ridgway) did so as ornithologists. Resentment of the AOU came from two directions: the bird taxidermists, collectors, and natural history museum dealers on the one hand, and "sentimentalist" bird watchers and protectionists on the other.

Joseph M. Wade, editor and owner of *Ornithologist and Oologist*, fired off right away, calling the AOU "not American, and it is formed too much on the principle of our city social clubs where each member carries a Yale key. It won't work in science, gentlemen; all nature belongs to all men."[9] One hears in Wade's grievance not only a very American appeal to democratic ideals, but also a snub directed at Ivy League, East Coast pretension. The latter would prompt complaints from avian colleagues in other regions: "among our western ornithologists there is a strong feeling that we, as a whole, should be better and more fairly represented in that [the AOU] body."[10]

A separate challenge came from the "opera glass students," citizens who were not only prone to report their bird sightings as factual information in various periodicals of the day, much to the annoyance of gun-toting professionals, but also an alliance that was increasingly ready to criticize outright any scientific collecting, that sacred cornerstone of early American ornithology. In 1915, an editorial challenging the research collection of birds appeared in *Science*. Not insignificantly, it exempted any purposes of frontier exploration from its justifications for why more restrictions on bird collection were needed otherwise:

> *[T]he usefulness of such collectors except in remote and little explored regions has largely gone by ... [collectors] assistance to real science is rarely more than very slight and oftener nothing at all; and that their destructiveness is very great.... The scientific value of the average bird collection, or even of one made with far more than average care, is greatly overrated....*[11]

Barely more than a decade into the new twentieth century, then, ability to procure ivory-billed woodpeckers via scientific collecting was already under harsh social approbation.[12] At this juncture, too, dealer-based, commercial collecting was legally proscribed[13] and entirely compromised in an economic sense (Chapter 3). By 1918 the Migratory Bird Treaty Act made it unlawful to pursue, hunt, take, capture, kill or sell listed birds unless one was exempted through special permit. Although it is doubtful that this brand-new regulatory environment halted killing of ivory-bills in the early twentieth century, it is not unreasonable to consider that it slowed down the unregulated take of the woodpecker.

Whether these measures came soon enough to make a conservation difference is not at issue here. Rather, within the first few years of the 1900s, state officials had already begun widespread restrictions on the scientific collection of birds and their eggs.[14] Our tally of ivory-bill specimens as revealed in the historical archive thus fell sway to these profound social and cultural changes. More directly, this specimen-based record was not an accurate temporal proxy for the bird's abundance, even within the narrow temporal window of 1880–1920 (Chapter 14).

During its first half century, the AOU hardly would be mistaken for a conservation-oriented body. Although the union early established a Committee on Protection of North American Birds, it would lie almost entirely dormant until the 1930s. One exception was when in 1895 AOU treasurer William Dutcher was appointed head of the protection committee. His activism led to substantial innovations in bird protection, including facilitation of the National Committee of the Audubon Societies of America in 1905. But Dutcher resigned from the AOU protection committee that same year, after the union's Council complained about undue restrictions placed on scientific bird collecting.[15]

Not again until W. DeWitt Miller, Willard G. Van Name, and Davis Quinn issued their 1929 call to the public, *A Crisis in Conservation: Serious Danger of Extinction of Many North American Birds,* would the AOU be shaken out of its complacency over avian conservation. Only at its annual meeting in 1937 (when

Tanner was studying ivory-bills in Louisiana) did the AOU pass a resolution that opposed some types of scientific collecting, and even then only in the special case of "seriously depleted species."[16]

Sierra Club

Philosophical foundations that were embodied in the Sierra Club gave the ivory-bill its key storylines for land protection, solitude, and wilderness as designated place. Formed in part to keep tabs on a newly created Yosemite National Park (1890), the Sierra Club included some scientists in its founding members. On May 28, 1892 in San Francisco, 182 charter members, including professor Joseph LeConte and Stanford University president David Starr Jordan, met to "enlist the support and cooperation of the people and government in preserving the forests and other natural features."[17]

Set-asides of land were an early, prominent conservation achievement for the Sierra Club. The Club's first victory was defeating a proposal to reduce boundaries to Yosemite National Park. A year after its founding, support from the Club helped establish a 13-million-acre Sierra Forest Reserve. By 1897 the Club was pushing addition of even more national parks, including Grand Canyon. Congress established Mt. Rainier National Park in 1899 in response to public support generated by the Sierra Club. These victories, plus designation of Glacier National Park in 1910 and the National Park Service in 1916, attracted national attention to the merits of land set-asides as a conservation tactic. This approach had shortcomings, a prominent one being the inability by John Muir's preservationist campaign in 1913 to block the Hetch Hetchy Reservoir, an effort that lost out to a commodity style of natural resource management promoted by Gifford Pinchot.[18]

Similar set-asides for a nascent National Wildlife Refuge system began during this same era by Theodore Roosevelt.

In 1903, Roosevelt established the first unit for protection of migratory birds: Pelican Island National Wildlife Refuge, in Florida. During the earliest years of the twentieth century, America's conservation movement thus came to view land preservation as a vital strategy to save our wildlife. When the presumed final haven for ivory-bills in the Singer Tract was threatened by logging in the 1930s, it was only instinctual for the bird's supporters to fall back again on this stratagem to protect the species. As with Hetch Hetchy, however, the preservationist sentiment lost out badly to the commodity rationale (Chapter 10).

National parks and other preservation land set-asides were founded on wilderness precepts.[19] Parks and wilderness areas were viewed as remnants of our lost frontier.[20] Wilderness[21] had colossal (though entirely misplaced) clout in fashioning our perceptions of the ivory-bill's living space. Descriptions of the woodpecker as a resident of ancient, mystical forest long have been a staple of American writing. William Faulkner's fiction weaved death, wilderness legend, and a heard-but-never-seen ivory-bill in his 1942 work *Go Down, Moses*.[22] Naturalists since Audubon also portrayed in colorful prose the woodpecker as a denizen of isolated and primordial places (year given for quote):

- 1831 "the favourite resort of the Ivory-billed Woodpecker ... the extent of those deep morasses, overshadowed by millions of gigantic dark cypresses, spreading their sturdy moss-covered branches, as if to admonish intruding man to pause and reflect on the many difficulties which he must encounter, should he persist in venturing farther into their almost inaccessible recesses, extending for miles before him.... Would that I could give you an idea of the sultry pestiferous atmosphere that nearly suffocates the intruder during the meridian heat of our dogdays, in those gloomy and horrible swamps!"[23]

- 1885 "It is always rare, inhabiting the most solitary places remote from mankind."[24]
- 1885 "Of all our wild American birds, I have studied no other one which combines all of the elements of wildness so perfectly in its character as does the ivory-billed woodpecker."[25]
- 1907 "ivory-billed woodpeckers ... seemed to me to set off the wildness of the swamp...."[26]
- 1914 "In Florida this splendid Woodpecker is now confined to the wildest and remotest swamps."[27]

A sense of the ivory-bill requiring large wilderness set-asides was made even more explicit by Tanner in his recommendations to conserve the few presumed survivors still living in northeast Louisiana during the late 1930s. There could be no qualms as to the sort of landscape that Tanner envisioned as required for this bird to persist:

> *More than that, the entire area with all its life, plant and animal, would be a primitive or wilderness area, an example in perpetuity of the North American wilderness. That is where the Ivory-bill fits; a denizen of the tall trees in thick forests and swamps, he belongs in ... our southern swamp wilderness. The most will be accomplished by preserving an area of virgin and primitive forest that will be a suitable habitat for the Ivory-billed Woodpecker ... a permanent monument of native trees, plants, birds and other animals inhabiting that wilderness area....*[28]

Steinberg reflected on this wilderness fixation too: "While Tanner was convinced the woodpeckers he studied were the last, I wonder if this belief wasn't in part driven by nostalgia for the majestic Singer Tract that had been destroyed."[29] And as Snyder and colleagues accentuated: "The image of the ivory-bill given to us by Tanner is possibly the *most powerful image of species endangerment and disappearing wilderness that we have* (emphasis supplied)."[30] We see foreshadowed just how much metaphoric weight we would expect this woodpecker to carry for us, in this instance for protection of wilderness tracts as an essential underpinning to conservation practice (Chapter 5).

Appeals to our allegiance to wilderness became even more frank during the drumming up of political support to protect the Singer Tract.[31] The erudite Aldo Leopold, in justifying a new national park to preserve the presumed last few ivory-bills, actually managed to tie elegantly the bird's long-term prospects to the third leg of frontier essentialism,[32] that is, a location noted for its closeness and conflict.[33] Leopold penned his plea:

> *the Ivory-billed Woodpecker – a bird inextricably interwoven with our pioneer tradition – the very spirit of that "dark and bloody ground" which has become the locus of the national culture. It is known that the Ivory-bill requires as its habitat large stretches of virgin hardwood.*[34]

Seventy years later, the ivory-bill was yet again linked emphatically to wilderness. When media reported on a putative rediscovery of ivory-bill in Arkansas, its coverage fell back on descriptions that placed the bird firmly within a setting of untrammelled land. The *Los Angeles Times* announced grandly:

> *The ivory-billed woodpecker, a symbol of the fading American wilderness that was thought to have been pushed into extinction, has been spotted by scientists for the first time in 60 years, taking wing in the wild swamplands of Arkansas.*[35]

Hill even selected a primeval backdrop ("flooded wilderness") for the title to his 2007 book recounting that team's searches for the ivory-bill in north Florida.[36]

Wilderness was an upshot of a Euro-American conception of

frontier as a place of vacant space.[37] But the continent was not empty. Moreover, its original inhabitants had already played a potent hand in persecuting the ivory-bill. Because homage to wilderness is one of our socially fabricated abstractions,[38] our insights into the ivory-bill's ecological needs would become overly constrained (Chapter 1) and prone to extensive bias (Chapter 7) due to this slanted depiction of the natural world.

> **Box 4.2.** The ***myth of the pristine*** holds that landscapes can be classified by whether or not they are corrupted by human influence, a sticky proposition given millennia-long occupation and profound impacts by earlier indigenous peoples in North America. Using "pristine" as a benchmark can easily mislead conservation to regard human absence as an overly simplistic solution, or human presence as an invariable threat to the protection of native species and habitats. This conceptual bias is known as ***tainted-nature delusion***.

"With the closing of the frontier, many Americans searched for their national identity in the primitive.... The wilderness cult represented a nostalgic attempt to regain America's lost past in the wilderness."[39] As a result, our views about the ivory-bill's conservation needs and its required living space were unduly shaped by a fundamental and deep-running pessimism rooted in the environmental ***myth of the pristine***.[40] That myth made it virtually impossible to imagine any sort of living ivory-bill surviving in modified landscapes given the potent narrative power of this "tainted nature delusion."[41]

National Audubon Society

As a non-game bird species, the ivory-bill was early influenced by the bird protection movement. Market hunting and lack of legal protections in America led to the unsustainable slaughter of many native wildlife, several species of which had vanished to near extinction by the late nineteenth century.[42] A sophisticated millinery industry was also exploiting millions of wings, quills and feathers of non-game birds for use in decorative hats and other fashion accessories. As early as 1886, J. A. Allen wrote a blistering critique of this wanton greed in *Science,* warning his compatriots: "The fate of extermination … if a halt be not speedily called by enlightened public opinion, [will] overtake scores of our song-birds, and the majority of our graceful and harmless, if somewhat less 'beneficial' sea and shore birds."[43]

Over-exploitation for economic, scientific, and sporting purposes began to shift American thought about birds, especially non-game species. George Bird Grinnell, editor of *Field & Stream,* issued a call in 1886 that led to formation of the very first Audubon society in the U.S. Even though this society grew so rapidly that the magazine was unable to keep up with its hungry demands, and it finally disbanded, Grinnell contributed prominently to bird preservation by creating openings for citizens to involve themselves directly in solving a conservation problem linked back to society's fashion tastes.[44]

Two Massachusetts women heard Grinnell's call. Harriet Lawrence Hemenway and her cousin Minna Hall, both part of Boston's aristocracy, founded the first state Audubon Society in 1896. Advocacy by these socialites and others prompted a citizen revolution in bird conservation. Within 15 years, 35 states had established Audubon groups. Boycott of the feather trade led to bans on interstate transport of birds via the Lacey Act (1900), formation of what was to be the National Audubon Society (1905), and the nation's first Migratory Bird Act (1913). Murders of three Audubon game wardens between 1905 and 1908 soured the nation against plume hunting for good.[45]

How successful were citizen efforts in changing social

outlook, including the overall shooting threats posed to non-game birds like ivory-bill? Several anecdotes illustrate how rapidly the new attitudes swept the nation. Despite his bully appetite for hunting and taxidermy, during a game excursion to rural Louisiana in 1907 Teddy Roosevelt merely watched and admired three fine examples of the large woodpecker.[46] Legal protections for ivory-bill came as early as 1901 in Florida, so that by 1905 a commercial collector was arrested for shipping specimens of the woodpecker out of that state.[47]

By 1915, social backlash against collecting bird specimens even for limited scientific purposes was so fervent that *Science* granted Joseph Grinnell more than three pages to complain about how ornithology was now "lapsing wholly into the status of a recreation or hobby, to be indulged in only a superficial way by amateurs and dilettantes."

Grinnell's alarm was such that it had become necessary now for the nation to instead *Conserve the Collector*. He vented:

> *It is with considerable apprehension that I have observed an unmistakable decrease in the number of collectors during the past six or eight years ... precision and accuracy in the field of ornithology are ... suffering as a consequence of this forsaking of the "shotgun method".... Ornithology as a science is threatened ... the present tendency toward extermination of the collector bears close relationship to the increasing number of extreme sentimentalists ... further ... inconsistency ... in the intemperance with which the reservation idea has been put into effect within the last few years.*[48]

From Grinnell's grandiose lecturing, there can be no wonder as to how professional ornithology and popular bird watching diverged so early, and why neither enterprise was (or is) predisposed instinctually towards land conservation. Tension between amateurs and professionals played out again in the 1940's campaign to protect the Singer Tract (Chapter 10), and later still over the rancorous debate about the identity of a large woodpecker seen in twenty-first-century Arkansas (Chapter 12).

Protections levied early in the 1900s nevertheless transformed radically America's attitude towards the killing of non-game birds. Between the 1880s and 1920, the publicly sanctioned, widespread shooting of non-game birds went from everyday norm to increasingly rare and contentious exception. Thus, a rise and then fall in numbers of ivory-bills collected during this same interval cannot possibly be held as a credible sample or an unbiased proxy for the woodpecker's actual population status across just these few decades, never mind a full century (see Chapter 14).

Transformation of American values through new public interest in bird watching had yet another consequence for the future perceptions cultivated about the ivory-bill. At a time when collecting birds was on the wane, a corresponding rise in publication of sight records of birds in such outlets of that day as *Bird-Lore* was being staunchly resisted by professional ornithology. Specimens were indisputable proof of a species occurrence, but what someone claimed to see by naked eye was inevitably open to dispute.[49]

A wide rupture on what constituted reliable evidence was born, with sight records never fully accepted as sufficient confirmation in professional ornithology. Yet by the early 1900s, the ivory-bill's status was no longer being detected adequately via collecting, either. Coupled with a lack of consensus over the doubtfulness of sight records, the woodpecker's status became increasingly veiled thereafter.

Ecological Society of America

Disciplinary ecology gave the ivory-billed woodpecker a robust conceptual model for describing the bird's relationship to the American environment writ large. From this interdisciplinary view, it was hoped that the necessary conservation measures could be identified in order to save the bird from extinction. Methodology used for and the content reported in Tanner's monograph[50] on the ivory-bill were fundamentally ecological in its orientation.

Ecology as an organized, interdisciplinary profession would not take hold in the United States until 1915,[51] the result of 50 attendees at the American Association for the Advance of Science meeting in Columbus, Ohio, splitting off to form a separate Ecological Society of America. Victor Shelford was nominated as its first president. Shelford later articulated the motives for integrating various disciplines of study within their new profession:

> *The development of modern ecology has received its greatest impetus in a few localities where there are physiographic conditions giving diversity of habitats in which environmental dynamics are apparent ... the workers in the field have thus far derived, and probably always will derive, their inspiration from ... the diurnal, seasonal and secular changes in the environments of single species or of communities of species ... botanists and zoologists – separated by the present-day organization of science but with much common ground which demands fusion.*[52]

Here now was a venue where animal and plant science could be integrated. Studies of animal ecology in the U.S. traced back to 1870, a line of scientific inquiry most notable in universities of the upper Midwest.[53] Stephen A. Forbes (1844–1930), probably most remembered for his 1887 lecture "The Lake as a Microcosm," was among the first to make systematic studies on the food habits of birds and fishes. Another early ecologist, C. Hart Merriam (1855–1942), led the Division of Ornithology and Mammalogy at the U.S. Department of Agriculture. Merriam's use of both animal and plant data to map seven life zones in his "Provisional Biological Map of North America Showing the Principal Life Areas" was extraordinarily innovative in establishing biogeography as a line of study. Another early American ecologist, Victor Shelford (1877–1968), was acclaimed for his work studying succession in both plant and animal communities.

> **Box 4.3.** ***Myth*** need not consist of strictly unproven or totally false belief used out of convenience to justify our collective social and cultural values. At times ***mythology*** also may consist of stories that dramatize history and distill human experience into compelling metaphor that is more easily absorbed in this narrative form. Elements of truth, then, can be embedded in ***myth***, including even those constructed to explain the enigmatic ivory-bill.

By 1913, now well into an era when ivory-bill was increasingly difficult to find, the field received its first comprehensive text on animal ecology, authored by Charles C. Adams (1873–1955).[54] Early advances in animal ecology during these years, especially a fusion of diet studies, biogeography, habitat use, and forest succession, were to become highly influential to Tanner's thinking about the ivory-bill two decades later. Ecological study was notably absent from most ornithology practiced during and prior to this era, because the latter field was still focused heavily on anatomy and taxonomy.

As the century turned, land set-asides became a staple and arguably the primary means arrayed to conserve American wildlife. Ecological study of avian habitat now had a disciplinary home. Ornithology began exerting itself as a profession, setting guidelines for what constituted a valid body of facts about America's birds. Due to concerted efforts by amateur bird devotees, killing ivory-bills and other non-game species became legally regulated and socially censured. A fissure arose between bird-watching and professional ornithology over the evidentiary protocols used to differentiate valid knowledge. These changing social mores about what constituted proper standards of evidence contributed substantially to the woodpecker's increasing invisibility in the public eye.

Four movements from the Progressive Era thereby melded how our early knowledge about the ivory-bill would be ***framed*** as mythological dialogue about this species well into the future. The nature of that ***framing*** carried immense power[55] for how we came to perceive the woodpecker's life history and evaluate its presumed fate. Yet as decisive as each of these four progressive crusades were to sweeping the woodpecker along in the fickle tides of history, two disciplinary ***frames*** yet to arise in American life would also greatly constrain what we did not (and could not) know about the ivory-bill.

Endnotes

1 Hays, S.P. 1999. *Conservation and the Gospel of Efficiency: The Progressive Conservation Movement, 1890–1920*. University of Pittsburgh Press, Pittsburgh, PA, p. 261.

2 Theodorakea, I.T., and E. von Essen. 2016. Who let the wolves out? Narratives, rumors and social representations of the wolf in Greece. *Environmental Sociology* 2: 1–12.

3 Social and individual preferences and outlooks hinge critically on framing. The relative appeal of options or explanations change when the same decision problem is framed in different ways. Tversky, A., and D. Kahneman. 1981. The framing of decisions and psychology of choice. *Science* 211: 453–458.

4 American Ornithologists' Union. 1892. The Code of Nomenclature Adopted by the American Ornithologists' Union, New York.

5 Huxley, J. 1916. Bird-watching and biological science: some observations on the study of courtship of birds. *The Auk* 33: 142–161.

6 Dwight, J. 1918. Correspondence, editor of "The Auk." *The Auk* 35: 262.

7 Barrow, M.V., Jr. 1998. *A Passion for Birds: American Ornithology after Audubon*. Princeton University Press, Princeton.

8 *Ibid.*, pp. 52–53.

9 *Ibid.*, p. 54; this passage was part of a suppressed editorial, "Plain English," to have been released by Wade in *Ornithologist and Oologist* 8, in 1883. It was never formally published, however.

10 Daggett, F.S. 1900. Concerning the active membership of the A.O.U. *The Condor* 2: 68–69.

11 Van Name, W.G. 1915. Bird collecting and ornithology. *Science* 41: 823–825.

12 Well more than a century later, professional ornithology still found it necessary to defend its reasons for the valid scientific collection of birds. Winker, K., J.M. Reed, P. Escalante, R.A. Askins, C. Cicero, G.E. Hough, and J. Bates. 2010. The importance, effects, and ethics of bird collecting. *The Auk* 127: 690–695.

13 An economic panic in 1893 curtailed financial returns on bird taxidermy such that, along with state regulations and public outcry, bird collections began to be reduced by the late 1890s. See the overview described by Barrow, pp. 134–146.

14 Barrow, p. 128.

15 *Ibid.*, pp. 128–134.

16 *Ibid.*, p. 153.

17 Cohen, M.P. 1988. *The History of the Sierra Club: 1892–1970*. Sierra Club Books, San Francisco.

18 Miller, C. 2001. *Gifford Pinchot and the Making of Modern Environmentalism*. Island Press, Washington, DC. In a speech delivered in 1903, Pinchot made clear: "The object of our forest policy is not to preserve the forests because they are beautiful – or because they are refuges for the wild creatures of the wilderness – but the making of prosperous homes – every other consideration becomes secondary." This commodity-favored approach, also known as the Resource Conservation Ethic, ultimately swayed the

1940s logging of the Singer Track, then perceived to be the last redoubt for the woodpecker.

19 Henderson, N. 1992. Wilderness and the nature conservation ideal: Britain, Canada, and the United States contrasted. *Ambio* 21: 394–399. Wilderness as an essential underpinning to the management of preserved lands like national parks comes very close to being a uniquely American paradigm. Guha, R. 1989. Radical American environmentalism and wilderness preservation. *Environmental Ethics* 11: 71–83.

20 Hall, C.M. The changing cultural geography of the frontier: national parks and wilderness as frontier remnants. Pp. 283–298 *in* S. Krakover and Y. Gradus, eds. *Tourism in Frontier Areas*. Lexington Books, Oxford, England.

21 Although the Sierra Club's John Muir popularized wilderness preservation at the end of the nineteenth century, its roots as a rationale for conservation go back earlier to such American Romantic and Transcendental writers as William Cullen Bryant, Ralph Waldo Emerson, and Henry David Thoreau. For more on this Romantic-Transcendental Preservation Ethic, see: Callicott, J.B. 1993. A brief history of American conservation philosophy. Pp. 10–14 *in* W.W. Covington and L.F. DeBano, eds. *Sustainable Ecological Systems: Implementing an Ecological Approach to Management*. Rocky Mountain Forest and Range Experiment Station, Fort Collins, CO.

22 Aiken, C.S. 2009. *William Faulkner and the Southern Landscape*. University of Georgia Press, Athens, p. 164. Cheatham, G., and J. Cheatham. 1988. The ivory-billed woodpecker in Faulkner's *Go Down, Moses*. *American Notes & Queries* 1: 61–63. Faulkner's story is also considered to be a classic representation of a frontier chronicle. Gleeson-White, S. 2009. William Faulkner's *Go Down, Moses*: an American frontier narrative. *Journal of American Studies* 43: 389–405. Consequently, and much like the woodpecker itself, Faulkner's literary world consisted of imagined landscapes where legend, truth, and fiction all merged seamlessly. Travis, C. 2011. William Faulkner and the Southern Landscape. *Journal of Cultural*

Geography 28: 225–232.

23 Audubon, J.J. 1832. *Ornithological Biography, Or an Account of the Habits of the Birds of the United States of America*. Volume 4. James Kay, Jun. & Co., Philadelphia, p. 215.

24 Clarke, S.C. 1885. Ivory-billed woodpecker in Florida. *Forest & Stream* 24: 367.

25 Thompson, M. 1885. A red-headed family. *The Elzevir Library* 4: 12.

26 Roosevelt, T. 1908. *In the Louisiana Canebreaks*. Louisiana Wild Life and Fisheries Commission, 400 Royal Street, New Orleans. 8 pp.

27 Phelps, F.M. 1914. The resident bird life of the Big Cypress Swamp. *Wilson Bulletin* 26: 86–101.

28 Tanner, J.T. 1942. *The Ivory-billed Woodpecker*. Dover Publications, Inc., Mineola, NY, p. 89.

29 Steinberg, M.K. 2008. *Stalking the Ghost Bird: The Elusive Ivory-billed Woodpecker in Louisiana.* Louisiana State University Press, Baton Rouge, p. 119.

30 Snyder, N., D.E. Brown, and K.B. Clark. 2009. *The Travails of Two Woodpeckers*. University of New Mexico Press, Albuquerque, p. 20.

31 Out of homage to our frontier heritage, The Wilderness Society (of the United States) was founded in 1935, just two years prior to the launch of Tanner's study of the ivory-bill in northeast Louisiana. Wildlife ecologist and author Aldo Leopold was one of the society's eight original founding members.

32 As cited in Jackson, J. A. 2006. *In Search of the Ivory-billed Woodpecker*, 2nd ed. HarperCollins, New York, p. 290.

33 Gouge, C. 2007. The American frontier: history, rhetoric, concept. *Americana: The Journal of American Popular Culture (1900 to present)* 6: 16.

34 Leopold, A. 1936. A Proposal to the Wildlife Conference for an Inventory of the Needs of Near-extinct Birds and Animals. P. 515 *in* J.B. Callicott and M.P. Nelson, eds. Selections from Aldo Leopold, *The Great New Wilderness Debate*. University of Georgia Press, Athens.

35 Maugh, T.H., II. 2005. A bellwether of the wild, the ivory-billed

woodpecker, is back. *Los Angeles Times,* The Nation, April 29.

36 Hill, G.E. 2007. *Ivorybill Hunters – The Search for Proof in a Flooded Wilderness.* Oxford University Press, London.

37 "Wilderness ... is so heavily freighted with meaning of a personal, symbolic, and changing kind as to resist easy definition." Nash, R.F. 2014. *Wilderness and the American Mind,* Fifth Edition. Yale University Press, New Haven, CT, p. 1.

38 Cronon, W. 1995. The trouble with wilderness: or, getting back to the wrong nature. Pp. 69–90 in W. Cronin, ed. *Uncommon Ground: Rethinking the Human Place in Nature.* W.W. Norton & Co., New York.

39 Hall, pp. 293–294.

40 Clement, C.R., and A.B. Junqueira. 2010. Between a pristine myth and an impoverished future. *Biotropica* 42: 534–536.

41 Shell, D., and E. Meijaard. 2010. Purity and prejudice: deluding ourselves about biodiversity conservation. *Biotropica* 42: 566–568.

42 Hornaday, W.T. 1913. *Our Vanishing Wild Life: its Extermination and Preservation.* C. Scribner's Sons, New York.

43 Allen, J.A. 1886. The present wholesale destruction of bird-life in the United States. *Science* 7: 191–195.

44 Merchant, C. 2010. George Bird Grinnell's Audubon Society: bridging the gender divide in conservation. *Environmental History* 15: 3–30.

45 McIver, S.B. 2003. *Death in the Everglades: The Murder of Guy Bradley, America's First Martyr to Environmentalism.* University of Florida Press, Gainesville.

46 Roosevelt, T. 1908. *In the Louisiana Canebreaks.* Louisiana Wild Life and Fisheries Commission, 400 Royal Street, New Orleans. 8 pp.

47 Tanner, p. 56.

48 Grinnell, J. 1915. Conserve the collector. *Science* 41: 229–232.

49 Barrow, pp. 154–155. Expressing concern that the authority of their discipline was threatened by dubious, unverifiable observations, professional ornithologists argued that the proliferation of bird watching's "sight records" was a case of "a little knowledge

[being] a dangerous thing."

50 See especially part V of Tanner's monograph, which employed a sophisticated ecological rationale for describing the "requirements and general policy of a conservation program" for the woodpecker, pp. 87–97.

51 For an expanded history of early twentieth century American ecology, see: Mitman, G. 1992. *The State of Nature: Ecology, Community, and American Social Thought, 1900–1950*. The University of Chicago Press.

52 Shelford, V.E. 1917. The ideals and aims of the Ecological Society of America. *Bulletin of the Ecological Society of America* 1: 1–8.

53 Egerton, F.N. 2014. History of ecological sciences, part 49: formalizing animal ecology, 1870's to 1920's. *Bulletin of the Ecological Society of America* 95: 59–81.

54 Adams, C.C 1913. *Guide to the Study of Animal Ecology*. Macmillan, New York.

55 Fairhurst, G.T. 2010. *The Power of Framing: Creating the Language of Leadership*. John Wiley & Sons, San Francisco, CA.

Chapter 5

Unconsummated Rescue

Never look back unless you are planning to go that way.
—Henry David Thoreau

Despite wildlife extinctions looming over turn-of-the century America, the country still lacked sophisticated means to ameliorate or even detail accurately the declines in the ivory-bill. Certainly, few such tools existed in the decade of the 1910s. Protections that were at hand, like land reserves and legal regulations on take, were too untested or recently implemented to be truly effective for the woodpecker. Instead, the necessary tools to save the nation's most imperiled wildlife would have to be proposed, verified, and implemented through future decades of trial-and error. This process of testing began in earnest during the 1930s, but out of supreme necessity such experimentation for how to best rescue endangered species still continues through the present day.

Therefore, even had we been able to fathom the ivory-bill's critical ecological needs prior to 1915,[1] we lacked key specialties that would have been indispensable for an effective response to safeguard the woodpecker from threats to its existence. Those absent salvations included the principles and practices of wildlife management,[2] and the theoretical insight and applied tools furnished today by conservation biology.[3] The first discipline did not really possess substantial capacity until the 1930s; the second profession did not even arrive onto the national scene until much later, the 1980s.[4]

No reserve cavalry, then, would charge to the bird's rescue. Both wildlife management and conservation biology provided too little and came far too late to do any practical good for the ivory-bill. Nonetheless, the particular outlooks generated from each of these natural resource professions would unquestionably dictate the intangible expectations that we would place on the woodpecker during the bird's real or putative appearances in the late twentieth and early twenty-first centuries.

Before dissecting any further what we didn't know or couldn't do about the ivory-bill in past eras, we should acknowledge some peculiar challenges to this kind of hypothetical tour. Overlaying any modern frame on an earlier epoch can be prejudicial, a device akin to chastising slow eighteenth-century road travel for relying on horses instead of the internal combustion engine. Still, in order to accentuate the vast uncertainty that controlled very early the boundaries of reliable knowledge about the woodpecker, it helps to portray just how much it was that we did not and indeed could not know at the very time when the ivory-bill became increasingly hidden to us.[5] Ultimately, this check-up will show that the uncertainties generated in earlier times predisposed us in future eras to cultivate massive distortions in our critical thinking concerning the woodpecker.

In applying a retrospective analysis like this we must also be mindful of two specific cautions. First, it would be a blunder to use ***presentism***[6] to garner any false sense of intellectual superiority as a result of what we might know or do better today. There is no virtue in holding the past accountable for historical norms that had not yet evolved. Second, it would constitute a serious thinking bias to expect those past events as being more predictable than they were at the time for the people living in those eras. This just pushes us into a cognitive error of ***hindsight bias***.[7] With these caveats in mind, then, how was it that the absences of wildlife management and applied conservation biology magnified uncertainty over the woodpecker into such fallacious proportions?

Had we actually known in 1915 that the ivory-bill needed rescue,[8] and there is no evidence from literature of the day

> **Box 5.1.** ***Presentism*** refers to our penchant to rely only on present-day attitudes, an inclination to interpret history only through filters of modern values or paradigms. ***Hindsight bias*** signals an "I-knew-it-all-along" error in our thinking whereby we come to see some event as more predictable that it actually was. In matters of the ivory-bill, a device of ***presentism*** helps us discover how evolving social mores on the evidentiary standards used to document birds in the early twentieth century shrouded the woodpecker from easy detection. Criticizing ornithologists or bird watchers of that era for not realizing that these altered detection methods might be responsible for the bird's *seeming* disappearance reveals a ***hindsight bias***.

that such belief was prevalent,[9] modern wildlife management surely would have expressed confidence that it was up to the task. Key attributes offered by wildlife management include a mechanism of direct intervention itself, especially for setting maximum sustained yield under conditions of exploitation, as well as victories achieved with the restorative North American Model of Wildlife Conservation (hereafter Model).[10] The Model rests on two major assumptions: fish and wildlife are a public trust for non-commercial use,[11] and they should be managed in perpetuity at optimum population levels to maintain numbers for public availability.

Ironically, it was hunting that first drove American wildlife to the brink (commercial market hunting for profit) but then later helped recover or prevent many species from outright extinction (recreational sport hunting).[12] The Model possesses seven key elements.[13] First, wildlife and natural resources are held in public trust, having no single owner, ensuring that both current and future generations always have access. Second, commercial hunting and sale is prohibited in order to improve prospects for sustainability. Third, access to wildlife via hunting relies upon democratic rule of law. Fourth, every citizen (not just landowners or nobility) has opportunity to hunt or access wildlife. Fifth, wildlife can only be taken for non-frivolous purposes, eliminating the casual and wasteful killing for body parts.[14] Sixth, whenever appropriate wildlife is managed cooperatively as an international resource. And seven, wildlife management is premised on the use of sound science to inform policy and regulatory decisions.

Had these measures been in place decades to a century earlier, thc ivory-bill certainly might have benefited from any implementation of numbers 1–3, 5, and 7 of the Model's primary tenets. The Model worked splendidly over the long term for restoring the nation's once-lost populations of deer, elk, bison, waterfowl, wild turkey, and other game that had been so depleted by the rampant market hunting practiced through the late nineteenth century. The Model's successes arose because recreational hunters had earlier banded together to take responsibility for stewarding the nation's wildlife.

Theodore Roosevelt and George Bird Grinnell founded the Boone and Crockett Club for this very purpose in 1877, setting up a uniquely American process that advocated for wildlife conservation, in part by advancing principles of fair chase carried out in authentic pioneer spirit.[15] Roosevelt in particular used his rhetorical inspiration to call for action and link a need for wildlife conservation to our past frontier values.[16] He tailored the conventional frontier myth by substituting a farmer-hero for the traditional conqueror-hero, and thereby fostered an American brand of conservation that had more to do with moral imperative than legislative proposal.[17]

Although the motivation behind recreational hunting was a continued access to game species for social pursuit, this inspiration would have been extraordinarily helpful to the ivory-bill had it been possible to implement the Model for this once heavily exploited species. In an ideal world, managing the ivory-bill for a sustained yield that could have supplied the taxidermists, museum collectors, and scientists with their cravings of that day, though doing so without compromising any continued viability in the woodpecker's population, theoretically would have been far more preferable to the blind exploitation that took place in fact.

Derivation of wildlife management principles was very prolonged, however, making assimilation of such informed protection measures for the ivory-bill completely unfeasible. For example, the Model's beginnings were rather modest, even primitive. During the 1910s, active wildlife management consisted largely of setting up game farms in order to restock depleted wildlife species back into vacated landscapes.[18] Other practices in this early era were limited to reserve establishment, law enforcement, predator control, season and bag limits, and basic population surveys. These practices were implemented usually without much confirmation from science.

Although Theodore Roosevelt called commissioners from Canada, Mexico, and the U.S. to a North American Conservation Congress as early as 1909, wildlife management as a distinct profession would not truly consolidate in this hemisphere until the 1930s. Jay N. "Ding" Darling was a conservationist and chief of the Biological Survey, a precursor of today's U.S. Fish and Wildlife Service. In 1934 Darling facilitated the passage of the Migratory Bird Hunting Act, a measure that required all waterfowl hunters to purchase a Federal Duck Stamp. Sales proceeds from these stamps were used to purchase wetlands and protect habitat. By 1937 sportsmen successfully lobbied congress to pass the Pittman-Robertson Wildlife Restoration Act, a mechanism that placed an excise tax on sporting arms and ammunition, ultimately generating more than $2 billion for habitat acquisition. Wildlife conservation now had steady national financing.

Wildlife management would foster both an applied and a research wing, each guided by the sterling leadership of Aldo Leopold.[19] From 1929 to 1932, Leopold served as chairman of the Committee on Game Policy for the American Game Association, forerunner of today's Wildlife Management Institute and National Wildlife Federation. In December of 1930, Leopold delivered a report to the 17th American Game Conference giving a direction to the wildlife profession that would last for decades. Leopold outlined what was to become the Model for conserving the nation's wildlife:

> *Demand for hunting is outstripping supply. If hunting as a recreation is to continue, game production must be increased.... Irreplaceable species may be destroyed before these forces become operative ... the majority adhere to the deep-rooted American pioneer tradition that hunting is a free privilege, and insist that it can be kept so.... We submit that public hunting under the license system is workable for game species inhabiting cheap land which the public can afford to own (or lease) and operate, but that compensation to the landowner in some form or other is the only workable system for producing game on expensive private farm land.... We believe, in short, that experiment, not doctrine or prophecy, is the key to an American Game Policy.*[20]

Notice some rather stark implications for ivory-bill embedded in this vision. Wildlife management would focus its attention on species for which a licensed, public hunting system was workable. Non-game species were nowhere to be found: they did not have a shelter in what is otherwise a highly practical system aimed at boosting and then sustaining population sizes for select species. So, even though an experimental science rather than some untested philosophy would influence the

Model used to steward the nation's wildlife, the Model would come close to neglecting entirely those species for which there was no sporting demand.[21]

A roll-up-the-sleeves, can-do self-confidence has always been a lofty strength of the American variant of wildlife management. .But the Model handicapped (and unwittingly delayed) prospects for preserving the ivory-billed woodpecker and other imperiled non-game wildlife. Wildlife management had scarcely arrived as a profession when the woodpecker had already shrunk to what was then believed to be its final refuge, the Singer Tract in northeast Louisiana during the late 1930s (Chapter 7). Had the Model been in place when we first surmised the woodpecker's impending fate, the ivory-bill still would have fallen through the crack of inattention. Leopold's concept of irreplaceability made no disciplinary home yet for endangered non-game species in this otherwise exceedingly practical discipline. The ivory-bill was basically invisible to the very profession that arguably might have been most effective at intervening directly to manipulate its population back into recovery.

Despite its imperfect scope, Leopold's vision for American wildlife management was implemented stunningly fast. Herbert L. Stoddard, Sr., who was involved in several ivory-bill debates during later decades, published in 1931 *The Bobwhite Quail*, a template for future texts on game management. During the 1930s, the wildlife management profession began certifying credentialed biologists, offering stable, equitable funding for wildlife projects, and building university training programs for new students. By 1933, the University of Michigan and University of Wisconsin had established the first curricula in game management. Leopold's classic 1933 textbook *Game Management* "recognized science as a tool for discharging [the profession's] responsibility."[22] Cooperative Wildlife Research Units for wildlife study were also placed on the nation's university campuses to facilitate collaboration with federal resource agencies.

By 1937 The Wildlife Society was founded to serve as the professional trade association and publication sponsor for the field, much like the AOU and ESA had functioned when they were established decades earlier (Chapter 4). North American practitioners would have the *Journal of Wildlife Management* and *Wildlife Society Bulletin* as outlets to share with their peers any successes and failures from experimental work on game management. Wildlife biology had arrived with a flourish, and would remain a vibrant field within the natural resource disciplines.

Yet the overlooked needs of non-game species like the ivory-bill contributed greatly to a disciplinary void. Eventually this oversight incentivized a rise in standing for conservation biology as a new, crisis-oriented field that broke wide open during the mid-1980s. Conservation biology in the U.S. was abetted notably by earlier passage of the Endangered Species Act (1973). Along with other landmark environmental laws passed during the same decade,[23] the table was set for a significantly heightened societal awareness and concern over the diminishing ranks of our natural heritage. Whatever the past neglects from the field of wildlife management, the nation's less studied non-game and especially its imperiled species needed our dedicated attention too.

Recounting the precise advent and essential features of conservation biology is fraught with more difficulty than unraveling the histories for other environmental disciplines that shaped modern constructs for the ivory-billed woodpecker. Even though some of conservation biology's philosophical underpinnings in the U.S. can be traced back to 1864,[24] the discipline as an organized profession did not arrive until an entire half-century after wildlife management arose. Consequently, conservation biology has "formed too recently to be viewed with historical detachment, and the trends shaping it are still too fluid to be easily traced."[25]

Moreover, conservation biology was criticized almost instantly as not substantially different from either ecology or wildlife biology. Critics claimed that it consisted of merely extending more widely the practice of applied wildlife ecology,[26] it restated an old split in philosophy of natural resource management noted by Leopold,[27] or even that it just placed an old wine into a new bottle.[28] A disgruntled Teer declared that "what the [Society for Conservation Biology] proposes to be, the profession of wildlife ecology and management has been for all of its history."[29]

Despite some initial heckling, conservation biology did not much listen to these distractions. Instead, the field plunged ahead, pulling in experts from all over to forge a new, eclectic profession having a common denominator that was centred around protecting all biological variety, whether expressed at molecular, population, species, community, or ecosystem levels. Fundamentally, then, conservation biology relied upon "description, explanation, appreciation, protection, and perpetuation of biological diversity."[30] This primary focus on biodiversity also distinguished conservation biology from older movements dedicated to preserving wilderness and other natural lands.[31]

In substantial part, conservation biology was a response to "a failure of traditional academic ecology and the natural resource disciplines to address modern conservation problems adequately."[32] In bringing new syntheses to an already strongly interdisciplinary field, conservation biology melded such themes as systematics, genetics, ecology, evolutionary biology, and, especially, quantitative population ecology.

What were the earlier failures in other disciplines, and how did any essential qualities of conservation biology set it apart from other environmental fields in shaping our perceptions about the ivory-billed woodpecker? First and foremost, conservation biology was and remains motivated over alarms about the irreversible nature of extinction. Concern over species extinctions was the primary worry of conservation biologists during the 1970s and 1980s.[33] No other environmental discipline has so embraced the calculated prevention of and rescue of highly imperiled species. In contrast, wildlife biology was perceived as concentrating its focus too much on single-species management for wildlife that had direct economic or recreational interest.[34]

Box 5.2. Under conditions of uncertainty, conservation practice generally favors application of the ***precautionary principle*** ("better safe than sorry"). Because this stance errs on the side of taking no risk(s) that might harm some species or habitat, the true probability of (and benefits gained by) taking on risk are not properly evaluated. This fear-based outlook stems from ***loss aversion***, a tendency that leads us to uncritically commit yet other decision mistakes from ***neglect of probability*** or succumbing to ***availability cascades*** driven by undue influence from the ***availability heuristic***.

To illustrate, if a rare and once-thought extinct species like ivory-bill were ever captured (cf. Chapter 13), the ***precautionary principle*** might justify a prohibition on radio-tracking out of concern for an individual woodpecker's welfare. But rigid fidelity to the ***precautionary principle*** in this case might blind us to a more complete risk assessment (***neglect of probability***) that would even-handedly weigh the conservation gains to the species made possible from knowing the woodpecker's spatial needs through that radio-tracking.

Several key guidelines steer the practice of conservation biology.[35] Most could be readily applied to what we wish we knew or had known about the ivory-bill. First, had ivory-bill been well distributed instead of confined to small portions of its original range, its prospects would improve. Second, large blocks of habitat containing larger populations of the woodpecker would represent an improvement over small patches with small populations. Third, habitat patches spaced closer together would be chosen over those dispersed far apart. Four, contiguous, interconnected habitat would have been better for the species than highly fragmented habitat. Five, habitat blocks that were road-less or otherwise less accessible to humans would be ideal. Six, given large uncertainties about the ivory-bill, conservation planning might emphasize less reduction and disruption of natural habitat than we might otherwise propose in instances where we possessed greater knowledge about other species. Herein the precautionary principle would be applied.[36] Seven, maintaining overall viability of ecosystem(s) for which the woodpecker was adapted, like age-diverse bottomland hardwood and other southeastern U.S. forest types, would over the long term be more efficient, economic, or effective than a piecemeal approach aimed at mitigating in isolation the various individual threats to that species.[37]

One of the major distinctions of conservation biology is its profound sense of urgency. It is first and foremost a crisis-driven discipline aimed at the need to act promptly, even before knowing all facts that we might wish were available to us.[38] This mission element gives the discipline a strong connection to the social sciences, including economics,[39] as well as a near-unavoidable ethical dimension wherein normative value is placed on preservation and restoration of the natural world for its own sake. For the ivory-bill, it would have been preferable to have this sense of resolve aimed at its protection prior to the 1930s. And no doubt it would have also been preferable to have had a better understanding of how social values both threatened and motivated any recovery prospects for the woodpecker. Instead, we entered the twentieth century mostly sightless about the significance of human values for this bird.

Principles and norms embedded in conservation biology that might have saved the ivory-bill had they been thoroughly known and applied earlier can be subsumed under the twin precepts of spatially explicit patch dynamics[40] and population viability.[41] In linking population survival of a species to its spatial needs, both concepts become vital to the practical application of conservation biology when the purpose is to recover most imperiled wildlife. Of course, the state of our knowledge a century ago was nowhere near advanced enough to be able to deliver that brand of technically detailed salvation to the big woodpecker.

> **Box 5.3.** ***Omniscience bias*** signals a false confidence from thinking we are able to know, detect, or understand something fully if enough resources are devoted to the problem. For example, the internet can easily make us think we already know or could know everything we need to about some topic, feeding an illusion that we are well-informed when in fact we are not. In matters of the ivory-bill, ***omniscience bias*** can skew the legitimate certainty we should have about the bird's life history traits or extant status.

Nevertheless, for those of us today who are even casually aware of the woodpecker's history and its presumed fate, we cannot escape carrying around certain expectations from these very modern sets of awareness. Whether we are mindful of them or not, these contemporary frames will inevitably creep into our present-day deliberations about the ivory-bill. As much as we might wish that either wildlife management or conservation biology could have ridden to the rescue of ivory-bill, however, they could not and they did not. To expect otherwise, we would only predispose ourselves to cognitive delusion, whether one of ***omniscience bias***,[42] ***hindsight bias***, or maybe both.

If we could truly strip away our restrictions on living only in the present tense, yet go back in time armed with today's knowledge, we would see just how well concealed the ivory-billed woodpecker was becoming to us during the 1910s era. The primal myth of enigmatic ivory-bill was birthed inside a crypt formed by vast technical and scientific ignorance. Witlessness over the bird's true providence would arise from a dearth of reliable information. This illiteracy would grow and metastasize over the decades until mythology about the woodpecker reached epic proportions, indeed until it touched the realm of the absurd.

Endnotes

1 1915 marked the beginning of the five-year interval in which historical specimen records for ivory-bill plummeted lastingly, the level of this decline reaching almost 400%: see Chapter 3. Numbers of geographic locations that had at least one report of ivory-billed woodpecker also declined 30% in the decade 1910–1920 compared to the prior decade (C. Hunter, *pers. comm.*).

2 Possibly or likely to include forest management as well, to the extent that stand- and landscape-level forest conditions were necessary to preserve or enhance the long-term prospects for the ivory-bill.

3 For historical and disciplinary contexts behind the development of conservation biology, see Meine, C., M. Soulé, and R.E. Noss. 2006. "A mission-driven discipline": the growth of conservation biology. *Conservation Biology* 20: 631–651.

4 Soulé, M.E. 1987. History of the Society for Conservation Biology:

how and why we got here. *Conservation Biology* 1: 4–5.

5 Hidden is not equivalent to irreversibly diminished, never mind extinct. Rather, new and changing social constructs would severely obstruct our view of what happened to the ivory-bill beginning in the late nineteenth century, with these limitations taking permanent root by about 1915 (see Chapter 6).

6 "In the war over presentism, there are few neutrals." See: Schonberger, H. 1974. Purposes and ends in history: presentism and the new left. *History Teacher* 7: 448–458. Using ***presentism*** as a device can be contentious in historical scholarship, particularly if accompanied by any "should have known betters" to justify our reshaping social policies and actions in the present tense. Here I limit the use of ***presentism*** for a widening of perspective. It can be exceedingly difficult to rise above one's own time and place when crafting historical viewpoint, anyway, leading some historians to just acknowledge that bias. See Loewenberg, R.J. 2007. "Value-free" versus "value-laden" history: a distinction without a difference. *Historian* 38: 439–454. Moreover, a common sense view provides that, necessarily, it is always true that only present objects and present outlooks exist. See Markosian, N. 2004. A defense of presentism. Pp. 47–82 *in* D.W. Zimmerman, D.W. (ed.). *Oxford Studies in Metaphysics*, Vol. 1, Oxford University Press.

7 Humans have very strong motivations to see the world as more orderly and predictable than it really is, and to avoid being blamed for failings that arise from very substantial levels of uncertainty. This tendency is a major drive behind formation in us of the ***hindsight bias***. See Roese, N.J., and K.D. Vohs. 2012. Hindsight bias. *Perspectives on Psychological Science* 7: 411–426.

8 The nineteen-teens were the first era in which range-wide extinction was seriously contemplated for the ivory-bill. Prior to that decade, extinction was rarely used universally in connection with this species. Instead, commentary typically pointed to local or regional extirpation. For example, speaking of his hope to

study and photograph ivory-bill in the Okeechobee area of Florida in 1912, Baynard referred to it as "this almost extinct bird." See: Baynard, O.E. 1914. Two months in the Everglades. *The Oologist* 31: 36.

9 For example, Snyder and colleagues synthesize numerous citations in which the ivory-bill is claimed by various authors as being common or even abundant up until about 1905. See: Snyder, N., D.E. Brown, and K.B. Clark. 2009. *The Travails of Two Woodpeckers*. University of New Mexico Press, Albuquerque, p. 23.

10 The first detailed description using modern terminology for the Model was apparently articulated by Geist, V., S.P. Mahoney, and J.F. Organ. 2001. Why hunting has defined the North American model of wildlife conservation. *Transactions North American Wildlife and Natural Resources Conference* 66: 175–185. However, the Model was essentially being implemented in practice almost a century beforehand.

11 Tober, J.A. 1981. *Who Owns the Wildlife?: The Political Economy of Conservation in Nineteenth-century America*. Greenwood Press, Westport, CT.

12 Heffelfinger, J.R., V. Geist, and W. Wishart. 2013. The role of hunting in North American wildlife conservation. *International Journal of Environmental Studies* 70: 399–413.

13 Organ, J.F., V. Geist, S.P. Mahoney, S. Williams, P.R. Krausman, G.R. Batcheller, T.A. Decker, R. Carmichael, P. Nanjappa, R. Regan, R.A. Medellin, R. Cantu, R.E. McCabe, S. Craven, G.M. Vecellio, and D.J. Decker. 2012. *The North American Model of Wildlife Conservation*. The Wildlife Society and The Boone and Crockett Club Technical Review 12–04.

14 Market hunting for wildlife parts continues to be a substantial threat to wildlife conservation, even today. See: Geist, V. 1988. How markets in wildlife meat and parts, and the sale of hunting privileges, jeopardize wildlife conservation. *Conservation Biology* 2: 15–26.

15 With a closed frontier and growing urban populace, Roosevelt and

Grinnell believed that recreational sport hunting would cultivate the country's pioneer skills and maintain the nation's fundamental moral character. See: Morris, E. 2010. *The Rise of Theodore Roosevelt*. Modern Library Press, Random House, New York; Brinkley, D. 2009. *The Wilderness Warrior – Theodore Roosevelt and the Crusade for America*. Harper Collins, New York; Miller, N. 1994. *Theodore Roosevelt: a Life*. Harper Collins, New York; Cutright, P.R. 1985. *Theodore Roosevelt, the Making of a Conservationist*. University of Illinois Press, Urbana.

16 Buehler, D.O. 1998. Permanence and change in Theodore Roosevelt's conservation jeremiad. *Western Journal of Communication* 62: 439–458.

17 Dorsey, L.G. 1995. The frontier myth in presidential rhetoric: Theodore Roosevelt's campaign for conservation. *Western Journal of Communication* 59: 1–19.

18 "It is time for us to unite in creating, through our own industry, a wealth of wild life to take the place of that which nature gave us, and we so thoughtlessly destroyed." Taken from: Huntington, D.W. 1915. *Game Farming for Profit and Pleasure*. Hercules Powder Co., Wilmington, DE.

19 Meine, C. 1988. *Aldo Leopold: His Life and Work*. University of Wisconsin Press, Madison.

20 Leopold, A. 1991. *The River of the Mother of God and Other Essays by Aldo Leopold*. Flader, S.L., and J.B. Callicott, eds. The University of Wisconsin Press, Madison.

21 Overly-narrow usage of the Model is acknowledged as problematic today, with one potential solution perceived to be broader funding sources, and a wider inclusion of other primary stakeholders and advocacy positions. Organ et al., p. 28.

22 Leopold, A. 1933. *Game Management*. Charles Scribner's Sons, New York, p. 18.

23 Other pivotal laws from the 1970s that influenced social consciousness behind the rise of conservation biology included the National Environmental Policy Act (1970), Clear Air Act

(1970), Clean Water Act (1972), Coastal Zone Management Act (1972), Federal Land Policy and Management Act (1976), Fisheries Conservation and Management Act (1976), Marine Mammal Protection Act (1972), National Forest Management Act (1976), Surface Mining Control and Reclamation Act (1977), and Resource Conservation and Recovery Act (1976).

24 "All nature is linked together by invisible bonds, and every organic creature, however low, however feeble, however dependent, is necessary to the well-being of some other among the myriad forms of life." This remarkably prescient and ecosystem-like view was given by: Marsh, G.P. 1864. *Man and Nature: Or, Physical Geography Modified by Human Action*. Harvard University Press, Cambridge, MA.

25 Meine, C. 2010. Conservation biology: past and present. Pp. 7–26 in *Conservation Biology for All*, Oxford University Press, London.

26 Wagner, F.H. 1989. American wildlife management at the crossroads. *Wildlife Society Bulletin* 17: 354–360.

27 Aplet, G.H., R.D. Laven, and P.L. Fielder. 1992. The relevance of conservation biology to natural resource management. *Conservation Biology* 6: 298–300.

28 Jensen, M.N., and P.R. Krausman. 1993. Conservation biology's literature: new wine or just a new bottle? *Wildlife Society Bulletin* 21: 199–203.

29 Teer, J.G. 1988. Conservation biology: the science of scarcity and diversity. *Journal of Wildlife Management* 52: 570–572.

30 Meine et al., p. 632.

31 Sarkar, S. 1999. Wilderness preservation and biodiversity conservation – keeping divergent goals distinct. *BioScience* 49: 405–412.

32 Noss, R. 1999. Is there a special conservation biology? *Ecography* 22: 113–122.

33 Noss, p. 116.

34 Temple, S.A., E.G. Bolen, M.E., P.F. Brussard, H. Salwasser, and J.G. Teer. 1988. What's so new about conservation biology? *North*

American Wildlife and Natural Resources Conference 53: 609–612; Edwards, T.C., Jr. 1989. The Wildlife Society and the Society for Conservation Biology: strange but unwilling bedfellows. *Wildlife Society Bulletin* 17: 345–350.

35 Adapted from Noss, p. 116.

36 See for example: Braunisch, V., J. Coppes, S. Bächle, and R. Suchant. 2015. Underpinning the precautionary principle with evidence: a spatial concept for guiding wind power development in endangered species' habitats. *Journal for Nature Conservation* 24: 31–40. The ***precautionary principle*** can be critiqued on grounds of faulty logic and cognitive bias. Implementation of the ***precautionary principle*** is prone to ***loss aversion***, our tendency to dislike losses far more than we like corresponding gains. This arises because an unfamiliar risk typically produces far more concern in us than a familiar risk, even if the latter is more probable (thereby leading us to ***neglect of probability***). Some risks are more cognitively "available" to us whereas other risks are not, so certain hazards will stand out prominently whether or not they are actually highly probable (thus causing us to succumb to the ***availability heuristic***). Consequently, "the problem with the precautionary principle is not that it leads in the wrong direction, but that – if taken for all it is worth – it leads in no direction at all." Sunstein, C.R. 2002–2003. The paralyzing principle. *Regulation* Winter 2002-2003: 32–37.

37 A piecemeal approach to ivory-bill conservation, for example, might focus only on remediation of disturbances from humans (e.g., protection from shooting) or only on habitat protection and enhancement. Conservation biology would seek first to identify all relevant threats, then rank these in importance, and finally direct efforts at those threats that would act to most reduce the future risk of extinction.

38 Soulé, M.E. 1985. What is conservation biology? *BioScience* 35: 727–734.

39 Halpern, B.S., C.J. Klein, C.J. Brown, M. Beger, H.S. Grantham,

S. Mangubhai, M. Ruckelshaus, V.J. Tulloch, M. Watts, C. White, and H.P. Possingham. 2013. Achieving the triple bottom line in the face of inherent trade-offs among social equity, economic return, and conservation. *Proceedings of the National Academy of Sciences* 110: 6229–6234.

40 Pickett, S.T.A., and P.S. White, eds. 2012. *The Ecology of Natural Disturbance and Patch Dynamics*. Academic Press, Orlando, FL.

41 Pe'er, G., Y.G. Matsinos, K. Johst, K.W. Franz, C. Turlure, V. Radchuk, A.H. Malinowska, J.M.R. Curtis, I. Naujokaitis-Lewis, B.A. Wintle, and K Henle. 2012. A protocol for better design, application, and communication of population viability analyses. *Conservation Biology* 27: 644–656.

42 Leslie, I. 2014. Omniscience bias: how the internet makes us think we already know everything. *New Statesman: Science & Tech*. http://www.newstatesman.com/internet/2014/06/omniscience-bias-how-internet-makes-us-think-we-already-know-everything (accessed 25 April 2016).

Chapter 6

Buried Alive

The report of my death was an exaggeration.
—Mark Twain

As America entered the 1910s, three factors converged to drag the ivory-bill even deeper into myth. The first contribution came from cessation of the shotgun method as a fully accredited and effective means to track the abundance and whereabouts of the big woodpecker. This transformation kept the bird largely out of our sight, misleading us to regard it as already extinct merely because we could no longer track it easily.[1] A second conspicuous influence was potency from tethering the woodpecker to broadly publicized and very real extinctions witnessed during the same decade. Losses of the passenger pigeon (1914) and Carolina parakeet (1918), especially, transfixed the nation as two of our showiest feathered icons became forever extinguished.

These two factors merged to produce a third, even more weighty influence: formation of a recurring archetype in which the ivory-bill was lost and then found again, declared extinct but later rediscovered, reckoned as deceased only to be miraculously revived. Death and rebirth forged an exceedingly potent ***meme***,[2] one certain to seize and then bind the human imagination. A ***meme*** of resurrection went on to arrange uncritically our intangible beliefs about the woodpecker. A recurring blueprint was launched, one that lent special weight to casting, however involuntarily, the ivory-bill as some sort of mythical rather than corporeal organism.

During waning years of the bird's accessibility to public view, the inventions of top-flight binoculars, GPS units, telephoto zoom lens, and high-speed film were far off into the future, making it

Box 6.1. ***Memes*** are pervasive ideas or thought patterns that replicate via our cultural and social connections; ***memes*** are fundamental yet also virtually parasitic units of information that transmit vertically (e.g., inherited within families) or horizontally (e.g., carried across social identities).

still necessary by the accepted science standards of that day to shoot birds in order to conclusively verify their identity, diet, and other habits in nature.[3] To illustrate the mindset, Ridgway's manual from 1911 on the practice of scientific bird study provided 26 pages of instruction on proper firearms, dissection equipment, packing, cleaning, and preparation methods for securing bird specimens.[4] Yet he gave merely two paragraphs on how to keep field notes about the specimen's habits in life! That skewed balance in professional ornithology presages why we learned and knew so comparatively little about the real ivory-bill.

Nevertheless, by 1915 substantial restrictions were already in place on the social practice of shooting birds (Chapter 4). A new national sensibility about avian life had set in. As early as 1912 even some scientists were remarkably blunt about what gunning down non-game birds had by then come to represent in the public eye:

> *The birds of Texas ... are on the rapid road to extermination, the result of ... the brutality of those inhuman beasts in the guise of men who annually get the 'blood lust' and go forth to slay for gain or so-called sport. To the true SPORTSMAN, lover of the woods and fields, who believes in obeying the laws, not only those of the State but also those of common decency and humanity and kills within the limit and then only such birds as can be considered*

> *GAME, I am willing to take off my hat and call brother. But for the man (?) who considers Robins, Larks and Mockingbirds fair game, I have absolutely no respect.... The Agriculturists are beginning to suffer their sin of omission in not shot-gunning the lunk-headed louts who formerly overran their fields and wooded pastures, slaughtering for the sake of killing, every living wild thing in sight....*[5]

Newfound censure over bird killing rarely factors into a serious appraisal of the ivory-bill's early conservation history. But the social disdain heaped on the discredited shotgun method cannot help but have swayed the quality and quantity of information generated as the woodpecker's prospects seemed to dim in the early twentieth century.[6] To not revise our perspective after the kind of evidence made available to us shifted so profoundly

Box 6.2. Failure to suitably revise our positions after receiving new, dispositive information leads to the mental blunders from ***Bayesian conservatism***. Our being overly careful in changing some view or belief is associated with a cognitive under-reaction to the diagnostic power of valid evidence. This bias might arise if we are overly ***anchored*** to (or vested in) some pre-existing opinion, or if we have predispositions toward questioning the reliability of the new data. ***Bayesian conservatism*** is one of many forms of bias that can issue from the ***neglect of probability***. ***Bayesian conservatism*** (also known as conservative bias) is one of several thinking errors that can befall persons who depreciate every single ivory-bill report (however much disputed) from the modern era (i.e., all those reports made after about 1944) as having any likelihood at all of being true.

risks our falling into thinking errors caused by ***Bayesian conservatism***.[7]

While the shotgun method was on the wane, however, sight records were not getting the same respectability as standing for reliable evidence inside ornithology. Instead, the professionals fretted constantly over giving in to a flimsier standard for documenting the identity and habits of birds. Avian scientists in the 1910s remained intensely distrustful, whether such information appeared in the top flight or more humble journals of that day:

> *What to do with sight records and how to be consistent in the practice of any plan that may be adopted?...'The Auk' has questioned the accuracy of certain 'sight' records published elsewhere and has in turn been criticized for certain 'sight' records that have appeared in its own columns.... 'The Condor' recently contained a severe editorial criticism of the publication of 'sight' records by incompetent observers and scored authors who have not posted themselves on the previous literature of their subject.... Certain minor ornithological journals and independent publications of 'bird clubs' consider that all is grist that comes to their mill and publish any records that their members may hand in. These statements are made not in a spirit of criticism but simply to show the difficulty of consistency and also the nature of the condition that we face.*[8]

The Auk's editor Witmer Stone wrapped up his lengthy post with a tone of exasperated resignation: "All sorts of sight records, good, bad and indifferent are being published and will be published in increasing numbers."

In other instances, authors of that era divulge the scorn they expected to get for this method when they used diffident, almost apologetic language to publish any sight records at all. For his annotated list of the birds in Sac County, Iowa, Spurrell confessed:

> *[A]ll data are based on careful sight records.... I do not expect subspecific sight records to be accorded the same value that they would have had the specimen been taken.... I have not listed some of the vireos and flycatchers which undoubtedly occur, owing to the difficulty of sight identification.*[9]

Even other animal biologists piled on. Herpetologists of the same epoch, eager to keep their own house in credible order, proclaimed: "It is unfortunate that records based on insufficient or inaccurate observations find their way from time to time into faunal lists, even though reported by 'observing gentlemen' with 'quite a taste for natural history,' as it is much easier to get them in than to get them out. Sight records with reptiles have less claim for recognition than with birds owing to the lesser difficulties which capturing the former [snakes, lizards] entail."[10]

What was to be done about vetting all of these sight records submitted by the amateur crowd, the opera-glass students of birds? In 1902, Frank M. Chapman focused "A Question of Identity" at this very problem. He did so from the perspective of an editor faced with the dilemma of trying to make responsible decisions on what constituted valid evidence for reporting on wild birds in the official literature:

> *We are frequently in receipt of reports of the occurrence of rare birds or of birds far beyond the boundaries of their normal range, while sent in perfectly good faith, are obviously based on faulty observation, though it is difficult, in fact sometimes impossible, to convince the observer of his error in identification ... when they are recorded in print they become part of the literature of ornithology and cannot be ignored, even by those who feel assured of their incorrectness ... in refusing to use the gun, must the opera-*

glass student be denied the privilege of having his observations accepted without question?... gun or no gun, we must take into consideration the mental attitude of the enthusiastic bird student afield.[11]

Just as the shotgun method was fading, then, proper monitoring of the ivory-bill was held hostage to arbitrary standards *within* the sight observation method, too. Chapman here endorsed making judgments based on an observer's presumed mental outlook, an explicit call to rely on specious ***argument from authority***.[12]

> **Box 6.3.** ***Argument from authority*** is an appeal to believe or disbelieve a claim or conclusion asserted to be true based solely on the expertise (real or not) of some authority. ***Argument from authority*** is a fallacy of irrelevance when the authority cited does not actually have the relevant expertise.

Editors and authors were praised if they worked "against the acceptance of 'opera-glass' records of rare or unusual species."[13] Moreover, Chapman's essay portends other cognitive error that would arise over disputes about bird sight records. He correctly pinpoints a complication of ***observer expectancy bias***[14] in these sightings. Struggles between the shotgun and sighting methods inside ornithology also reflected a very evident ***system justification***.[15] Scientists of that day who insisted on ignoring sight records altogether set up a propensity to commit ***omission bias***.[16]

Abandoning the outmoded evidentiary standard (collecting birds with firearms) along with inconsistency over validity of sight records led to much less reliable evidence of *any* sort being gathered to evaluate the actual status of the living ivory-bill. Our sensors for detecting the big woodpecker deteriorated badly during the 1910s. That state of affairs then became naïvely equated with demise of the bird itself.

Box 6.4. ***Omission bias*** is a tendency for us to judge *action* as more harmful than *inaction*. Skewing our decisions from this cognitive error may reflect our tendency to make harsher moral judgments (based on subjective values) about "sins of commission" as opposed to "sins of omission."

Only reinforcing this information vacuum, the bird itself seemed to be retreating to more remote strongholds. Ivory-bill searches during this decade were fraught with formidable struggle, consisting of travel hindrances that we would think of as entirely prohibitive today. In February 1914, Frederic H. Kennard, teamster Peter Hogan, and guide Tom Hand set out to look for the woodpecker in Florida's distant expanses of the Big Cypress Swamp.[17] After leaving their auto at the edge of the swamp, they used a wagon shaped like a prairie schooner and two yoke of oxen to transit the vast wilds. Oxen had to be employed in the expedition because their spreading toes were less likely to get bogged down in the quagmire than those of mules or horses.

Besides cutting trees as levers to extricate themselves and the wagon from soft clay, the trio at times had to cut enough timber to build a miniature corduroy road in order to make any progress. "We waded through miles of swamp, crawled through miles of jungle, dodging snakes, and devoured by red bugs, our necks stiff from searching the tree tops for possible [ivory-bill] nests."[18] After further struggles with lost and sick oxen, they were rewarded with seeing (and of course, shooting) a lone female ivory-bill. Despite spending almost a month searching in this region of south Florida, and chasing down second-hand rumors of other ivory-bills as well, only one bird was ever found.

Kennard concluded with a downbeat summary of their expedition: "Our trip, so far as ivory-bills were concerned, had been pretty discouraging. We had secured one specimen, to be sure, but had found no nest, and had learned but little of the birds."[19] From this and other similarly frustrating missions to pursue a now rapidly waning species, we can see foreshadowed how inescapably a ***meme*** of the quest also would be foisted on future searches for the ivory-billed woodpecker.

Commentary during the 1910s began to state more assertively that the ivory-bill was either gone or had departed from much of its former range. "The large, handsome Ivory-billed Woodpecker has been exterminated in this State [South Carolina]."[20] Referring to Texas, and making clear parallels with genuine extinction, Strecker exclaimed: "many of these [other bird species] are soon to be numbered with the ivory-billed woodpecker and Carolina paroquet as birds which have recently become extinct!"[21] Here we see already presaged the linkage from other authentic extinctions now extended to the still living ivory-bill. Strecker admitted later that the ivory-bill's official status in his state was "now almost extinct." In nearby Arkansas, the woodpecker stood "on the verge of extinction" although "a few are believed to remain in the wilder parts of southeastern Arkansas."[22]

Between 1910 and 1920, other writers could only draw passing attention to the dwindling numbers of ivory-bill. In the Okefenokee region of Georgia, it was considered "rare, but still existent in small numbers in the northwestern part of the swamp."[23] By 1913 the ivory-bill was "very rare" in Alachua County, Florida.[24] In southwest Florida, the ivory-bill had joined several other bird species as "missing ... forever," even though it was deemed inexplicably to have been comparatively abundant just a decade earlier.[25]

Frank M. Chapman pronounced the ivory-bill as "common in but few localities," its home confined now to the "almost limitless cypress forests of our southern coasts and river valleys."[26] Beal warned: "Unfortunately, it appears to be rapidly becoming extinct ... and is nowhere numerous."[27] Yet after pointing out the woodpecker's value "as a guardian of the forest," Beal went on to state with somewhat more confidence:

> *Wise legislation, backed by intelligent public opinion, may retard, if not absolutely prevent, the present destruction and allow the bird to regain something of its former abundance. There is plenty of room for this splendid species and much need of its services in the great southern forests.*[28]

So, and despite the alarms, prospects for the ivory-bill were not seen as entirely hopeless during this era. Indeed, the messages transmitted about the woodpecker were decidedly mixed, a feature that seems often to have been the case with this inscrutable species.

Confusion over the fate of ivory-bill likely arose too by its having been swept up prematurely in the extinctions of other species, a natural proclivity during an era when the prospects for many species were still so dire. Earliest years of the twentieth century were characterized both by a rapid disappearance of much of America's native wildlife and somber warnings of even more extinctions to come. Some early speculation listed the ivory-bill as already extirpated or extinct in certain regions. The bird was gradually being buried alive.

William Temple Hornaday (1854–1937) gave one of the most thoughtful warnings and urgent set of predictions ever about American wildlife extinctions. Director of the New

York Zoological Park (today's Bronx Zoo), Hornaday's global travels had convinced him that nowhere was nature being destroyed so rapidly as in the United States. His 1913 opus *Our Vanishing Wild Life: Its Extermination and Preservation*[29] exhorted his fellow citizens to use every means at their disposal – sentimental, educational, and legislative – to engage in the cause of preservation. Considered among the most influential conservationists of the nineteenth century, Hornaday's treatise was far ahead of its time. Not only did his championing of wildlife draw attention to the excesses of market hunting, his audacious, pugnacious, and intrepid style[30] were engaged in an outright war for conservation that dispensed with "any handicaps of false modesty."[31]

Our Vanishing Wild Life described the former plenitude of American wildlife, summarized the nation's avian extinctions, and identified the country's likely next victims. Hornaday's book reveals important context for the ivory-bill because it unmasks what the nation then believed about animal extinction, including a check on how good we actually were at predicting such irreversible declines. Hornaday listed six birds as already extinct by 1913: great auk, Labrador duck, Eskimo curlew *Numenius borealis*, Pallas' (spectacled) cormorant *Phalacrocorax perspicillatus*, passenger pigeon (by then believed only to occur in captivity), and Carolina parakeet. Note that Hornaday hardly was perfect on this score of extinction. Despite his insistence that "the species is absolutely dead," the Eskimo curlew was certainly not yet gone.[32] Although for the Carolina parakeet he thought that "there is no reason to hope that any more wild specimens ever will be found," undocumented reports for this species continued in the Okeechobee region of south Florida until the 1920s.

In his chapter "The Next Candidates for Oblivion," Hornaday listed a "partial list" of 23 species of North American birds threatened with early extermination.[33] These included but one species that in truth went extinct, the heath hen. Ivory-bill was not even among that list. Although he also prophetically identified a few species that would remain threatened, endangered, or candidates for these protections at federal or state levels a century later (whooping crane *Grus americana*; trumpeter swan *Cygnus buccinator*; black-capped petrel *Pterodroma hasitata*; Gunnison's sage-grouse *Centrocerous minimus*; Attwater's prairie-chicken *Tympanuchus cupido attwateri*), the simple fact is that more than 70% of Hornaday's predictions were wrong, even absent the gross oversight for ivory-bill!

Incredibly, nowhere did Hornaday draw special attention to the potential or imminent extinction of the woodpecker. After querying 250 experts state-by-state, and acknowledging that data were incomplete, his breakdown for ivory-bill extirpations listed only Indiana, Missouri, North Carolina, South Carolina, and Texas as then missing the big woodpecker. Hornaday's assessment was entirely wrong for at least the three latter states.[34] His track record was only 5% accurate for predicting correctly our national bird extinctions by species. During an era in which the ivory-bill mostly escaped our attention, this check on human forecasting ability is not at all reassuring (see also Chapter 14).

During the 1910s, two real extinctions of prominent American birds unfolded in front of a captivated national audience. Losses of the charismatic passenger pigeon and the Carolina parakeet were extended to other rare species of the day. This fueled speculation that other persecuted birds like the ivory-bill were sure to soon follow these two birds into a spiral of oblivion.[35] Images of these last of their kinds languishing inside the Cincinnati Zoological Gardens evoke a sort of palliative ward for terminally ailing wildlife. One imagines a desperate clock ticking down, only the fickle longevity of individual rebels reminding us of what once were inestimable races of these wild birds.

Expirations in these lasts of their kind were public spectacles.[36] Lone survivors became national celebrities during their final days. After two male companions had died in the same zoo in 1910, "Martha," named for George Washington's wife, passed on September 1, 1914, at the advanced age for a passenger pigeon of 29 years. While alive, she attracted long lines of curious visitors. At one time the pigeons had darkened the skies by the billions, their migratory flights lasting days. But after ceaselessly being shot, netted, and burned for the commercial market, our hunting and clearing of eastern forests so disrupted the pigeon's social breeding system[37] that it was gone before we could react.

Less than four years later, the story repeated for the Carolina parakeet, the nation's only native, endemic species of parrot. The last parakeet, "Incas," would die on February 21, 1918, within a year of his mate, "Lady Jane." The birds expired publicly in the very same pagoda-shaped aviary at the zoo that had once showcased "Martha." The parakeet's bawdy screeches and gaudy yellow, orange, and green plumage had once painted the country from Florida and the Gulf coast north to Colorado, the upper Midwest, and New York. Now it too was gone.[38]

Watching these two elegant birds admitted to the intensive care ward of captivity, only to live out their last days in a slow decline, must have fueled a grave sentiment to those concerned with extinction. National vigils over these very prominent conservation failures were rather prolonged, rendering a forlorn and melancholy finale as the fitting closure for our greed. No wonder Hornaday tried so diligently to shock his compatriots into action about redressing quickly the problem of extinction.

Views of the ivory-bill were likely molded too by wider cultural forces at work shaping our outlooks in post-frontier America. Just a few years prior to Tanner's study of the ivory-bill, "Shadow Catcher" Edward S. Curtis published a seminal work photographing native peoples based on a project that he began at the turn of the century. Curtis and his contemporaries tended to portray American natives as members of a noble but vanishing race.[39] Michaels suggests that similar views were likely to have been propagated about the woodpecker: "Tanner's description of the Singer Tract and especially his ivorybills are reminiscent of Edward S. Curtis's Native Americans, magnificent, worthy of honor, but ultimately doomed and incapable of adapting."[40]

Images of calamity fed only half the resurrection meme being fabricated for the ivory-bill. Because we could scarcely see the big woodpecker during and after the 1910s,[41] surely it too must be shadowing the pigeon and parakeet along that dark path. It could be only a matter of time before the ivory-bill too was gone for good. Our legitimate concern over extinction, however, set us up for profound miscalculation. All that was required now to fix the woodpecker in myth was to complete the other half of resurrection[42] – a rebirth. We would indeed get that very rider, and not just once. Indeed, we would fall victim to this cycle over and over again.

A first resurrection from the grave occurred in 1924.[43] Just seven years prior, T. Gilbert Pearson had pronounced the ivory-bill "now nearly extinct ... unable to withstand the advance of civilization ... soon [to] be numbered with the lengthening list of species that have passed away."[44] In the winter of 1924, Arthur A. Allen, founder of Cornell Laboratory of Ornithology, spent his sabbatical time searching for rare birds in the southeastern U.S. Although the ornithological community had written off ivory-bill, Allen and his wife, Elsa, along with local guide Morgan P. Tindall, managed to find, photograph, and study briefly a breeding pair of the woodpeckers in central Florida.[45] Before they could return for more thorough research, however, local taxidermists shot both birds, apparently under permit,[46] eliminating what was then believed to be the entire then-known population of the woodpecker.[47] Despite the best of intentions, Allen may have agonized that he had inadvertently "helped to seal the fate of America's rarest bird."[48]

> **Box 6.5.** "Out of sight, out of mind" signals inability to perceive some item as still in existence if it is not immediately detectable. An inclination towards ***object impermanence*** (or ***lack of object constancy***) normally fades in childhood, but it can still persist in adults. Those persons who are more reflective in outlook allocate their attention to all accessible information. In contrast, the more impulsive among us are influenced by the incentive value of proximate stimuli, we overlook broader information, and consequently we get easily tricked by this thinking error. Subjective belief surrounding this ***object impermanence*** can even lead some individuals to conclude that recurring (but still real) objects possess a supernatural or magical essence.

Just like the fabled phoenix,[49] though, the ivory-bill was not yet finished. After wandering for a while in the netherworld of occluded ornithology,[50] it came back again, this time in 1932. Its first try at resuscitation was through a claim by Charles N. Elliott that it persisted still in the Okefenokee Swamp.[51] This assertion was roundly derived by noisy doubters who exacted a physical proof: "Show us or we cannot believe ... the King is dead."[52] A blustery, linen-suited attorney, Mason D. Spencer, did just that, boasting he could bring one in to the Louisiana Department of Conservation if they would only grant him a permit. He got that authorization, and Mason delivered on his promise with an ivory-bill corpse, launching yet another round of discovery and loss for the woodpecker (Chapters 7, 10). Even decades later on, the ivory-bill was still going "extinct again."[53]

So it was that a cycle of death and rebirth became entrenched in our idle musings over the ivory-bill. Through mere flukes of poorly corroborated natural history, we granted the woodpecker the most crucial attribute ever needed by deity – a dying and a rising. A motif for transcending mortality was hereby bestowed unwittingly on this bird, much like those granted to the Egyptian Horus, Old Norse Baldur, Hindu Krishna, Christian Jesus, and Aztec Quetzalcoatl. Each starting as an entity that lived among us, then died, then resurrected. Rebirth is such a potent symbol for our gullible species.[54] Little wonder that we went on to concede so much metaphorical authority to a mere bird.

Endnotes

1 Suddenly fewer detections might occur if the woodpecker was truly rarer, because it had become increasingly wary as a result of persecution (see Chapter 9), or because the shotgun "observation" method had by then become mostly obsolete (with no truly effective replacement). These three factors might have reinforced each other, fortifying a cryptic façade that fell quickly over the ivory-bill beginning in the 1910s, and persisting through later decades.

2 An idea, concept, theory, paradigm, style, behavior, object, or other attribute that spreads rapidly among individuals in a culture through written and oral communication. Erikson, T, and T. Ward. 2015. The Master Communicators Handbook. Changemakers Books, John Hunt Publishing. ***Memes*** are perceived as the cultural analogs to the transmission of and selection for genes in biological systems. See: Dawkins, R. 1989. *The Selfish Gene*. Oxford University Press, London.

3 See editing comments from Edgar Kinkaid for the ivory-bill account in Oberholser, H. C. 1974. *The Bird Life of Texas*. University of Texas Press, Austin.

4 Ridgway, R. 1911. *Directions for Collecting and Preserving Specimens*. Bulletin No. 39. Smithsonian Institution, United States National Museum, Washington, DC.

5 Strecker, J.K., Jr. 1912. *The Birds of Texas: An Annotated Checklist.* Baylor University Bulletin, Volume 15, No. 1. Baylor University Press, Waco, TX, pp. 3–4.

6 Changed social norms may have either slowed down the actual shootings of ivory-bill or reduced the reporting of such incidents. Regardless, the number of geographic locations that reported by sight alone at least one ivory-billed woodpecker plunged from 20 between 1900–1909 to only 6 in 1910–1919. By the 1920s decade, the number had fallen further, to just 2 locations (W.C. Hunter, *pers. comm.*).

7 A tendency to revise one's beliefs too little when presented with new, extra, or better evidence.

8 Stone, W. 1917. Correspondence: subspecific designations. *The Auk* 24: 373–376.

9 Spurrell, J.A. 1919. An annotated list of the land birds of Sac County, Iowa. *The Wilson Bulletin* 31: 117–126.

10 Babcock, H. L. 1920. Some reptile records from New England. *Copeia* 85: 73–76.

11 Chapman, F.M. 1902. A question of identity. *Bird-Lore* 4: 166.

12 An illogical premise in which a person who is perceived to be an expert affirms a proposition to a claim merely by stating the proposition to be true.

13 Swarth, H.S. 1914. Publications reviewed: Birds of Connecticut. *The Condor* 14: 97.

14 A cognitive bias in which observers or researchers expect a given result, and therefore unconsciously manipulate an experiment or misinterpret observations (or data) in order to find or reinforce it.

15 A tendency to defend and justify the status quo even if it is demonstrably contradicted, or even injurious. A theory from social psychology that proposes that humans need to maintain stability and order in their lives, even if the premises for that stability are flawed.

16 This asymmetric stance towards sight records indicates professional ornithologists were more concerned about acts of

commission (accepting as true sight records of birds that were erroneous) than acts of omission (not accepting legitimate sight records at all, or only some legitimate sight records).

17 Kennard, F.H. 1915. On the trail of the ivory-bill. *The Auk* 22: 1–14.

18 *Ibid.*, p. 10.

19 *Ibid.*, p. 14.

20 Williams, B. 1916. *The Decrease of Birds in South Carolina*. No. 47. University of South Carolina, Columbia. Of course, the ivory-bill was not yet extinct in that state, with accounts appearing from later eras and extending possibly up through the present day. See, for example, the account by Moskwik, M., T. Thom, L.M. Barnhill, C. Watson, J. Koches, J. Kilgo, B. Hulslander, C. Degarady, and G. Peters. 2013. Search efforts for ivory-billed woodpecker in South Carolina. *Southeastern Naturalist* 12: 73–84.

21 Strecker, p. 4.

22 Howell, A.H. 1911. *Birds of Arkansas*. Bulletin No. 38. U.S. Dept. of Agriculture, Biological Survey, Washington, DC, p. 46.

23 Wright, A.H., and F. Harper. 1913. A biological reconnaissance of Okefinokee Swamp: the birds. *The Auk* 30: 477–505.

24 Baynard, O.E. 1913. Breeding birds of Alachua County, Florida. *The Auk* 30: 240–247.

25 Phelps, F.M. 1912. A March bird list from the Caloosahatchee River and Lake Okeechobee. *The Wilson Bulletin* 24: 117–125.

26 Chapman, F.M. 1916. *Handbook of Birds of Eastern North America*. D. Appleton and Company, New York, p. 323.

27 Beal, F.E.L. 1911. *Food of the Woodpeckers of the United States*. Bulletin No. 37. U.S. Dept. of Agriculture, Biological Survey, Washington, DC, p. 62.

28 *Ibid.*, p. 63.

29 Hornaday, W.T. 1913. *Our Vanishing Wild Life: its Extermination and Preservation*. Charles Scribner's Sons, New York.

30 Although once an ardent taxidermist himself, Hornaday had been transformed in his later years into a zealous protectionist: "We are weary of witnessing the greed, selfishness and cruelty of 'civilized'

man toward the wild creatures of the earth. We are sick of tales of slaughter and pictures of carnage. It is time for a sweeping Reformation; and that is precisely what we now demand." Ibid, p. x.

31 Bechtel, S. 2012. *Mr. Hornaday's War: How a Peculiar Victorian Zookeeper Waged a Lonely Crusade for Wildlife that Changed the World.* Beacon Press, Boston, MA.

32 Another North American bird buried too soon in the mausoleum of extinction. A specimen was collected in 1963 from Barbados, and unconfirmed sight records for the bird continued from 1981 through 2006; BirdLife International. 2015. Species factsheet: *Numenius borealis*. Accessed from http://www.birdlife.org on 20 April 2015.

33 Hornaday, p. 18.

34 For several much later records from North Carolina, see: Hunter, W.C. 2010. *Interpreting historical status of the ivory-billed woodpecker with recent evidence for the species' persistence in the southeastern United States.* Appendix E, Recovery Plan for the Ivory-billed Woodpecker (*Campephilus principalis*), U.S. Fish & Wildlife Service, Atlanta, GA. For South Carolina, see: Moskwik et al. 2013. In Texas, there were at least 25 published records of ivory-bill after 1913. See Shackelford, C.E. 1998. A compilation of published records of the ivory-billed woodpecker in Texas: voucher specimens versus sight records. *Bulletin of the Texas Ornithological Society* 31: 35–41.

35 A major concern of avian conservationists during turn-of-the-century America was a troubling singularity wherein as the numbers of a birds dwindled, their value increased greatly to collectors, whose accelerated takings further diminished any relic populations. Barrow, M.V., Jr. 1998. *A Passion for Birds: American Ornithology after Audubon.* Princeton University Press, Princeton, p. 103; and also Appendix A.

36 Reinforcing a somber mood of that era, this decade in American history was also the very last in which a very few native peoples managed to survive free and outside the total domination of

European influence. See, for example: Kroeber, T., and K. Kroeber. 2004. *Ishi in Two Worlds: A Biography of the Last Wild Indian in North America*. University of California Press, Berkeley and Los Angeles.

37 Bucher, E.H. 1992. The causes of extinction of the passenger pigeon. *Current Ornithology* 9: 1–36. Springer Publishing Company, New York.

38 Snyder, N.F. 2004. *The Carolina Parakeet: Glimpses of a Vanished Bird*. Princeton University Press, Princeton, NJ.

39 Kennedy, M.H. 2000. Review of *Edward S. Curtis and the North American Indian, Incorporated* by Mick Gidley. *Great Plains Quarterly* Winter 2000: 79–81.

40 Michaels, M. 2014. How the ivory-billed woodpecker might have survived. http://projectcoyoteibwo.com/2014/08/05/how-the-ivory-billed-woodpecker-might-have-survived/ (accessed 11 February 2016).

41 Psychological inclinations towards perceptual errors caused by ***object impermanence*** are greater in adults who are impulsive and less reflective in their cognitive outlook. den Daas, C., M. Häfner, and J. de Wit. 2013. Out of sight, out of mind: cognitive states alter the focus of attention. *Experimental Psychology* 60: 313–320. When objects are manipulated to change or seemingly disappear, e.g., "the ivory-bill went extinct decades ago," individuals may either leap to some magical explanation or deny altogether upon re-seeing this "disappeared" object. See: Subbotskii, E.V. 1991. Existence as a psychological problem: object permanence in adults and preschool children. *International Journal of Behavioral Development* 14: 67–82.

42 "Would there be – could there be – another resurrection of the Lord God Bird?" Cokinos, C. 2001. *Hope is the Thing with Feathers: A Personal Chronicle of Vanished Birds*, Warner Books Inc., New York, p. 65.

43 The 1920s were not an optimistic time; extinction in ivory-bill might well have taken root as inevitable during this especially gloomy era. The harsh realities of World War I left many Americans

disillusioned about human progress. Kennedy, D.M. 2004. *Over Here: The First World War and American Society*. Oxford University Press, Oxford and New York. After horrors of poisonous gas and other combat "innovations" from humanity's new-found technical prowess, a notion that science could simply lead mankind to a beneficial relationship with nature was among the several casualties.

44 Pearson, T.G. 1917. *The Bird Study Book*. Doubleday, Page & Company, New York, p. 129.

45 Allen, A.A., and P.P. Kellogg. 1937. Recent observations on the ivory-billed woodpecker. *The Auk* 54: 164–184. See especially plate 11, apparently the first photograph ever of an ivory-bill, although one that would never be considered to serve as definitive evidence today.

46 Jackson, J. A. 2006. *In Search of the Ivory-billed Woodpecker*, 2nd ed. HarperCollins, New York, p. 118.

47 These shot ivory-bills were not necessarily the same birds that had been seen earlier by the Allens. Indeed, these two birds were the only firm evidence for the species persistence during the entire decade. Chuck Hunter, *pers. comm.*

48 Hoose, P. 2004. *The Race to Save the Lord God Bird*. Farrar, Straus, and Giroux, New York, p. 58.

49 The phoenix possesses striking parallels with at least the allegorical ivory-bill, including such qualities as longevity, resurrection, immortality, eternal youth, paragon (uniqueness), self-sufficiency, complete transmutation, heraldry, and in psychology, also dream, change, and living from moment to moment. See De Vries, A. 2004. *Elsevier's Dictionary of Symbols and Imagery*, Elsevier, Amsterdam, pp. 439–440. As embedded in Egyptian, Greek, Arab, Jewish, Russian, Japanese, Native American (thunderbird), and other cultural myths, the phoenix and its various analogs also represent our own aspirations for immortality, eternal life, destruction, creation, and fresh beginnings.

50 Another mythological parallel to our socio-cultural handling of

the ivory-bill can be made with the Greek goddess Persephone, who alternated her existence between earth and the underworld.

51 Elliott, C.N. 1932. Feathers of the Okefenokee. *American Forests* 38: 202–206, 253.

52 Bird, A.R. 1932. Ivory-bill is still King! *American Forests* 38: 634–635, 667. Thus began a snarky, condescending tenor used by doubters, skeptics, and deniers when levying criticism at future prospectors of the woodpecker's continued survival. Note in Bird's account a tone of derision, and use of the moniker "King," a prescient if entirely inadvertent connection with Elvis.

53 Achenbach, J. 2006. Ivory-billed woodpecker goes extinct again. *Washington Post*, March 16. See: http://voices.washingtonpost.com/achenblog/2006/03/ivorybilled_woodpecker_goes_ex.html (accessed 1 May 2015).

54 See for example: Jung, C. 1964. *Man and His Symbols*. Anchor Press, New York. Resurrection myths shape human experiences that range from social initiation practices to psychological developmental stages, while also symbolizing such rituals as purification and redemption, fertility and growth; see Henderson, J.L., and M. Oakes. 1990. *The Wisdom of the Serpent: The Myths of Death, Rebirth, and Resurrection*. Princeton University Press, Princeton, NJ.

Enforcing Myth: 1932–2004

Chapter 7

Consistent Contradiction

Contradiction is not a sign of falsity, nor the lack of contradiction a sign of truth.
—Blaise Pascal

Our reading room for the ivory-billed woodpecker is full of blatant contradictions. Beyer grasped this riddle back in 1900: "accounts of the habits of this species seem to be considerably at variance."[1] Hoyt was similarly adamant in 1905: "I can state some facts that do not agree with other writers."[2] Decades later Allen and Kellogg, too, underscored that "some of our observations are at variance with those published by others."[3] And by the time the ivory-bill was fading from its perceived last stronghold in the Singer Tract, Pough confessed "that the ivory-bill problem puzzles me exceedingly, and I do not feel that Tanner's report begins to explain the reasons for the drastic decline in this species."[4] Throughout its documented history, then, ivory-bill refused to be easily branded with our set conventions. Yet these weighty contradictions are a very reason to throw open fresh inspection of the woodpecker's cultural and ecological history.

Revelations disclosed here are limited to samples of this contrary reporting, mostly leaving for elsewhere the deeper explanations for such stark inconsistency. Despite stressing the many conflicts, it is not my goal to cast aspersions on particular authors, points of view, or theses offered about the woodpecker. To make room for larger perspective, it is necessary to show that contradictions are real, sizable, and widespread. This exercise seeks not to discount, dismiss, or discriminate among all incongruities anyway, but instead embrace as many of them as possible.

For convenience, I organize coverage of these inconsistencies into three parts: 1) the woodpecker's basic life history, 2) self-contradictions that originate from reversals in commentary by the same author, and 3) arbitrary rules for gauging historical reliability of the scholarly record. In some places the year that a quoted statement was made helps set the individual or collective statements into a fuller historical alignment. The examples presented are illustrative, by no means exhaustive.

Contradictions in life history traits

Identification and body size in life

Not even the ease of the bird's identity or apparent size in the field can be agreed upon. "It is readily distinguishable from the pileated woodpecker in flight by the large amount of white on the wings. Its call is quite different, too."[5] Some writers who saw the bird in life were quite clear that it was noticeably large. They also based that experience on direct comparison to the pileated woodpecker:

> *its extremely large size – longer and heavier than I would have expected....*[6]
>
> *The ivory-bill is decidedly larger than the pileated, and this difference in size is very apparent, as we had ample opportunity to observe, when by chance birds of both species fed at the same time on a tall decayed stump within 80 feet of our hiding place....*[7]

One ornithological dignitary confirmed that the sizes of the two big woodpeckers were alike, but that everything else was so different that it was next to impossible to confuse them with each other:

[I]ts color, its actions (particularly its manner of flight), and its notes are so totally different that once seen it need never be mistaken for that species [pileated], or vice versa. The pileated woodpecker is a noisy, active bird, always in evidence from its loud yelping or cackling notes or its restless movements. The ivory-bill, on the other hand, is comparatively quiet and secluded....[8]

Yet despite accounts of first-hand observers, supposed trials in separating the two large, North American woodpeckers in the field were (and still are) put forth as ample justification to denigrate practically any historically recent sighting, i.e., those after 1944. Tanner did not consider the size difference reliable unless the two species were seen together, so he used a putative confusion with the pileated to ignore from any serious consideration some 22% of the 45 regions in which he very briefly looked into ivory-bill accounts of the species' persistence through the late 1930s.[9]

Confusing the pileated with ivory-billed woodpecker continues to be arrayed to discredit visual observations up through the present day (see also Chapter 11). "The problem is that to the untrained eye, the common pileated woodpecker ... looks like an ivory-bill."[10] Referring to observers in the Arkansas search of ivory-bills during 2004–2005, Jackson opined:

[F]or an individual to be able to say that an ivory-billed woodpecker in flight, at 100 meters, is much larger than a pileated woodpecker implies an ability to easily distinguish between a meter stick and a yard stick down the length of a football field....[11]*I find it difficult to believe than such size judgments could be anything but wishful thinking.*[12]

Overstating similarity in size between the woodpeckers was deployed repeatedly to discredit the Arkansas reports. Yet this is impossible to rationalize on objective grounds because "on average, the ivory-billed woodpecker is 80% heavier, is 15–20% longer, has 15–20% longer wings and wingspan, has a 35–40% longer bill, and has a significantly longer neck that was described as noticeably distinctive in flight by virtually every naturalist who wrote about the species."[13]

What makes putative confusion between the two large woodpeckers even more illogical is not an utter lack of similarity between them, but rather the total lopsidedness with which this mix-up is ever allowed to occur. We are permitted, demanded even, to believe observers easily confuse(d) the common pileated for the rare ivory-billed woodpecker. But if the two species are or were so alike, where are the comparable warnings that caution observers about confusing the ivory-bill with the pileated? Why does a fancied resemblance between these two birds run in just the one direction?[14]

Flight

Audubon was first to describe the ivory-bill's flight, portraying it vividly as:

> *graceful in the extreme, although seldom prolonged to more than a few hundred yards at a time, unless when it has to cross a large river, which it does in deep undulations, opening its wings at first to their full extent and nearly closing them to resemble propelling impulse ... performed by a single sweep ... swinging itself ... forming an elegantly curved line....*[15]

The account here by Audubon differs so markedly from ivory-bill flight noted by other observers (but is quite similar to pileated) that one cannot help but question if he separated the two species correctly in all field conditions.

Allen and Kellogg took issue with Audubon's portrayal,[16] describing instead a direct, duck-like flight more like a red-headed *Melanerpes erythrocephalus* than pileated woodpecker.

Tanner echoed this description, pointing to its strong pintail-like flight with steady wing beats, and also emphasized the bird's long flight distances, typically a half mile or more.[17] But Tanner also thought the two woodpeckers at times shared flight characteristics, with the pileated able to "fly directly, in no way different from the flight of the larger bird."[18] On the other hand, Dennis' "recollection of the flight of the ivory-bill [was] that it was never undulating, but always in a straight line."[19]

Colloquial names

Ivory-billed woodpecker was given dozens of local European-language names by rural peoples, in at least four languages.[20] These include Indian hen, kate, log-cock,[21] wood-chuck, and woodcock. All five of these terms could refer to the pileated woodpecker as well. Moreover, Lord-God[22] has been pointedly reserved for the ivory-bill, yet it too was used for the pileated. Because researchers often relied upon colloquial names for the ivory-bill in their interviews of locals in order to evaluate sites or to narrow down search efforts in the field,[23] any dual use of names could lead easily to erroneously attributed identification, habits, location, or abundance.

Sounds and vocalizations

After reviewing a glut of inconsistent descriptions on the auditory behavior of ivory-bill, Jackson questioned: "Is one person right or another wrong with regard to the nature, frequency, and diversity of ivory-bill vocalizations?"[24] Mnemonics used to describe the bird's most common call note, whether or not the species called in flight, and noisiness of birds engaged in foraging behavior were all reported in contradictory fashion by naturalists. Moreover, the only unquestioned sound recording of ivory-bills came from the Singer Tract, and these are consequently limited solely to the behavioral context(s) in which those recordings were made. Also never measured in the ivory-bill was any diversity of vocalizations typically given by birds among individuals and across their once rather extensive range.

Distribution

Tanner concluded that by 1888 the woodpecker "had disappeared from ... all the region west of the Mississippi Delta except the very southeastern part of Texas...."[25] But Cooke's 1914 review of the winter status of birds in eastern Oklahoma 25 years later still remarked nonchalantly that ivory-bills were "not considered by the local hunters as any great rarity."[26] (See also **Self-contradictions**; both Wayne and Tanner provided contrary information about the presence of ivory-bills in the Wacissa and Suwannee drainages, respectively, of north FLorida.)

Local abundance

In March and April of 1890, William Brewster, Frank A. Chapman, and Slover Allen, accompanied by a savvy woodsman named Du Bose, traveled 120 miles down the Suwannee River from Branford, Florida to the Gulf of Mexico. In almost 3 weeks, they heard one and shot a second ivory-bill, both within 20 miles of the Gulf. That experience plus interviews with locals led them to regard the bird as rare.[27] Yet only two years later, leaving from the same town, taking an identical route, and traveling during the same months, Arthur T. Wayne took *13* specimens, and saw but thankfully spared an additional *10* ivory-bills.[28]

Historical abundance

Accounts for historical abundance in the ivory-bill are all over the place. Described as rare in the early part of the twentieth century up through Tanner's day, Snyder and colleagues instead postulated a "conspicuous early plenitude" that made it even more prevalent than pileated woodpecker in certain portions of Louisiana and Florida.[29] Yet in 1785 British naturalist Thomas

Pennant had described the ivory-bill "in North America a scarce bird."[30] Although Tanner too believed the ivory-bill "never was common,"[31] Snyder and colleagues cite 20 sources that they believed showed the species to be "formerly very common" or even "abundant" between 1837 and 1905.[32] Until the early 1920s, it was still listed as "uncommon" in the Pascagoula Swamp of Mississippi.[33] Tanner, however, quoted an anonymous author speaking as early as 1879 that "this bird is not at all abundant, and specimens may be regarded as good additions to one's cabinet."[34] In 1868 Elliott Coues also considered the ivory-bill as "rare."[35]

Paradoxes in these wide-ranging accounts about ivory-bill abundance dissolve substantially after abandoning the notion that the woodpecker resided in fixed areas and defended breeding territories. As Richard Pough remarked in 1944: "Such contradictory accounts would be just what one would expect of a long-lived, highly-mobile species, which probably did not breed except in areas where an abundant food supply was temporarily available within a small area, as a result of a fire, cyclone blow-down or other form of die-off."[36]

Range

At least since the time of Audubon, ivory-bills were deemed to have a confined range across limited portions of the southeastern United States. Audubon regarded the bird as "never having been observed in the Middle States within the memory of any person now living there."[37] Nonetheless, on March 7, 1780, Colonel William Fleming saw a pair of well-described ivory-bills, one of which was shot, near Stanford, Lincoln County, Kentucky, just west of the Cumberland Plateau.[38] Jackson, too, speculated that ivory-bill may have once had a wider range in the inland or upland "virgin forests" of the southeastern U.S.[39]

Population density

Based on his Louisiana study, Tanner gave a maximum density

of one pair of ivory-bills per 6 square miles, although he also gave even lower estimates of one pair per 10 to 17 square miles elsewhere.[40] Snyder and colleagues computed estimates with different data and assumptions. Using reports by Wayne from the late 1800s, numbers of birds seen and the area over which this avid collector roamed, they argue for ivory-bill densities in north Florida from 5 to 18 times higher than Tanner's. Considering earlier depletion by Wayne's collection of specimens in the same region, Tanner's estimate may well "have been more than an order of magnitude too low."[41] Because the ivory-bill was never known to defend territories like other woodpeckers, attempts to estimate density in this species using conventional means were fraught with flawed assumptions.

Habitat affinity

Other than a concession to wider habitat use in Florida, Tanner considered the ivory-bill as having had "almost always lived in virgin or primitive stands of timber" in "bottomland forests where sweet gums and oaks predominate."[42] Within this narrow set of conditions, Tanner placed these birds within stands having recently (2–3 years) dead trees with more large wood boring insect larvae. Tanner usually discounted ivory-bill use of upland pine forests in the southeastern U.S., although he allowed that originally "fire probably was the most important agent of timber death"[43] in creating favorable habitat conditions for ivory-bill.

Such narrow habitat specialization was questioned, however, and omissions were partially corrected by later scholars. Due to an earlier loss of this forest type to logging, Jackson regarded loss of old-growth pine forests as overlooked in so far as its potential role in contributing to ivory-bill decline.[44] He and both late nineteenth- and early twentieth-century writers provided substantive evidence that ivory-bills used pine woodlands for both nesting and feeding, including outside Florida in at least the Carolinas, Alabama, and Texas.[45]

Diet

Tanner asserted emphatically: "Wood-boring larvae are unquestionably the most important food of the Ivory-bill."[46] Along with scaling tight bark from recently dead hardwoods to get at these protein-rich larvae, this single conclusion became an assumed foundation for the woodpecker's ecological requirements, a marker of its habitat preferences, definitive sign of the species' ongoing presence, and a primary justification for its conservation needs. Putative foraging specialization via bark scaling continues to sway perceived success about current searches for the ivory-bill.[47]

Box 7.1. The human mind tends to be tricked into making faulty conclusions if odd behavior is observed inside a rare group. That cognitive defect causes us to make stereotypical associations when we should not. This false association, or ***one-shot illusory correlation***, is more readily created if both the behavior itself (foraging on large grubs) and group in which it is observed (rare ivory-billed woodpecker) are both uncommon. Consequently, a shared uniqueness leads us to entertain group membership as the explanation for some curious behavior.

Nevertheless, dietary specialization cannot be reconciled with other data about foods taken by the woodpecker. "Most of the ten stomach analyses conducted with the species in historical times revealed a preponderance of plant materials (fruits and seeds) in gut contents, not insects."[48] The bird's diet was remarkably broad. It was known to take blackberries, grapes, cherries, persimmons, hackberries, acorns, pecan nuts, hickory nuts, southern magnolia seeds, plant galls, and poison ivory berries, not to mention other insects, including carpenter ants or termites, as does the pileated.[49]

Still, Tanner insisted that plant food sources were not as important as large insect larvae.[50] Improbably, the numerical information available indicates that plant material could form the *majority* of the diet, as much as 55–62%![51] Tanner doubled down on his food specialization hypothesis in conservation recommendations, believing that "the quantity of woodpecker food could be artificially increased by killing certain trees."[52] (These discrepancies in dietary emphases, and management calls for tree girdling in Arkansas, were what so stumped my first reading of the bird's recovery plan in 2005.)

How large grubs[53] came to anchor rigid beliefs about specialization in the ivory-bill is traceable to our psychology. Seeing an exceedingly rare species engage in unusual behavior like transporting visibly-large prey items (themselves limited to restricted environmental conditions) triggered a common mental error known as ***one-shot illusory correlation***. "Rare-rare combinations receive more attributional processing" (i.e., dedicated attention) in us than do other kinds of random associations during our attempts to establish a correlation.[54]

Primary cause of endangerment

Tanner concluded: "This [destruction of large, southern swamp forests where the ivory-bill could find sufficient food] has been a more general and widespread factor than has shooting."[55] This hypothesis on a narrow foraging specialization would go on to monopolize our requiem of extinction for the ivory-bill.

Shooting pressure, including that exacerbated by inroads from people moving into newly open logged lands, is the primary alternative thesis. Human predation agrees with some of the historical record. Snyder and colleagues presumed lesser vulnerability of the other large woodpecker, the supposedly tamer pileated, to the same human depredations from shooting.[56] That hypothesis is utterly falsified by contemporaneous accounts that reveal the pileated also to have been "rather shy," and likely to seek refuge "long ere we came within gun range,"[57] traits identical to those described for ivory-bill. Pileateds were "very wary birds and if once they see you, it is useless to pursue them with a view to getting a shot at them."[58] Regardless of identity, there is evidence for both threats behind the ivory-bill's endangerment, although neither is without its deficiencies, especially when either is considered in total isolation from the other.

Behavioral wariness

Perhaps no other life history trait of the ivory-billed has received such disparate treatment by writers. When it comes to the bird's presumed habits of visibility and accessibility, texts could hardly be more conflicting. On the one hand we are informed that the woodpecker was conspicuous, tame, or incautious, easily located from its noisy, far-carrying calls. Several statements make this assertion or otherwise seem to cast doubt on the ivory-bill's propensity for wariness (year given for quote):

- 1888: "in northeastern Arkansas ... not very wild or wary..."[59]
- 1896: "as they are not generally shy..."[60]
- 1900: "if unmolested and not alarmed, they are certainly noisy, and by their oft-repeated cry we became accustomed to locate them...."[61]
- 1938: "in my experience the birds have not been particularly shy around the nest nor seriously upset by the presence of people or of blinds.... The ivory-bill is not usually wary of man nor seriously affected by man's presence...."[62]

- 1948: "At no time, however, would I say that they were shy."[63]
- 2005: "Yes, I'm afraid that we have no reliable evidence to indicate that the ivory-bill ever stopped being half-tame."[64]
- 2006: "I know of no evidence that suggests anything more than individual wariness as a result of negative interaction with humans...."[65]
- 2007: "these woodpeckers were evidently ill-prepared to survive well-armed humanity because of their fairly large size, conspicuous and relatively unwary habits ... both visually and aurally, and as such, they were not easily missed by rural residents inhabiting their ranges...."[66]

Because he was a genuine spectator, Tanner's assertions stand out as irreconcilable with other accounts. Yet beginning with Audubon, we are informed over and over again that the woodpecker did in fact possess an uncanny vigilance at close approach from those who might cause it harm. Indeed, accounts drawing attention to elusive behavior that made so difficult finding and studying the woodpecker are in the vast majority (year given for quote):

- 1831: "I observed that in two instances, when the Woodpeckers saw me thus at the foot of the tree in which they were digging their nest, they abandoned it forever."[67]
- 1877: "we saw a single pair of the rare Ivory-bills.... They were very shy, restlessly swinging from tree to tree, and taking good care to keep beyond gun-range...."[68]
- 1879: "it is very shy and hard to approach."[69]
- 1882: "Very shy and not easy to approach."[70]
- 1887: "They are now comparatively rare and very shy."[71]
- 1891: "its extreme wildness and desire for seclusion ... surely a bird as wild, as wary, as this would not remain

in the vicinity of where man was constantly to be met!"[72]

- 1892: "I saw and heard four Ivory-bills ... but could not get a shot as they were too wild, and couldn't be approached nearer than 300 or 400 yards."[73]
- 1895: "the Ivory-billed Woodpecker is an exceedingly wild and suspicious bird, and as the country becomes more settled it retires from the advance of civilization to the more inaccessible swamps, where it is not so liable to be molested."[74]
- 1895: "It is not rare, but is rarely met with."[75]
- 1895: "to be found ... nowhere within sound of the guns of the game butchers who infest Florida in winter. It is always rare, inhabiting the most solitary places remote from mankind."[76]
- 1895: "very rare and hard to get a shot at...I tried to get her [a female ivory-bill] but failed as she was exceedingly wary."[77]
- 1898: "the female kept out of sight in the thick, leafy growth further back in the swamp ... she was too sagacious to appear...."[78]
- 1905: "Mr. Brown told me that these birds are the shyest and most cunning of anything that wore feathers; he would sit concealed for hours and watch them, yet they must have known of his presence as they would go in and out of every old nest in the swamp, but steer wide of the new one."[79]
- 1905: "I saw six in one day ... two of these I secured, but the others were so wild that I got only a long distance shot at one flying above the pine tops...."[80]
- 1906: "In disposition it is wild and wary ... they are a wild and suspicious bird.... These Woodpeckers are very silent at all times so far as their voices are concerned. Especially is this true during the breeding season."[81]
- 1907: "those which most interested me were the great

ivory-billed woodpeckers. Of these I saw three.... They were noisy but wary...."[82]

- 1912: "Our guide heard an Ivory-bill within the swamp..., but the rest of our party ... were not fortunate in either seeing or hearing it...."[83]
- 1913: "now confined to the wildest and remotest swamps... good fortune to see and hear it, the reward of hours of laborious wading...."[84]
- 1916: "the Ivory-bill is a wild, shy bird...."[85]
- 1930s: "the bird was skittish and called only when disturbed."[86]
- 1931: "they were very wild and wary...."[87]
- 1937: "failure to find the birds in a given area is no proof that they are not there, for they are not noisy except when disturbed; their voice doesn't carry nearly as far as that of the pileated woodpecker.... We had great difficulty in following them ... senior author at one time stood under a giant oak and caught in his hand chips of bark and wood that an ivory-bill was scaling from a dead branch high in the tree without either one being able to see the other...."[88]
- 1941: "This bird, although naturally shy of man...."[89]
- 1942: "Both F.H. Kennard ... and Robert Ridgway ... had difficulty finding Ivory-bills in the Big Cypress, so that my failure to find any birds does not eliminate the possibility of their being there...."[90]
- 1942: "the difficulty of securing one even with a gun...."[91]
- 1950: "the nearest bird coming to within a hundred feet of me but very carefully staying out of my sight, although I am sure that I did not stay out of the bird's sight...."[92]
- 1960s: "They were shy, retiring and easily spooked."[93]
- 1971: "we finally concluded that the bird became aware of our presence before we could see it ... evidently having deserted the roost after the morning scare...."[94]
- 1987: "it was extremely wary and would not let us

approach closer than about 150 feet."[95]

- 1988: "From the time of its discovery early in the eighteenth century, it has been an elusive, retiring bird, becoming if anything warier and harder to find over the years."[96]
- 1999: "they were hard to find and harder to follow."[97]
- 2006: "Inherent in the biology of the ivory-bill is its wariness of humans…."[98]
- 2007: "Ivorybills … are very shy, quiet, reclusive birds."[99]
- 2007: "Ivorybills along the Choctawhatchee River are shy and skittish birds that flee from people."[100]
- 2008: "the Arkansas bird was extremely shy and quiet."[101]

Given the sweeping extent and historical duration of these behavioral accounts, it is irresponsible for modern-day biologists to dispute the fact that ivory-bill was wary. And yet this insistence continues: "there is no indication given by past writers or naturalists that these birds were exceedingly shy."[102]

Box 7.2. Because Tanner studied a pair of woodpeckers acclimated earlier to human presence, yet disregarded or downplayed scores of previous historical accounts that emphasized the ivory-bill's extreme wariness, his descriptions about the species' supposed tolerance to human disturbance were unduly shaped by a ***recency effect***. This ***recency effect*** went on to fix later beliefs that described the ivory-bill as a conspicuous species, creating a nearly intractable ***anchoring bias*** for a so-called "tame ivory-bill."

Most accounts that so blatantly contradicted Tanner took place before his study was ever conducted. Likely due to a ***recency effect***, Tanner's intimation of an "approachable ivory-bill" went on to affix and then distort human reason about the bird's distinguishing behavior. In trying to reconcile a stubborn error from that ***anchoring bias*** with the modern-day actuality of sparse detections, our writing about the woodpecker would become contorted to the point of nonsense: "loud, mobile ivory-bills, scaling downed dead trees are a creature of the past, replaced by relatively quiet, reclusive, canopy-dwelling denizens."[103]

Self-contradictions

Ivory-bill researchers, scholars, and other commentators not infrequently contradict themselves. One account indicates the woodpecker was very rare in 1895, even to the point of complete disappearance within a portion of the Florida panhandle: "once very common ... where it was in a large measure secure, but it is now rapidly becoming extinct on the Wacissa [River, Florida] ... every one is shot."[104] Nevertheless, a decade later in 1905 the very same author, Arthur T. Wayne, referring to the same river basin, stated: "birds of the Aucilla and Wacissa rivers [Florida].... I know that I left more than 100 birds in a radius of 20 square miles."[105]

In the 1960s, John V. Dennis made multiple searches for the ivory-bill, including those conducted in eastern Texas that received such a very public dig from Tanner himself. Dennis obviously thought that the woodpecker was alive during that decade. Yet almost 20 years earlier, Dennis had himself claimed in print that "the American bird is now extinct except for a few stray individuals."[106]

In speaking about the Suwannee River area further east in north Florida, Tanner blamed Wayne's collecting in 1892 and 1893 for the birds being "wiped out," with no reliable reports subsequently from that region. But oddly, in the same reference, Tanner lists records of ivory-bill from the upper Suwanee (Okefenokee's Craven Hammock and Minnie Lake Island) between 1910 and 1915. Tanner lists two more records from the lower portion of the river, both at Suwanee Hammock, in 1917 and 1925.[107]

Despite his insistence on the ivory-bill's specialization for bottomland hardwood forests of a certain age, tree composition, and topographic setting, Tanner contradicts that weighting elsewhere within his own monograph. Summarizing his review of habitat use across the bird's range, he notes: "There is no one type of forest that is the habitat of the ivory-billed woodpecker.... In ... Florida the ivory-bill was not confined to any particular type of forest, nor was there any one characteristic of the woods throughout the bird's habitat there that appeared to have an important influence on its distribution."[108]

Even reviewing evidence for the woodpecker's very existence can prompt a stunning reversal. In reacting to a presentation of the Luneau video by Cornell Laboratory of Ornithology staff on April 28, 2005, Jerome A. Jackson first exclaimed with unabashed enthusiasm:

> *My God, it is an ivory-bill! Look at that beautiful big white butterfly. I felt a surge of adrenalin as I recalled my encounter, and a bit of envy, too, as I rued not having captured my encounter with a possible ivory-bill along the Noxubee River in Alabama ... the sightings were so similar.*[109]

And yet Jackson would go on to become one of the Arkansas episode's most persistent critics, calling that search for the ivory-bill an episode of faith-based ornithology, one laden with motivational politics and other disrupting agendas, admonishing his colleagues for their alleged breaches with scientific process.[110]

Reliability of historical sourcing

A veritable chasm divides the assumed reliability given to ivory-bill sight reports before around 1915 versus those made after 1944. Anecdotal accounts prior to 1915 were (and still are) usually accepted entirely at face value, especially as to accuracy of the woodpecker's identity and habits. Yet incidents of ivory-bills reported after 1944 were nearly always depreciated.[111] The era 1915–1944 was a transition period in which the species was presumed extinct on several occasions, then "re-discovered," such that for the first time some but not every one of the records were strongly doubted (see Chapter 6).

Some scholars seemed to realize they were making a large leap into outright supposition concerning this early historical veracity. Over and over again, Snyder and colleagues admit to a reliance on blind trust to advance their argument about the bird's relative abundance, a major tenet used to build out a thesis for direct killing as the major factor propelling the bird towards extinction, starting in the late 1800s:

> *We struggle onward, relying on the basic assumption that most early naturalists reporting observations of these species were probably not liars and frauds ... unless one might wish to dismiss the honesty or competence of an impressive array of naturalists commenting on status of the ivory-bill in early times.... Assuming these hunters were not mistaking pileateds for ivory-bills ... ignoring these claims is equivalent to dismissing them without justification.*[112]

Well indeed, certainly. But banking on such feeble ***argument from authority*** is disquieting. Note what a dichotomy in conjecture about reliability of the historical sourcing demands from us.[113] In the context of ultimately accepting the species' likely extinction date, we are asked to completely trust in *all* of the pre-1915 accounts of the ivory-billed,[114] and most of those prior to the early 1940s, yet essentially fall back to complete suspicion concerning *all* of the post-1944 accounts.

This conjecture is even worse than it appears. We are

implored to believe that all early pioneer travelers, writers, and naturalists never had reason to embellish their stories, relied only on direct experience, not secondary sources, and always distinguished accurately the ivory-billed from the pileated woodpecker during the strenuous journeys that were typical of frontier America. We must assume that early writers always correctly identified large woodpeckers without quality optics, definitive field guides, or the adversarial interrogation of their identification skills by an army of skeptics.

Conversely, we are entreated to depreciate all incidents of ivory-billed woodpeckers reported after 1944. This obliges us to dismiss 100 or more such incidents made up of written reports, sightings (including multiple birds, by multiple observers), photographs, audio recordings, and video.[115] Curiously, we are asked to discount these incidents even though all but the most guileless during this suspicious era knew of the bird's presumed extinction, so had to expect that their reporting would be ruthlessly dismembered. Their daring also meant surmounting the cognitive discord that arises from seeing a pinioned apparition. Finally, we must believe that persons reporting incidents since 1944 were less honest *and/or* worse at field identification than their nineteenth-century counterparts. The latter entails that more recent observers were innately inferior at separating ivory-billed from pileated woodpecker despite access to excellent optics, numerous bird references, and a cohesive birding fraternity unafraid to levy ruthless verdicts from doubting peers.

Placing our fealty in this kind of historical caprice is a shocking petition. From where, exactly, did a chronological upsurge in misleading or biased identification of these two large American woodpeckers come from? This is not a lucid outlook. Taken literally, holding such an opposing credence in the historical reliability of field ornithology can only be seen as an appeal to raw faith, its most obstinate version, too. Such arbitrary confidence is patently illogical.

Endnotes

1 Beyer, G.G. 1900. The ivory-billed woodpecker in Louisiana. *The Auk* 17: 97–99.

2 Hoyt, R.D. 1905. Nesting of the ivory-billed woodpecker in Florida. *The Warbler* 1: 53–55.

3 Allen, A.A., and P.P. Kellogg. 1937. Recent observations of the ivory-billed woodpecker. *The Auk* 54: 164–184.

4 Pough, R.H. 1944. Report to the executive director, National Audubon Society: present condition of the Tensas River forests of Madison Parish, Louisiana and the status of the ivory-billed woodpecker in this area as of January, 1944, p. 4.

5 Phelps, F.M. 1914. The resident bird life of the Big Cypress Swamp region. *The Wilson Bulletin* 26: 86–101.

6 Crompton, D.H. 1950. My search for the ivory-billed woodpecker in Florida. *Bulletin of the Massachusetts Audubon Society* 34: 235–237.

7 T. Gilbert Pearson, upon seeing six birds in the Singer Tract, northeast Louisiana, May 1932.

8 Ridgway, R. 1898. The home of the ivory-bill. *Osprey* 3: 35–36.

9 Tanner, J.T. 1942. *The Ivory-billed Woodpecker*. Dover Publications, Inc., Mineola, NY, p. 26.

10 Gallagher, T. *The Grail Bird: The Rediscovery of the Ivory-billed Woodpecker*. Houghton Mifflin Harcourt, Boston, MA, p. 8.

11 J.A. Jackson, as quoted in "Scientist decries 'faith-based ornithology' in Ivory-bill search," by Chuck Hagner, 17 February, 2006, *Birdwatchers Digest*; see http://www.birdwatchingdaily.com/featured-stories/scientist-decries-faith-based-ornithology-in-ivory-bill-search/ (accessed 15 March 2015).

12 Jackson, J. A. 2006a. *In Search of the Ivory-billed Woodpecker*, 2nd ed. HarperCollins, New York, p. 290.

13 Fitzpatrick, J.W., M. Lammertink, M.D. Luneau, Jr, T.W. Gallagher, B.R. Harrison, G.M. Sparling, K.V. Rosenberg, R.W. Rohrbaugh, E.C. Swarthout, P.H. Wrege, and S.B. Swarthout.

2006. Clarifications about current research on the status of ivory-billed woodpecker (*Campephilus principalis*) in Arkansas. *The Auk* 123: 587–593.

14 At least one observer searching Arkansas' Big Woods during the 2004–2005 incident (Chapter 12) did in fact misidentify a woodpecker as a pileated because he was unaware that the female ivory-bill possessed a black crest, a diagnostic feature for this species. Chuck Hunter, *pers. comm.*

15 Audubon, J.J. 1842. *The Birds of America,* Volume 4, pp. 214–226.

16 Allen and Kellogg, p. 183.

17 Tanner, p. 58.

18 *Ibid.*, p. 1.

19 Dennis, J.V. 1948. A last remnant of the ivory-billed woodpeckers in Cuba. *The Auk* 65: 497–507.

20 Trapp, J.L. 2007. Alternative names for the ivory-billed woodpecker. *Birds Etcetera,* 27 January; see http://birdstuff.blogspot.com/2007/01/alternative-names-forivory-billed.html (accessed 17 March 2015).

21 At least during the era of Alexander Wilson, however, ivory-bill may have been designated *greater logcock,* with *lesser logcock* given to the pileated. Jackson 2006a, p. 107.

22 *Ibid.*, p. 2.

23 Tanner, p. 21.

24 Sources for and diversity of accounts about vocalizations and other sounds produced by ivory-bill are contrasted extensively in Jackson, 2006a, pp. 17–21.

25 Tanner, p. 18.

26 Cooke, W.W. 1914. Some winter birds of Oklahoma. *The Auk* 31: 473–493.

27 Brewster, W., and F.M. Chapman. 1891. Notes of the birds of the lower Suwanee River. *The Auk* 8: 125–138.

28 Wayne, A.T. 1893. Notes on the birds of the Suwanee River. *The Auk* 10: 336–338. Differences in search effort between the two expeditions may explain the difference in the ivory-bill numbers

that were detected. Whatever the reason, however, such disparities in detection rates cast doubt on our beliefs about the ivory-bill's true habits.

29 Snyder, N., D.E. Brown, and K.B. Clark. 2009. *The Travails of Two Woodpeckers*. University of New Mexico Press, Albuquerque, pp. 24–25.

30 Pennant, T. 1785. *Arctic Zoology, Volume II, Class II, Birds*. Henry Hughs, London, p. 268.

31 Tanner, p. 31.

32 Snyder et al., p. 23.

33 Corrington, J.D. 1922. The winter birds of the Biloxi, Mississippi region. *The Auk* 39: 530–556.

34 Tanner, p. 55.

35 Coues, E. 1868. Synopsis of the birds of South Carolina. *Proceedings of the Boston Society of Natural History* 12: 119.

36 Pough, p. 4.

37 Audubon, J.J. 1832. *Ornithological Biography, Or an Account of the Habits of the Birds of the United States of America*. Volume 1. James Kay, Jun. & Co., Philadelphia, p. 341.

38 Schorger, A.W. 1949. An early record and description of the ivory-billed woodpecker in Kentucky. *Wilson Bulletin* 61: 235.

39 Jackson, 2006a, p. 50.

40 Tanner, pp. 32–33.

41 Snyder et al., pp. 27–30.

42 Tanner, pp. 46, 99.

43 *Ibid.*, p. 53.

44 Jackson, 2006a, pp. 53–57.

45 *Ibid.*; Nehrling, H. 1882. List of birds observed at Houston, Harris Co., Texas, and in the counties Montgomery, Galveston and Ford Bend. *Bulletin of the Nuttall Ornithological Club* 7: 166–175.

46 Tanner, p. 41.

47 See for example the photos and discussion of woodpecker scaling of trees with tightly-adhered bark in Louisiana, available here: http://projectcoyoteibwo.com/2015/02/25/trip-report-feb-20-24-

2015-still-more-smoke (accessed 22 March 2015).

48 Snyder et al., p. 36.

49 Bent, A.C. 1939. *Life Histories of North American Woodpeckers*. Smithsonian Institution National Museum Bulletin 174 (1964 Dover edition); Higley, W.K. 1906. *Birds and Nature*, Volume III, A.W. Mumford and Company, Chicago, p. 123; Allen and Kellogg, p. 167; Jackson, 2006a, p. 197; Snyder et al., p. 36.

50 Tanner, p. 45.

51 Bent, pp. 9–10.

52 Tanner, p. 100.

53 Tanner emphasized importance of all grubs to the ivory-bill, but it was later writers who would go on to selectively focus on the larger grubs as the principal reason for the woodpecker's dietary specialization and this scarce food source as the major cause behind its decline.

54 Risen, J.L., T. Gilovich, and D. Dunning. 2007. One-shot illusory correlations and stereotype formation. *Personality and Social Psychology Bulletin* 33: 1,492–1,502.

55 *Ibid.*

56 Snyder et al., p. 66.

57 Wright, A.H., and F. Harper. 1913. A biological reconnaissance of Okefinokee Swamp: the birds. *The Auk* 30: 477–505. See p. 496.

58 Pleas, L. 1891. The pileated woodpecker. *The Oölogist* 8: 236–237.

59 Cooke, W.W. 1888. Report on bird migration in the Mississippi Valley in the years 1884 and 1885. *Bulletin of the Division of Economic Ornithology* 2: 127–128. U.S. Department of Agriculture.

60 Maynard, C.J. 1896. *The Birds of Eastern North America*. C.J. Maynard & Co., Newtonville, MA.

61 Beyer, p. 98.

62 Tanner, pp. 77, 100.

63 Dennis, p. 503.

64 Tom Nelson, Minnesota birder, blogger, climate change and ivory-bill skeptic, commenting here in 2005, conspicuously ignorant to the extensive historical literature that describes exactly

contrary behavior in this woodpecker http://www.birdforum.net/showthread.php?t=33968&page=1-&pp=25 (accessed 19 March 2015).

65 Jackson, J.A. 2006b. Ivory-billed woodpecker (*Campephilus principalis*): hope, and the interfaces of science, conservation, and politics. *The Auk* 123: 1–15.

66 Snyder et al., p. 9.

67 Audubon, J.J., 1832, p. 344.

68 Brewster, W. 1881. With the birds on a Florida river. *Bulletin of the Nuttall Ornithological Club* 6: 38–44.

69 Krider, J. 1879. *Forty Years Notes of a Field Ornithologist*. Joseph H. Weston Press, Philadelphia, PA, p. 17.

70 Nehrling, p. 170.

71 Scott, W.E.D. 1888. Supplementary notes from the Gulf coast of Florida, with a description of a new species of marsh wren. *The Auk* 5: 183–188.

72 Hasbrouck, E.M. 1891. The present status of the ivory-billed woodpecker (*Campephilus principalis*). *The Auk* 8: 174–186.

73 Field notes of Arthur T. Wayne from April 22, 1892, as given in Tanner, p. 63.

74 Bendire, C. 1895. Life histories of North American birds, from the parrots to the grackles. *Special Bulletin of the U.S. National Museum* 3: 42–45.

75 *Field & Stream* 14: 407, a comment by "Yell" when referring to the tendency of the species to inhabit "unfrequented swamps" in Arkansas.

76 *Field & Stream* 14: 367, a comment from S.C. Clarke, Marietta, Georgia.

77 *Field & Stream* 14: 427, a comment from W.A.D., Hawkinsville, Florida, who also revealed he and his two brothers had shot between 20 and 25 ivory-bills in the vicinity during the previous 10 years.

78 Ridgway, p. 36.

79 Hoyt, p. 52.

80 Account by Vernon Bailey, field naturalist with the U.S. Biological Survey, in describing his experience with collecting the species in Texas; see Oberholser, H. C. 1974. *The Bird Life of Texas*. University of Texas Press, Austin.

81 Higley, p. 122.

82 Roosevelt, T. 1908. *In the Louisiana Canebreaks*. Louisiana Wild Life and Fisheries Commission, 400 Royal Street, New Orleans. 8 pp.

83 Wright and Harper, p. 504.

84 Phelps, p. 99.

85 Chapman, F.M. 1916. *Handbook of Birds of Eastern North America*. D. Appleton and Company, New York.

86 J.J. Kuhn, as quoted in Bales, S.L. 2010. *Ghost Birds: Jim Tanner and the Quest for the Ivory-billed Woodpecker, 1935–1941*. University of Tennessee Press, Knoxville, p. 126.

87 Nice, M.M. 1931. *The Birds of Oklahoma*. University of Oklahoma Press, Norman.

88 Allen and Kellogg, p. 165.

89 McIlhenny, E.A. 1941. The passing of the ivory-billed woodpecker. *The Auk* 58: 582–584.

90 Tanner, p. 27.

91 *Ibid.*, p. 55.

92 Crompton, p. 236.

93 Gallagher, p. 229, referring to the birds that Dennis observed during this decade in the Big Thicket region of eastern Texas.

94 Agey, H.N., and G.M. Heinzmann. 1971. The ivory-billed woodpecker found in central Florida. *The Florida Field Naturalist* 44: 46–47, 64.

95 Jackson, 2006a, p. 185, describing the behavior of a bird responding to and moving towards taped calls of ivory-bill that he and Malcolm Hodges broadcast on March 29, 1987, near the Yazoo River, Mississippi.

96 John V. Dennis, as quoted in Bales, p. 165.

97 Tanner, J.T. 2000. A postscript on ivorybills. *Bird Watcher's Digest* July/August 2000: 52–59.

98 Jackson, 2006a, p. 239. Note that Jackson contradicts himself, claiming in one reference that individual ivory-bills were only wary after negative interactions, but also stating this wariness was an *inherent* feature of their biology in a second reference that appeared the same year.

99 Hill, G.E. 2007. *Ivorybill Hunters: The Search for Proof in a Flooded Wilderness*. Oxford University Press, London, p. 245.

100 *Ibid.*, p. 159.

101 Steinberg, M.K. 2008. *Stalking the Ghost Bird: The Elusive Ivory-billed Woodpecker in Louisiana.* Louisiana State University Press, Baton Rouge, p. 56.

102 Keith Ouchley of The Nature Conservancy, as reported in Steinberg, p. 42.

103 Cyberthrush. 2016. Explaining the inexplicable… http://ivorybills.blogspot.com/ (accessed 1 May 2016). Misconstruing the ivory-bill's cautious habits is rife. "Perhaps all the noisy and approachable ones were killed off a century ago." Gallagher, p. 248. No evidence exists that the species was *ever* approachable; the relevant time line for gauging this trait is not a single century (exposing a Euro-American chauvinism), but rather many centuries or even several millennia (matching the ivory-bill's long persecution by native peoples).

104 Wayne, A.T. 1895. Notes of birds of the Wacissa and Aucilla River regions of Florida. *The Auk* 12: 362–367.

105 Letter of Arthur T. Wayne to Frank M. Chapman, sent from Mount Pleasant, South Carolina, October 12, 1905, as reproduced in Snyder et al., p. 136.

106 Dennis, p. 497.

107 Tanner, 1942, pp. 19, 5–6.

108 Tanner, 1942, p. 16.

109 Jackson, 2006a, p. 257.

110 Jackson, 2006b.

111 "For more than fifty years, every time someone reported a seemingly credible sighting of this species, the skeptics slammed

it." Gallagher, p. 257.

112 Snyder et al., pp. 10, 26, 45.

113 This capricious trust on anecdotal sourcing is usually not acknowledged forthrightly. The discrepancy is embedded tacitly from evidentiary criteria deployed by ornithological or birding communities, however, when they judge reliability of modern-day ivory-bill reports. Whether supported by physical evidence or not, written accounts from earlier eras are not disputed, at least overtly. No ivory-bill records since ca. 1944 are entirely undisputed, however, such that a knee-jerk conviction of complete skepticism holds utter sway, e.g., "no conclusive records of their [ivory-bill] existence in recent decades," Snyder et al., p. 1, and "we do not yet have confirmation of the existence of a single live ivory-bill in the twenty-first century," Jackson, 2006a, p. 264. This same refrain is then repeated by media as an established fact, e.g., "last seen: 1944," *Extinct birds land at 'Once There Were Billions' at the National Museum of Natural History*, S. Dingfelder, Washington Post, June 26, 2014, http://www.washingtonpost.com/express/wp/2014/06/26/extinct-birds-land-at-once-there-were-billions-at-the-national-museum-of-natural-history/ (accessed 30 March 2015).

114 "Scientists ... generally accept the earlier sight records of ornithologists who knew the birds as a result of previous experience with them." Jackson, 2006a, p. 268. This reasoning is presumptive and circular, reinforcing an indiscriminate belief in the veracity of the historical record as perceived through ***argument from authority***.

115 Hunter, W.C. 2010. *Interpreting historical status of the ivory-billed woodpecker with recent evidence for the species' persistence in the southeastern United States*. Appendix E, Recovery Plan for the Ivory-billed Woodpecker (*Campephilus principalis*), U.S. Fish & Wildlife Service, Atlanta, GA. Hunter's synthesis is grouped into "incidents," some of which constitute multiple birds or as seen by several observers. Thus, the 2004–2005 Arkansas sightings by

"many" observers, plus the audio recordings and a video of a bird from the Bayou de View, Cache River National Wildlife Refuge, represent only a single incident in this treatment.

Chapter 8

Curse of Small *n*

What makes [a hypothesis] ... more 'vampirical' than 'empirical' – unable to be killed by mere evidence – is that [it] seems so logically compelling that it becomes easy to presume ... true.
—Jeremy Freese[1]

Blessing can bear a remarkable likeness to curse. When the two arrive as conjoined twins, the latter may be disguised so cunningly that the former's malice catches us wholly off guard. Nowhere is this more true than when arbitrating the credibility of insights to be gleaned from the 1937–1939 study of ivory-bill in the 82,000 acre Singer Tract of Madison Parish in northeast Louisiana.[2] James T. Tanner's study had a sample size (n) of just one, flouting any confidence whatsoever in its universality. Perils for drawing conclusions arise not so much from *incorrect* material garnered by the Tanner study, but rather from its *incomplete* quality in serving as sturdy evidence for ascertaining

> **Box 8.1**. Extending unwarranted conclusions from samples to populations, or exaggerating the importance of a sample that is too small to faithfully represent the claims made, exemplify the cognitive error of ***insensitivity to sample size***. Larger samples justify more confidence because they are closer to the truth. However, many aspects of the ivory-billed woodpecker's life history were represented by a sample of a *single* pair studied in just *one* geographic region, compromising greatly this essential science and statistical requirement.

the woodpecker's true habits.[3]

Given the ivory-bill's dire prospects entering the twentieth century, and the assumed extinction(s) repeatedly assigned to the bird, there is no question we are privileged to have Tanner's study at all.[4] But that is not the question posed here. Instead, we must ask if the Tanner study was *characteristic* of the ivory-bill, that is, did it reflect the norm or the range of habits expected for this species across conditions under which it originally lived, or could live? If not, how have the Tanner findings inhibited what we think we know? How strongly should we continue to clutch the hypotheses or theories put forth by and based entirely on the Tanner report?

To help answer, three issues merit a vetting. The first evaluates behaviors in the unique population that Tanner investigated, especially how those birds reacted to researchers given levels of past human activity in Madison Parish, Louisiana. The second entails reappraisal of the bird's habitat affinities, including a supposed feeding specialty on large beetle larvae in virgin forests of remote bottomland hardwoods. The last concerns the unrestrained sociological and cognitive anchoring caused by clutching at Tanner's study, the unconscious steering of interpretation that resulted from a singular study of the woodpecker.

No other researcher ever duplicated such prolonged surveillance of nesting and feeding ivory-bills. Tanner did so intermittently for three consecutive years, beginning in 1937. However, during this entire interval, his efforts were limited mostly to studying *a single pair*, one that nested repeatedly in the vicinity of John's Bayou. Even that pair was pinned down by a local, resident game warden hired to protect the birds. J. J. Kuhn was a gifted woodsman who "knew where to look."[5] Moreover, Tanner's work was confined to the breeding season (January through May) when the woodpecker's conspicuous activity, ties to a nest cavity, and a less-vegetated forest canopy made it far easier to detect.

Box 8.2. Sampling was so sparse that it seriously impeded what we truly know about the ivory-bill. This ***insensitivity to sample size*** extends to the most basic traits about the bird. For example, although Cornell researchers were able to record some (but not all) vocalizations of a single pair they studied during the 1930s, those samples remain wholly inadequate to describe the acoustic repertoire for the ivory-bill or to faithfully represent individual variation in the species' vocalizations across its entire range.

Despite benefit of local expertise and confining searches to "ivory-bill areas,"[6] daily pursuit of the ivory-bill was anything but easy, even in the known hotspots. Tanner's field notes are replete with examples of supreme toil in tracking the birds. On May 4, 1937, he spent "the rest of the day searching the region of John's Bayou ... finding scattered fresh feeding sign but no birds. Kuhn searched the Greenlea Bend area without seeing a thing."[7] Later in the same week, and exploring further south from John's Bayou closer to Spring Bayou, "neither [Tanner or Kuhn] found evidence of ivory-bills.... Buck Guthrie, a local resident ... had no news of recent ivory-bill sightings."[8]

As summer's heat descended on northeast Louisiana the next month, within "an area that was still hospitable to the ghost birds, they were proving most difficult to find."[9] Expanding their search zone even further afield, to the west side of the Tensas River, Tanner interviewed a sharecropper named Levi Jefferson for any clues to the birds' whereabouts. Although Jefferson had lived in the same vicinity for *two decades*, this local resident "had never seen an ivory-bill west of the river."[10]

During his last year of study at the Singer Tract, Tanner and new assistant Jesse Laird continued to search for ivory-bills beyond the single pair they found consistently at John's Bayou. But in April 1939, "Tanner's frustration over Titepaper and Mack's Bayou continued ... Tanner and Laird ... found fresh feeding sign but nothing else.... He was putting in the time but with diminishing returns."[11] When Tanner combed the southerly reaches of the Singer Tract "hoping to find signs of other ivory-bills. He found none."[12] Wrapping up his fieldwork on the ivory-bill, Tanner described plainly the immense challenge of tracking his quarry:

> *Hunting for localities where ivory-bills were, and in these localities trying to find the birds, was like searching for an animated needle in a haystack.*[13]

Other ivory-bills present in the Singer Tract were not easy to detect, never mind study. Although Tanner set the record for consecutive time following a single pair (fifteen weeks),[14] repeated observations of an identical subject still represents a sample size of only one. With judicious use of both tree and ground blinds as cover, that single pair gradually acclimated to observers.[15] Aside from birds tracked to their roosting or nesting cavities, however, all other ivory-bills present in the Singer Tract were elusive.

Land-use history for the Singer Tract could hardly be said to embody typical or representative conditions for the ivory-bill. Not only had at least a few individual woodpeckers acclimated to human presence, the Singer Tract and surrounding areas of Madison Parish had a long history of relative isolation from the heaviest blights of human settlement. Compared to elsewhere within the woodpecker's range, this section of northeast Louisiana may have been incomparable as a place where the ivory-bill had not been exploited continuously for centuries.

Prior to 1800, only the native Taensas people permanently inhabited the region, although even they mostly lived just south of Madison Parish. Sedentary, agricultural, and canoe-faring, the Taensas numbered about 1,200 scattered in seven villages.[16] After first European contact with the La Selle expedition in 1682, their numbers were decimated by a smallpox outbreak in 1698. Due to warfare with adjacent tribes, the Taensas were relocated in 1740 by the French to the Tensas River near Mobile, Alabama.[17] Madison Parish then remained largely unoccupied for the next 90 years.

Not until the 1830s, when southern farmland to the east had become over-tilled, did substantive European settlement begin. Between 1838 and the onset of the Civil War, Madison became one of Louisiana's wealthiest parishes by virtue of its rich alluvial land, steamboat access to New Orleans, and a plantation economy based on cotton.[18] The parish had five small towns: Tallulah, Richmond, Milliken's Bend, Delta, and New Carthage. Settlement was clustered along the eastern side of the parish near the Mississippi River, leaving mostly vacant the vast swampy forestlands of the delta to the west (including the large Singer Tract).

Given proximity to Vicksburg, Madison Parish experienced bitter fighting and a harsh occupation during the Civil War. By war's end, parish buildings had been burned, its livestock confiscated, and its plantations reduced to shambles.[19] Parish seat Richmond was burned to the ground. Similarly, New Carthage and Milliken's Bend were never rebuilt from ruins of the conflict. The devastated parish became a no-man's land, with the population sinking 39% and total area under cultivation falling by a staggering 82%. Much of this abandoned farmland gradually reverted back to forest.[20]

After a second, smaller depopulation occurred between the 1890s and 1920s, Madison Parish would reach its pre-Civil War population of 14,000 by the time of Tanner's study. Nevertheless, farmland under cultivation when Tanner arrived was not even half that existing prior to the war. A catastrophic Mississippi River flood in 1927 only reinforced the inaccessibility of some delta forests surrounding the Singer Tract. Consequently, Tanner researched the ivory-bill in a landscape that had remained heavily forested but also lightly populated (if not depopulated) for almost two centuries.

Whatever number of ivory-bills persisted in Madison Parish by the 1930s, they had faced relatively light contact with humans. No evidence exists that woodpeckers had been recently persecuted. Western Madison Parish escaped the frenetic zeal by the most insatiable nineteenth-century taxidermists and wholesale museum dealers who worked over portions of Florida, for example. The earliest record for ivory-bill in the parish was of a single bird taken by E. M. Haskell in 1891. Others included a male and female in December 1908, and four in March–April 1909.[21] Only one other specimen is known, the one permitted to Mason Spencer in 1932.[22] Game warden Kuhn thought ivory-bills had actually increased in the Singer Tract since 1925, as it had been three or four decades since locals had shot any ivory-bills there out of mere curiosity.[23]

Carpet-bagging companies rushed in to scoop up the South's cheap forest land after the Civil War ended. Prior to arrival of the Cornell team, the Singer Tract had changed owners several times due to defaults on the parcel's taxes. Singer Manufacturing Company finally acquired the property in 1913. By 1920 the Singer Tract became an inviolate sanctuary managed by the state of Louisiana. Local wardens were hired to enforce wildlife protections. Hence, when Tanner began, woodpeckers had been shielded from human persecution for at least an entire ivory-bill generation.[24] These auspicious circumstances for a study of the rare woodpecker were peculiar to the Singer Tract: as far as we know, they could not have been duplicated anywhere else.[25] "In terms of ivory-bill research, John's Bayou was like the mythical

El Dorado, the Lost City of Gold."[26]

Tanner's model for ivory-bill can be condensed as "the huge woodpecker with a huge bill needed huge trees."[27] The model was slanted inevitably by what Tanner was constrained to discover at one site in northeast Louisiana.[28] In ecological and conservation terms, his model had the following key elements: 1) a remote or wilderness setting; 2) bottomland hardwood habitat dominated by sweet gum *Liquidamber styraciflua*, oak *Quercus*, and hackberry *Celtis*; 3) old-growth condition, and 4) recent mortality (within 2–3 years) of trees containing 5) abundant, large wood-boring insect larvae, especially Cerambycidae (long-horn beetles), Buprestidae (jewel beetles), and Elateridae (click beetles).[29]

Primitive or wilderness features are entirely a human construct (see Chapter 4). Yet it was obvious that Tanner let that construct slant his beliefs about where the ivory-bill might still be found. He gave considerable plausibility to the "virgin and primitive condition of the Big Cypress, one of the last real wildernesses of the country."[30] On the other hand, he depreciated any prospect of the ivory-bill still surviving in east Texas because, in his view, that region was "greatly over-rated as a wilderness area."[31]

Each of the ecological elements in his model, however, can be reviewed for putative accuracy in representing how ivory-bills actually lived:

Habitat type

Ivory-bills were never restricted to bottomland hardwoods, although by the time of Tanner's study, it might have seemed so. Gauged by its occupied range, the woodpecker's habitat affinities were likely under a double illusion. After heavy, centuries-long persecution by Native Americans,[32] the initial distribution witnessed by Europeans was already some trace of its former extent. The ivory-bill was mostly absent from interior regions long occupied by native peoples who had settled, hunted, and cleared prime alluvial land for crops. Yet several eighteenth and very early nineteenth-century ivory-bill records from interior Kentucky,[33] western Virginia or eastern West Virginia,[34] and piedmont western South Carolina[35] all suggest a once larger extent to the ivory-bill's interior range.[36]

As European settlement expanded, the woodpecker's range shrunk to areas where large blocks of multi-aged forest still persisted and where human persecution was less intense due to the complex terrain posed by swampy, seasonally flooded woodlands located fortuitously next to other forest types. When Tanner started, the woodpecker was no longer widely distributed (or easily seen) anywhere, even in these conditions, leading him out of pure necessity to focus on the sweet gum-oak-hackberry association as crucial for the bird. But whenever explanations are formed merely with the first association that we happen to make (***primacy effect***), our decisions are highly error prone, a bias stemming fundamentally from our over-reliance upon the ***availability heuristic***.[37]

Box 8.3. In seeking a ***cognitive closure***, we tend to make conclusions in haste with the most convenient or easily recalled information available to us (***availability heuristic***). When Tanner linked the ivory-bill to the first (and only) forest habitat type that he ever studied, he was likely over-influenced by a ***primacy effect***, one of several biases that can arise from a serial position effect, i.e., the order in which we receive or absorb information.

Ivory-bills in fact lived in varying forest types. Woodpeckers fed, roosted, or nested in such diverse tree species as bald cypress *Taxodium distichum*; red maple *Acer rubrum*; red bay *Persea borbonia*; sweet bay *Magnolia virginiana*; Nuttall *Quercus texana*, overcup *Q. lyrata* and willow oaks *Q. phellos*; American *Ulmus americana* and cedar elms *U. crassifolia*; tupelo *Nyssa aquatic* and black gum *N. silvatica*; green *Fraxinus pennsylvanica* and pop ash *F. caroliniana*; hackberry or sugarberry *Celtis*; pecan *Carya illinoinensis* and hickories *Carya* spp.; American persimmon *Diospyros virginiana*; honey locust *Gleditsia triacanthos*; and cabbage palm *Sabal palmetto*.[38] In southwestern Florida, ivory-bills even inhabit(ed) regions with subtropical tree species like royal palm *Roystonea regia* and black mangrove *Avicennia germinans*.[39]

Upland southern pine (*Pinus*) forests were also exploited by the big birds in at least the Carolinas, Florida, Alabama, and Texas.[40] In Florida "the birds apparently did most of their feeding in the dead pines at the edge of the swamp, scaling off the bark … or actually getting down on the ground like Flickers."[41] Jackson too suggested that ivory-bills were once more common in southern pines prior to that forest type's extensive removal during the process of European settlement.[42]

Stand age

A persistent contention was that the ivory-bill was an old-growth obligate. This belief traces directly to Tanner: "ivory-billed woodpeckers have almost always lived in virgin or primitive stands of timber."[43] More than a half century later, this notion was so deeply ingrained that it appeared reflexively and obligatorily in virtually every account written for the species, e.g., "ivory-billed woodpeckers … inhabit … particularly old-growth forests supporting healthy populations of beetles."[44] I am as guilty as the next in blindly repeating this central but unsubstantiated assertion.[45]

Certainly, the woodpecker fed on old trees and in late successional stands where feeding conditions were favorable, as Tanner showed. Yet evidence suggest the bird did not require either virgin forests or large trees (Appendix B). In 1935 the Cornell team found one of four nests at the edge of a natural clearing, hidden behind poison ivy and catbrier,[46] conditions indicating the stand was recently disturbed. When he returned to northeast Louisiana in 1937, Tanner found the John's Bayou pair in slash, a stand of woodland that had been previously logged.[47] In the Greenlea section of the Singer Tract, ivory-bills fed in an area ranging from "second growth in the old fields, to big solid oaks and gums."[48] In Florida's Apalachicola swamp, Tanner noted that "three of the places where ivory-bills have been reported are cut-over, showing that the birds can find a living somewhere in the large swamp there."[49]

In central Florida, ivory-bills fed by "scaling off the bark of those small and medium-sized pines that had been killed by fire."[50] Gazing at Allen and Kellogg's photograph, one is struck profoundly both by the relatively small diameters and the sparse density of the pines in which the woodpecker was foraging. Towards the end of logging in the Singer Tract, ivory-bills even fed on residual trees too small to have any commercial value. Tanner witnessed them feeding in the cutover slash itself. Richard Pough found a lone female foraging on "relatively small Nuttall oaks."[51]

Tree mortality

Tanner gave us incontrovertible evidence (though still with a sample size of only one) that ivory-bills fed selectively on trees of larger diameter.[52] Mature trees as found in the Singer Tract would be more likely to suffer from crown die-back. Because all woodpeckers benefit from tree mortality, that assumption to Tanner's ivory-bill model is sound. What was never demonstrated, however, was that the ivory-bill was restricted to trees that had died within a 2- to 3-year window, feeding only on trees where bark-scaling abilities facilitated access to the largest beetle larvae. Declaring an intractable reliance of the woodpecker on this very narrow set of conditions nurtured an ***illusory correlation***.[53]

Feeding specialization on large beetle larvae

"Wood-boring larvae are unquestionably the most important food of the Ivory-bill."[54] Although Tanner indicated that he never saw the woodpecker feed on anything else, by their very nature large items would be more visible and identifiable to a human observer. Smaller prey, including lesser wood-boring beetle larvae, termites, ants, or other insects, would not be as detectable.[55] Here Tanner was exposed to ***confirmation bias***,[56] and almost certainly by ***omniscience bias***,[57] calling into question any putative specialization by this bird on a single food item. Dietary specialization requires a contrast made between the food items that are available and accessible versus those actually consumed, an analysis that was never undertaken anywhere for the ivory-bill.

Other considerations cast incontrovertible doubt on narrow dietary specialization in the ivory-bill. On May 10, 1935, the Cornell team watched parent woodpeckers arrive at a nest cavity to feed nestlings that were apparently quite young. Adults "had trouble feeding large grubs to the nestling."[58] Although protein-rich items like large larvae are clearly advantageous from an energy delivery perspective, provisioning young with prey that are too big to consume is exceedingly maladaptive. Ivory-bills must have been capable of finding and delivering small items. Indeed, one adult female was observed to feed a nestling 38 times in fifteen minutes,[59] a rate that would have been impossible to achieve by excavating only the large wood borers.

Across its range and through varying seasons, the ivory-bill's diet was diverse. Notably it included fruit and mast: grapes, cherries, persimmons, hackberries, acorns, pecan nuts, hickory nuts, magnolia seeds, plant galls, and poison ivory berries. It also consumed other insects, including carpenter ants or termites, similar to the pileated woodpecker.[60] In fact, the 1935 Cornell team specifically pointed out how in Louisiana a female ivory-bill dug a foraging hole into a hackberry tree just like the pileated.[61] In Florida, Cornell biologists "watched them digging for borers exactly like hairy or pileated woodpeckers."[62]

Despite so much evidence to the contrary, accounts would go on to claim falsely that large beetle "grubs were almost the only thing that ivory-billed woodpeckers ate.... The birds were [otherwise] starving to death."[63] A supposed reliance on large larvae was then embellished further to conjecture about other traits in ivory-bill: "the ... inability to exploit [ants, termites, and so on] could have been a major limiting factor on population and fecundity."[64] As the only North American woodpecker able to "pry away the tight-fitting bark and get at the beetle grubs inside," the woodpecker's unique ability was extended rashly to a belief that it "had carved out a slim, precarious niche for itself."[65] Such heedless extrapolation launched a tenacious ***illusory correlation***: an unverified link between the ivory-bill's real decline and a single foraging method and dietary item.

Vanishing habitat and a highly specialized feeding style went on to shape an elegant, sophisticated proposal to explain the ivory-bill's disappearance. Like any thesis it was only as reliable as its basic assumptions. Yet this hypothesis fails under the scrutiny of evidence-based critical thinking. Intuitionist hypotheses alone can never confirm how the world operates. Intuition is specious, leading us to engage in an off-the-cuff ***truthiness***, a fickle gut attachment to opinions because they just *feel right*.[66] Faulty decisions then result.[67] Despite its entirely plausible appeal, the Tanner hypothesis (and its later corollaries) about the ivory-bill are not truth, they are merely ***truthy***.

Tanner studied a single pair, at a single location, in a single forest type having a peculiar vegetative composition and distinctive history of human activity. Any scientist of that era probably would have drawn quite similar conclusions, especially given such sparse data. Yet Tanner himself warned clearly of the stumbling blocks that were posed by his own study's limitations:

> *The chief difficulty of the study has been that of drawing conclusions from relatively few observations, necessary because of the extreme scarcity of the bird. My own observations of the birds have been entirely confined to a few individuals in one part of Louisiana, and although these observations covered a large percentage of all the Ivory-bills living in the country, the conclusions drawn from them will not necessarily apply to the species as it once was nor to individuals living in other areas ... one must draw conclusions carefully and with reservations.*[68]

Box 8.4. ***Anchoring bias***, also known as ***focalism***, results when we become so strongly attached to a view, opinion, or belief that even in the face of new information, we can only revise our position in tiny increments away from the anchor. This bias is often but not invariably caused by first impressions (first found, first stated, first published, and so on). In the routine conduct of science, such overly-rigid attachment to a hypothesis or a theory can be so stubborn that it blinds our awareness entirely to other explanations, even those that better account for the actual facts (***congruence bias***).

Despite his very pointed cautions, Tanner's advice was patently ignored by almost everyone who came along later.[69] Others went on to grant his hypotheses a sacrosanct prestige. "Notions that Tanner himself often recognized as tentative became hardened into unchallenged dictums without a good basis for doing so."[70] Misapplications from that hypothesis became pervasive through the tenacity of ***anchoring bias***.[71] Later researchers fell to evaluating potential ivory-bill habitat "only in what appeared to be the best available habitat."[72]

Even worse, findings from the original hypothesis were never verified, only presumed. In vile desecration of the scientific method, an $n = 1$ acted as both a hypothesis and its own confirmation. A reckless bias from ***insensitivity to sample size*** deceived us into thinking that an extreme value from such a narrow example, the Singer Tract study, could faithfully represent the woodpecker's real habits. In fact, small samples like this are even more likely to be extravagant and deviant from the norm.[73]

Box 8.5. Tanner's study was so plagued by ***insensitivity to sample size*** that it exposed him and others to yet more cognitive error. The stereotypical associations that we make between two rare, odd, distinctive, or unusual phenomenon make us prone to ***one-shot illusory correlation***. Rarity of the ivory-billed woodpecker itself acted as the first variable in setting us up for that bias. The ivory-bill's alleged "tameness," foraging on large grubs, use of bottomland hardwoods, occurrence in old-growth, and links to "wilderness," all factors derived solely from the Singer Tract study, then served as a second variable that steered us into ***one-shot illusory correlations***.

So it was that the vampirical first stalked and then devoured the empirical. With the ivory-bill's rarity or extinction barring ever again gathering much additional evidence, a rogue thesis went on to attack, destroy, and consume every challenge that ever dared cross its path. In a bizarre chain of cognitive farce, the death of evidence became equated with mortality of an entire species. A scourge from small *n* held spellbound our ideas about the woodpecker.

Endnotes

1 Freese, J. 2008. The problem of predictive promiscuity in deductive applications of evolutionary reasoning to intergenerational transfers: three cautionary tales. Pp. 145–178 in A. Booth, A.C. Crouter, S. Bianchi, and J.A. Seltzer (eds.). *Caring and Exchange Within and Across Generations*. Urban Institute Press, Washington, DC.

2 Tanner, J.T. 1942. *The Ivory-billed Woodpecker*. Dover Publications, Inc., Mineola, NY.

3 "Tanner's attempt at a one-man inventory of all ivory-billed woodpeckers and ivory-bill habitat in the United States was perhaps the greatest folly in the history of [the species'] conservation and one of the greatest follies in the history of U.S. bird conservation." Hill, G.E. 2007. *Ivorybill Hunters: The Search for Proof in a Flooded Wilderness*. Oxford University Press, London, p. 66.

4 In addition to the 1942 Tanner monograph itself, circumstances leading up to this field study by Tanner and the Cornell Laboratory of Ornithology, one sponsored by the National Association of Audubon Societies, can be found elsewhere. A succinct summary about Tanner's study is given by Jackson, J.A. 2006. *In Search of the Ivory-billed Woodpecker*, 2nd ed. HarperCollins, New York, pp. 125–137. For an expanded history of Tanner's direct experiences working in the South during his study, as amplified by his detailed, daily field notes, see Bale, S.L. 2010. *Ghost Birds: Jim Tanner and the Quest for the Ivory-billed Woodpecker, 1935–1941*. The University of Tennessee Press, Knoxville.

5 Bale, p. 106.

6 Tanner, p. 37.

7 Bale, p. 115.

8 *Ibid.*, p. 117.

9 *Ibid.*, p. 120.

10 *Ibid.*, p. 122.

11 *Ibid.*, pp. 206–208.

12 *Ibid.*, p. 203

13 Tanner, p. 20.

14 Bale, p. 125.

15 Even the acclimated pair of ivory-bills was skittish. Tanner and the Cornell crew remarked on this frequently: "Tanner had to leave the safety of the canvas sheltering the wagon to cover the microphone. 'This alarmed the birds.' ... 'The camera bothered him some and he flew' ... 'The activity seemed to make both birds nervous, especially the female'.... 'They called and acted nervously when I approached to within 70 yards'.... 'If they are aware of me, they call quite a bit, nervous yaps.' ... Tanner was suddenly uncovered, fully visible to the ivory-bill family. The woodpeckers quickly showed their agitation." Bale, pp. 44, 45, 138, 204.

16 Mooney, J. 1912. Taensa Indians. In *The Catholic Encyclopedia*. Robert Appleton Company, New York. From: http://www.newadvent.org/cathen/14429c.htm (accessed 4 May 2015).

17 Swanton, J.R. 1911. *Indian Tribes of the Lower Mississippi Valley and Adjacent Coast of the Gulf of Mexico*. Smithsonian Institution, Bureau of American Ethnology, Bulletin 43. Government Printing Office, Washington, DC, p. 271.

18 Sevier, R.P. 2003. *Images of America: Madison Parish*. Arcadia Publishing, Charleston, SC.

19 *Ibid.*, p. 29.

20 Economic geography of large farms modified the Southern landscape twice after the Civil War. See Aiken, C.S. 1998. *The Cotton Plantation South since the Civil War*. John Hopkins University Press, Baltimore, MD. Emergence of tenant farming first altered the spatial arrangement of large Southern plantations. Decline of tenant farming during the Great Depression then produced a second, major re-arranging of farmland in the South. See Kirby, J.T. 1987. *Rural Worlds Lost: The American South 1920–1960*. Louisiana

State University Press, Baton Rouge.

21 C. Hunter, *pers. comm.* These takings may have represented, however, but a single pair and family group, respectively.

22 George G. Beyer's collection in 1897 of seven ivory-bills took place further west in Franklin Parish, and was aided both by a keen local guide and an extended drought that facilitated access into the normally flooded swamp forest. See Beyer, G.G. 1900. The ivory-billed woodpecker in Louisiana. *The Auk* 17: 97–99.

23 Bale, p. 106. Kuhn had informed Tanner that it had been 25–40 years since herdsmen had shot ivory-bills "simply because they were wild." *Ibid.*

24 If assuming that fifteen years is the maximum longevity of this large-bodied woodpecker; see Fitzpatrick, J. W., M. Lammertink, M.D. Luneau, T.W. Gallagher, B.R. Harrison, G.M. Sparling, K.V. Rosenberg, R.W. Rohrbaugh, E.C.H. Swarthout, P.H. Wrege, S.B. Swarthout, M.S. Dantzker, R.A. Charif, T.R. Barksdale, J.V. Remsen Jr., S.D. Simon, and D. Zollner. 2005. Ivory-billed Woodpecker (*Campephilus principalis*) persists in continental North America. *Science* 308: 1,460–1,462. Jackson, p. 240, allows that it is possible that the species' longevity may reach 20 or more years.

25 The Cornell Laboratory of Ornithology had been trying since the 1920s with very little success to study ivory-bills, including repeated attempts in Florida. Only in the Singer Tract did conditions ever permit a detailed opportunity for prolonged research.

26 Bale, p. 201.

27 Hill, G.E. 2008. An alternative hypothesis for the cause of the ivory-billed woodpecker's decline [review]. *The Condor* 110: 808–810.

28 Tanner's emphasis on logging and forest loss driving the decline of the ivory-bill may have been profoundly shaped by views of E.A. McIlhenny, with whom both he and the original Cornell team visited at Avery Island, Louisiana, on several occasions. McIlhenny also regarded cutting of virgin timber as the reason for

the woodpecker's disappearance. See McIlhenny, E.A. 1941. The passing of the ivory-billed woodpecker. *The Auk* 58: 582–584.

29 Tanner, pp. 99–100.

30 *Ibid.*, p. 27.

31 *Ibid.*, p. 25.

32 Cultural trading in prized ivory-bill parts and artifacts was widespread among North American native peoples; these hunting practices took place over at least centuries, if not millennia. See Jackson, pp. 78–89.

33 Schorger, A.W. 1949. An early record and description of the ivory-billed woodpecker in Kentucky. *Wilson Bulletin* 61: 235.

34 Prior to 1810 while he was based briefly in Shepherdstown, West Virginia, Alexander Wilson collected an ivory-bill between Martinsburg and Winchester, Virginia. An ivory-bill may also have been shot as recently as 1900 in Doddridge County, West Virginia. Jackson, pp. 281–282.

35 Both specimens and eggs, now lost. See Hunter, W.C. 2010. *Interpreting historical status of the ivory-billed woodpecker with recent evidence for the species' persistence in the southeastern United States.* Appendix E, Recovery Plan for the Ivory-billed Woodpecker (*Campephilus principalis*), U.S. Fish & Wildlife Service, Atlanta, GA, p. 89.

36 Conversely, or simultaneously, these interior records could represent a predisposition by the ivory-bill to engage in itinerant wandering.

37 The ease with which instances or associations come to mind; a mental shortcut in which the first thing to occur or to be recalled is used to frame or interpret an observation, usually of something seen or heard that can be accessed most readily from memory. See Schwarz, N., H. Bless, F. Strack, G. Klumpp, H. Rittenauer-Schatka, and A. Simons. 1991. Ease of retrieval as information: another look at the availability heuristic. *Journal of Personality and Social psychology* 61: 195–202.

38 Jackson, p. 32; Tanner, p. 40–45.

39 Rohrbaugh, R., M. Lammertink, and M. Piorkowski. 2009. Final Report March 2009 surveys for Ivory-billed Woodpecker and bird counts in the Fakahatchee Strand State Preserve, Florida. Cornell Laboratory of Ornithology Report submitted to Fakahatchee Strand State Preserve. http://www.birds.cornell.edu/ivory/folder.2010-04-20.2993097079/2009%20Fakahatchee%20Final%20Report1.pdf (accessed 8 May 2015).

40 Jackson, pp. 53–57; Allen, A.A., and P.P. Kellogg. 1937. Recent observations of the ivory-billed woodpecker. *The Auk* 54: 164–184; Nehrling, H. 1882. List of birds observed at Houston, Harris Co., Texas, and in the counties Montgomery, Galveston and Ford Bend. *Bulletin of the Nuttall Ornithological Club* 7: 166–175.

41 Allen and Kellogg, p. 167.

42 Jackson, pp. 52–57. As late as the early 1930s, observers noted that Florida "ivory-billed woodpeckers were attracted to recently burned areas as foraging sites." *Ibid.*, p. 64.

43 Tanner, p. 46.

44 Ibid., p. 55.

45 Haney, J.C., and C.P. Schaadt. 1996. Functional roles of eastern old-growth in promoting forest bird diversity. Pp. 76–88 in M.B. Davis (ed.). *Eastern Old Growth Forests: Prospects for Rediscovery and Recovery* (M.B. Davis, ed.). Island Press, Washington, DC. Contributing to fallacy from an ***availability cascade***, I also merely restated uncritically other ***memes*** that are routinely asserted for the ivory-bill's ecological habits and fate, e.g., "it is almost certainly extinct in North America," p. 77.

46 Bale, p. 62. The Cornell team also described the first-year nest site at John's Bayou as being located in an area with second-growth trees, many downed tree trunks, and thick, tangled, jungle-like undergrowth of briers and poison ivy. *Ibid.*, p. 41.

47 *Ibid.*, p. 106.

48 *Ibid.*, p. 114.

49 *Ibid.*, p. 212.

50 Allen and Kellogg, p. 167.

51 *Ibid.*, p. 253.

52 Tanner, p. 43. His figure 20 shows ivory-bills fed more frequently than expected from availability alone on those trees in the larger size classes: the categories of 12–24, 24–36, and greater than 36 inches diameter at breast height.

53 A bias from an erroneous conclusion about the relationship between two events. An observer draws an association where there is none. ***Illusory correlation*** may arise from the co-occurrence of events that are in fact statistically infrequent. See: Hamilton, D.L., and R.K. Gifford. 1976. Illusory correlation in interpersonal perception: a cognitive basis of stereotypic judgments. *Journal of Experimental Social Psychology* 12: 392–407. ***Illusory correlation*** can also explain dogged persistence of stereotypical thinking in the absence of any confirming evidence; see Hamilton, D.L., and T.L. Rose. 1980. Illusory correlation and the maintenance of stereotypic beliefs. *Journal of Personality and Social Psychology* 39: 832–845.

54 Tanner, p. 41.

55 In other large *Campephilus*, as much as 20% of dietary items are too small to be identified at all. Ojeda, V.S., and M.L. Chazarreta. 2006. Provisioning of Magellanic woodpecker (*Campephilus magellanicus*) nestlings with vertebrate prey. *The Wilson Journal of Ornithology* 118: 252–254. Similarly, prey was never visible in the bills of parent crimson-crested woodpecker *Campephilus melanoleucos* feeding nestlings up to three weeks of age, indicating that large larvae are not essential for provisioning young in this relative of the ivory-bill. Kilham, L. 1972. Habits of the crimson-crested woodpecker in Panama. *Wilson Bulletin* 84: 28–47. Moreover, adult crimson-crested woodpeckers were seen feeding on wood riddled with tunnels made by termites, further rebutting the notion that *Campephilus* foraging is necessarily and narrowly specialized compared to other woodpeckers.

56 ***Confirmation bias***, the seeking or interpreting of evidence in ways that are partial to existing beliefs, expectations, or a hypothesis at hand, is exceedingly pervasive and a very strong influence

on decision-making by humans; see Nickerson, R.S. 1998. Confirmation bias: a ubiquitous phenomenon in many guises. *Review of General Psychology* 2: 175–220. This bias is strongly inferred by Tanner's not finding ivory-bills in what otherwise seemed to be suitable habitat: "At times, reading Tanner's journal, one gets the sense that he returned to the John's Bayou to reassure himself that ivory-bills indeed remained in the region." Bale, p. 119. Tanner gave fair warning about the risks of giving his ecological insights from the Singer Tract undue weight. Rather, his particular ***confirmation biases*** were most conspicuous and therefore detrimental to the future thinking about the ivory-bill when he spent too few days searching for the bird elsewhere (but concluded they were still absent), or discounted as unsuitable those habitats that were notably different than the ones he researched in northeast Louisiana; see especially Tanner, pp. 20–29.

57 A mistaken belief that we have access to the right information to form accurate opinions or to make informed judgments; the illusion that everything we need to know to be well-informed is accessible to us.

58 Tanner, p. 76.

59 Bale, p. 154.

60 Bent, A.C. 1939. *Life Histories of North American Woodpeckers*. Smithsonian Institution National Museum Bulletin 174 (1964 Dover edition); Allen and Kellogg, p. 167; Jackson, p. 197; Snyder, N., D.E. Brown, and K.B. Clark. 2009. *The Travails of Two Woodpeckers*. University of New Mexico Press, Albuquerque, p. 36.

61 Bale, p. 48.

62 Allen and Kellogg, p. 166.

63 Hoose, P. 2004. *The Race to Save the Lord God Bird*. Farrar, Straus, and Giroux, New York, p. 112.

64 Michaels, M. 2016. Digging deeper into Tanner, part 2 of 3 – foraging substrates. https://projectcoyoteibwo.com/2016/03/19/digging-deeper-into-tanner-part-2-of-3-foraging-substrates/ accessed 21 March 2016).

65 Bale, p. 221. Upon being shown feeding signs of ivory-bills during Singer Tract logging, an executive of the Chicago Mill and Lumber Company, S.C. Alexander, is said to have remarked to Tanner: "They ought to learn to feed on something different." Bale, p. 241. That statement reveals how powerful the meme of feeding specialization had already taken hold, a meme that only grew stronger and more resistant to criticism with passing time.

66 Narvaez, D. 2010. Moral complexity: the fatal attraction of truthiness and the importance of mature moral functioning. *Perspectives on Psychological Science* 5: 163–181. Intuitively-appealing (but still misleading) ***truthiness*** can arise when any of the following are present: 1) it feels familiar, 2) it doesn't require difficult mental processing, 3) the 'story' flows smoothly, 4) the source seems familiar and/or trustworthy, and/or 5) the evidence is easy to recall. Schwarz, N., and E.J. Newman. 2017. How does the gut know truth. *Psychological Science Agenda*, 31(8).

67 When people make judgments about the truth of a claim, non-probative information that is transmitted concurrently strengthens their tendency to believe the claim – an effect labeled as "truthiness." See: Newman, E.J., M. Garry, D.M. Bernstein, J. Kantnor, and D.S. Lindsay. 2012. Non-probative photographs (or words) inflate truthiness. *Psychonomic Bulletin & Review* 19: 969–974; Newman, E., and N. Feigenson. 2013. The truthiness of visual evidence. *The Jury Expert* 25:1–6; and Newman, E.J., M. Sanson, E.K. Miller, A. Quigley-McBride, J.L. Foster, D.M. Bernstein, and M. Garry. 2014. People with easier to pronounce names promote truthiness of claims. *PloS One* 9: e88671.

68 Tanner, p. XVI.

69 "Tanner's assertion that ivorybills can persist only in fully mature forests with abundant recently killed trees has shaped the perceptions of all subsequent ivorybill searchers, including the leaders of recent search efforts in the Big Woods in Arkansas and along the Choctawhatchee River in Florida." Hill 2008, p. 808. "Ivory-bill hunters, I think, put too much stock in Tanner's

assessment, and in so doing missed several obvious places to look for the birds." Hill 2007, p. 68.

70 Cyberthrush. 2016. Back to 2005. Wednesday, March 2, 2016. Ivory-bills live???! http://ivorybills.blogspot.com/ (accessed 30 April 2016).

71 "once ivory-bills had been pronounced extinct by the greatest ivory-bill expert in the world, it became virtually impossible for the species to be resurrected." Hill 2007, p. 66.

72 Jackson, p. 158.

73 ***Insensitivity to sample size*** is a bias when unwarranted generalities are extended from samples to populations. See: Reagan, R.T. 1989. Variations on a seminal demonstration of people's insensitivity to sample size. *Organizational Behavior and Human Decision Processes* 43: 52–57. Despite Jacob Bernoulli's terse optimism that even the "stupidest man" knows that the larger the sample, the more confidence one can have in being close to the truth, this adage was flouted brazenly with ivory-billed woodpecker from the 1930s through to the present. This cognitive oversight may arise from the fact that humans can more readily factor into account the sample size for *frequency distributions*, but are less able to do so for *sampling distributions*. Limitations imposed by the Tanner study are largely related to the latter. See: Sedlmeier, P., and G. Gigerenzer. 1997. Intuitions about sample size: the empirical law of large numbers. *Journal of Behavioral Decision Making* 10: 33–51.

Chapter 9

Poetic License

Tell them to be patient and ask death for speed; for they are all there but one – I, Chingachgook – last of the Mohicans.
—Last of the Mohicans (1992 film version)

Words bundle power. But it is image that ultimately forces the capitulation.[1] Tanner's report from the Singer Tract hardened a near immoveable set of tangible beliefs that were widely accepted about the ivory-bill. The bird's figurative death would be affirmed soon thereafter by a melancholy portrait commissioned to venerate the woodpecker's make-believe last survivor.[2] An artistic rendition of a solitary female ivory-bill winging her way over a ruined forest went on to fix a tenacious anchor of its own, one that had even more insidious effects on reinforcing myth about the woodpecker. That undisputed image and our careless turns of phrase further disparaged the real ivory-bill.

National Association of Audubon Societies' artist Donald Eckelberry is widely endorsed as the last "credible" naturalist ever to see an ivory-billed woodpecker. Audubon's president, John Baker, first sent the society's scientist Richard Pough in December 1943 down to Louisiana to scout around for any ivory-bills still left in the Singer Tract. The area was already years into relentless logging. After six weeks, Pough at last found one lone female in a small island of timber surrounded by cutover desolation. Eckelberry followed afterward, arriving in April of 1944. Guided to the bird by local game warden Jesse Laird, Eckelberry spent two weeks sketching this single ivory-bill as she left in the morning and returned during the evening to an isolated roost cavity inside an ash tree, it too set alone in the midst of cut-over ruin.

Eckelberry's depiction of that forlorn bird flying over a wooded wasteland is reproduced in Jackson's extensive history about the species.[3] Eckelberry also gave her a masterful requiem in haunting prose:

> *She came trumpeting in to the roost, her big wings cleaving the air in strong, direct flight, and she alighted with one magnificent upward swoop. Looking about wildly with her hysterical pale eyes, tossing her head from side to side, her black crest erect to the point of leaning forward, she hitched up the tree at a gallop, trumpeting all the way … bathed in rich orange light of the setting sun, she alternately preened and jerked her head about in a peculiar, angular way, quite unlike the motions of any other woodpecker I knew. I was tremendously impressed by the majestic and wild personality of this bird, its vigor, its almost frantic aliveness.*[4]

It is a touching and unforgettable scene, an artist bequeathing to posterity an illustrated and literary dénouement for a majestic species. Against hopeless odds, one final survivor displays such fierce will to live. "She was important because she was the last of the Mohicans."[5] Facts linked with fancy, the legend was now secured even more firmly.

For reasons never tolerably defended, Eckelberry's sight record in 1944[6] became the next (and some say, last) date certain of extinction put forth for the ivory-bill.[7] Later writers only reinforced that baffling declaration: "[Eckelberry] felt like he was staring at eternity. This single unmated female was all that remained of the Lord God bird that had commanded America's great swamp forests for thousands of years."[8] Media outlets just restated uncritically that "the ivory-bill … hasn't been definitively seen since World War II."[9] Without any tangible evidence, this sighting was then perpetuated as "the last universally accepted sighting of one of these birds in the United

States."[10] There is no proof to that reading. No evidence given in support whatsoever.

Box 9.1. "Repeat something often enough, and it becomes true," then "go along with the crowd," describes an ***availability cascade***. It is a self-reinforcing thinking gaffe caused by shared belief formation. A perception triggers a chain reaction, in turn giving that perception ever increasing plausibility due to its expanding availability in public discourse. Through its constant and durable repetition, "the last confirmed report of ivory-bill was in 1944" furnishes us a textbook example of the ***availability cascade***.

Never would oral reportage like this be respected as fitting testimony for a live ivory-bill today, so one must interrogate harshly why it was ever deemed an adequate standard for marking the absolute extinction of the entire species. The story's frequent reprise makes for a fine example of a wayward ***availability cascade***.[11] But repetition is not truth. In any event, the Eckelberry narrative served to bury the woodpecker. It fixed and then cornered on the defensive those who might later see the bird. For those who did not subscribe to this memorial service, myth narrative for ivory-bill nudged them unwillingly into one of two camps: quixotic searchers who mistook windmills for giants (the pileated for the huge ivory-bill), or naïve laypersons too clueless to abandon the bare witness of their own eyes and ears.

Storybook narratives about the ivory-bill have long thrived among us: "the ivory-billed woodpecker has always drawn attention to itself."[12] "It is a bird with pizzazz."[13] Our early naturalists were no less prone than we are to fall back on florid superlatives and linguistic excess when describing the huge woodpecker. But generally, they did so with depictions that focused on the bird's plumage and behavior, not some attribute burdened with antiquated myth, e.g., "a twelfth-century bandit nobleman might have gloried in such apparel as my ivory-billed woodpecker wore."[14]

Audubon compared the bird's physical appearance to the work of a Flemish Baroque painter of the seventeenth century:

> *in the plumage of the beautiful Ivory-billed Woodpecker, there is something very closely allied to the style of colouring of the great Vandyke. The broad extent of its dark glossy body and tail, the large and well-defined white markings of its wings, neck, and bill, relieved by the rich carmine of the pendent crest of the male, and the brilliant yellow of its eye, have never failed to remind me of some of the boldest and noblest productions of that inimitable artist's pencil ... whenever I have observed one of these birds flying from one tree to another, I have mentally exclaimed, 'There goes a Vandyke!'*[15]

At a minimum, Audubon here granted art and life coequal status. The comparison was at least fitting because it was grounded in a contrast between two tangible objects, not a real object and some fairy tale.

Although adoption of the Eckelberry narrative epitomizes the artistic license used in cultural positioning of the ivory-bill, it is by no means the only misleading instance of our fictional distortions. Indeed, a number of other abuses rooted in fable, legend, and cultural allegory have done nearly as much harm to real versions of the woodpecker as did the shotgun and cross-cut saw in earlier times. Our reckless semantics are a chief reason that so much cognitive distortion now plagues any logical discourse about the bird's destiny.

Various linguistic devices warrant criticism as to whether

they serve the best interests of a genuine bird. Among the more prevalent when applied to the woodpecker are the following:

Lord-God bird

Of this woodpecker it is said: "Sacred things are best shrouded by silence and solitude."[16] A spontaneous exclamation once blurted out by those astounded to stumble upon the big woodpecker in the wild, this label verifies how even well-intentioned praise can inhibit clear perspective. *Lord-God* appears in the titles of at least five books or short stories,[17] all of which have their plots centered in some way on the ivory-bill. Notably, four of these are complete works of fiction, only one is a real history.

Colorful descriptions are expected for this iconic species, certainly, but divinity is not among the labels that can be literally justified. Linking ivory-bill to deity only makes it seem larger than life. It was not; it is not. Alive or dead, it is still just a bird. *Lord-God* also removes the ivory-bill from its grounding in nature, the ultimate home to us all, and stations the woodpecker somewhere far away, remote and distant from reality. That realm may be sacred to some, but *Lord-God* conveys a sly invisibility about the bird itself. Something not quite real; an entity that is unattainable and inaccessible to us humble mortals. And, of course, the search for deity never ends.

Ghost bird

This moniker for the woodpecker is anything but subtle. Instead of a purely invisible object, we borrow a hackneyed title for representations of the dead. After seeing an enormous ivory-billed woodpecker one April morning while walking along the valley road, Aunt Maggie Cude goes into a swoon over what she believes was a sure omen of impending death: after all, "she's seen a ghost."[18] Commenting on the species putative rediscovery in 2004, the media could not resist this imagery either: "The ivory-billed woodpecker remains ghostly and mysterious."[19]

Ghost Bird titles a documentary film about the ivory-bill, a work reviewed quite fittingly as one that: "considers the ways in which collegial debate, intellectual rigor, and a collective desire for objective truth are in danger of extinction."[20]

Referring to ivory-bill as specter invariably denotes extinction. Dead things cannot be with us anymore. Dead things don't come back. Dead things do not appear to sophisticated, modern observers. Seeing dead things suggests that one is too imaginative and easily misled at best, a bit touched in the head at worst. Claims of finding a bird reputed to be long deceased lead inexorably to cognitive depreciation for any likelihood in the woodpecker's continued survival, whether in eras just after its putative extinction or in the present day. "Stalking the ghost bird,"[21] then, is likely to strike most of us as some foolish errand clouded by hallucination.

Lazarus bird

"Like Lazarus, the woodpecker had risen from the dead."[22] This might be the least objectionable of all of the many allusions heaped on to the woodpecker. Unlike the phoenix, Lazarus only died and resurrected one time. Still, the comparison was too much to leave alone. After the announcement of a rediscovery, the Milwaukee Journal Sentinel heralded: "Is the 'Lazarus bird' back from extinction?"[23]

A Lazarus species, or taxon, is one thought to have become extinct, only to reappear. Because the Lazarus effect was used originally to denote species that appear and then reappear geologically in the fossil record,[24] the allusion does have a science-based usage. Nevertheless, the term is misapplied in those cases wherein, as with the ivory-bill, the disappearance and re-appearance are mere human artifacts, whether due to our defective sampling (Chapter 13) or our cognitive aberrations (Chapter 12).

The Golden Fleece

One of the oldest myths to portray the hero's voyage, Jason and the Golden Fleece is a classic Greek tragedy containing betrayal, vengeance, and supernatural creatures of all description, including centaurs, sirens, hydras, dragons, and an island of husband-killing Amazons. It is a story that highlights dangers of selfishness, jealousy, and complex family loyalties. "If the Cornell expedition [of 1935] could be compared to the fabled quest of Jason and the Argonauts, then the ivory-bill was the Golden Fleece."[25] Uh-oh, the woodpecker gets linked to yet another charade. And, of course, that loaded verb "fleece": to deceive, to swindle, to con, and to dupe.

Elvis

No less an authority than National Geographic repeated this totally unfit association with a dead music entertainer. At its website, the Society proclaimed "it was hailed as the birding equivalent of finding Elvis alive"[26] and "the Elvis of the bird-watching world is alive in eastern Arkansas."[27] But *National Geographic* was hardly alone. The *Chicago Tribune* decreed: "Army of scientists searching for Elvis, researchers comb Arkansas' swampy woods hoping to find the elusive ivory-billed woodpecker affectionately named after the rock star."

Drawing a parallel between a rediscovered member of an authentic species and a single deceased performer is complete nonsense. It does not help that the word Elvis conjures up legions of lounge-lizard imposters, either, a thinly disguised link to deception giving us yet another reason to object to its use in comparison with the woodpecker. When the Cornell Laboratory of Ornithology used Elvis as code for their clandestine searches of the bird in Arkansas, they may not have grasped fully the unintended consequences that arose from that unfortunate choice:

Its meaning gradually morphed from 'Elvis found' to 'Elvis often seen but never confirmed' as sightings came in without evidence, akin to jokes that refer to 'Elvis' having 'entered the building.'[28]

Words matter. Extending Presley's name demeans the real woodpecker because it fosters an as-yet unsupported stance on the bird's demise. It belittles, too, our mathematical literacy to equate one entire species' extinction with a single individual death of another.

Grail bird

Grail is particularly dicey when directed at the ivory-bill. The imagery prompted by referencing the woodpecker to this object invokes both the sacred and the silly, not to mention pointless and never-ending quests for something that may or may not be there. "Every culture ... needs something perfect and unattainable just outside its reach."[29] We are obligated first to ponder whether woodpecker as grail conforms more to the medieval version of a divine chalice depicted in the stately Arthurian legends, or a Monty Python parody that depicts a bunch of bumbling idiots.[30] It is nigh impossible to take seriously a grail search if our association happens to be tainted by a "cult classic as gut-bustingly hilarious as it is blithely ridiculous."[31]

Only when grail is used to refer to "something that is earnestly sought after" might it apply suitably to the woodpecker. In the National Public Radio piece *In Search of Birding's Holy Grail*, the lede announces: "Obsessed birders searched the swampy woods of the southern United States for years, listening for the call of the Ivory-billed Woodpecker. How was the bird finally found?"[32] Even where usage of grail can handle the most irrelevant forms of semantic baggage, searching for an avian version can sound too much like credence in the power of wishful thinking:

In the culture of enchantment, the search itself becomes a sacred

> *ritual, what Emile Durkheim called an 'imitative rite,' meaning that the worshippers act in the hope that* 'like produces like.' *The Lord God bird in the heart might lead to a manifestation, a Lord God bird in the swamp.*[33]

Conjuring physical reality from prayerful meditation evokes the alchemic language found in sacramental rites, of fundamentalist religion, and around practices centered on unreasoned or unshakeable belief.

Portraying a search for the real ivory-bill inside the grail meme runs a legitimate risk of being perceived as cult. Psychological roots behind the blind pursuit of ivory-bill should be placed under bright lights of suspicion. Hunts for rare or elusive species may easily devolve into fixation.[34] In his *The Life of the Skies: Birding at the End of Nature,* Jonathan Rosen revealed interactions with various pilgrims who were steadfast in proving that the woodpecker still lives, a journey described from a fellow "denizen of a subculture of obsessive ivory-billed seekers."[35]

Even comparison with a guileless quest can easily degenerate into unintentional cheapening. Much like Cervantes' hero, the ivory-bill is not uncommonly framed by a meme centered on years of fruitless hunting. Despite its being a masterpiece of our world literature, one review of *Don Quixote de la Mancha* points to a: "hero's transformation from a 'clinical' case of literary insanity and chivalric folly, through the artifice of interlocking fictions..., to mythical status as redemptive hero of social satire."[36] Does all this simile really fit the description of a bona fide bird?

Damnation and redemption

In a context of American environmental history, the ivory-bill is commonly wielded as instrument of judgment over our own shoddy treatment of the environment. The woodpecker represented "a more environmentally pristine South."[37] And "nothing symbolizes what we have lost more than the ivory-billed woodpecker."[38] If it is extinct, the bird serves to remind us of our ignorance about species protection generally, or our faults in not setting enough (or the right types) of land aside for the big woodpecker. And if not really dead, then the ivory-bill represents "a message of forgiveness: maybe our environmental sins weren't so bad."[39] This "narrative arc" is just too compelling for us to pass up: "Only through the ivorybill's apocalypse—the utter destruction of its world, its believed extinction—was the groundwork laid for its subsequent resurrection and, therefore, for our subsequent redemption."[40]

City slicker teams with country bumpkin

A certain narrative convention recurs in stories about the refound ivory-billed woodpecker. Creative tension is achieved by juxtaposing an insider versus an outsider into what should be fact-based evaluation of woodpecker searches. To whit: "every time, these expeditions seemed to be remakes of the exact same buddy flick. The courtly individual from the Yankee Ivy League college gets taken into the woods by the joshing redneck who knows the ground.... This repetitive quality to the stories ... was American mythology."[41] Whether this is snide elitism, bigoted regionalism, or snobbish intellectualism, or just trying to be too cute by half, our own cultural narratives have no dispositive relevance to the actual identity or persistence of a wild bird.

The truth is out there

Cue the eerie music, a Cigarette Smoking Man lurks in dark shadow. The New Jersey devil, a crashed UFO, and alien abductions haunt two FBI inspectors. This phrase is the signature line of Jerome A. Jackson's decades-long search for the ivory-bill.[42] "Oh, look here. Look here, says the Fox Mulder [Jackson] of the Fakahatchee Strand."[43] In this instance, pop cultural links to *The X-Files* are, to put it mildly, ill-chosen. After all, the show's ever-diligent agent "Mulder, alienated from the truth that there are aliens, overcomes his alienation by using 'abduction' to infer the existence of aliens."[44]

Handicapping the woodpecker with ties to a series that revels in the supernatural could hardly create worse obstruction to clear thinking about a bird species native to North America. Neither paranormal nor fictional, the ivory-bill deserves better. In referencing this series and our relationship with scientific inquiry, Richard Dawkins noted that whenever *The X-Files* offers a rational theory and a paranormal theory to explain its plotline, "week after week, the rational explanation loses."[45] At least in this sense, the show parallels quite closely the abandonment of logical rigor so often manifest in the contentious debates over the ivory-bill's existence (Chapter 12).

Recurring cover-ups in *The X-Files* spawn distrust in what we are being told by those who represent "authority." Even the show's admiring critics divulge that it "fuelled the paranoia of those prone to adopt government conspiracies and in this way may have indirectly spawned various new unfounded theories about the government's covert involvement in our lives."[46] Connecting the woodpecker to *The X-Files* facilitates conspiracy views about the woodpecker itself, including the notion that sightings are fabricated for exerting tyrannical governmental rights to seize property (see Chapter 12).

Another reason for us to reject parallels with the show is its failure to ever reach a resolution. A recurring theme to *The X-Files* is a tension between the logical, rational Scully and the mystical, intuitive methods used by Mulder as they face various challenges of interpreting the bizarre evidence uncovered. Extending the premise to the ivory-bill, a clear inference is that ultimate resolution is never possible. Instead, we abide in "a society of illusions where only the gullible believe in anything except their own interests and where the powerful make policy decisions in secret."[47] We are forever doomed to reside in a limbo of not knowing whether the woodpecker is *still real*.

Extraordinary claims require extraordinary evidence

Popularized by Carl Sagan as diagnostic for the proper application of scientific method, this canard is irrelevant in the context of any modern-day existence of the ivory-billed woodpecker. There is no foundation for applying a standard of extraordinary evidence to a species authenticated already by more than 400 specimens, plus various still photographs, video footage, and audio recordings. Disputes about the sort of consensus needed to authenticate any continued existence of the ivory-bill *are* real. Indeed, disputes over evidence contribute substantially to muddled discussion about the actual bird. But raising the standards to meet some "extraordinary claim" yardstick is nothing but a ***straw man***.

Deploying Sagan's turn of phrase is intended to depreciate a present-tense ivory-bill with a false equivalency to a search for Bigfoot, the Loch Ness Monster, or unicorns, a textbook case

Box 9.2. Gross over-estimation of how much we actually know predisposes us to the cognitive bias of ***overconfidence effect***, a disparity between what we really know versus what we *think* we know. ***Overconfidence effect*** is a measure of the difference between objective accuracy and what we come to believe subjectively. Proclamations of certainty about the ivory-bill's extinction, whether made in the 1940s or in more modern times, illustrate this ***overconfidence effect***, in large part because humans are inclined to make wrong attributions about species' extinctions in a broader sense (i.e., the ***Romeo error*** that is triggered by ***base rate fallacy***; see Chapter 14).

of another fallacy, this one of the ***red herring*** variety. Neither extinction nor survival is an "extraordinary" phenomenon when it comes to the ivory-bill. Given ample uncertainty, both extinction and survival are very real likelihoods lying somewhere between the possible and the probable. Due to a sweeping ***neglect of probability,***[48] those who insist (with current levels of knowledge) in either the absolute extinction or the certain survival of the ivory-bill are under the influence of a weighty ***overconfidence effect.***[49]

Allusions to cryptozoology

A Google joint search for "Bigfoot" and "ivory-billed woodpecker" turned up 4,930 results; another search for the woodpecker and "Loch Ness Monster" revealed 2,650 more.[50] Scott Crocker, who wrote and directed *Ghost Bird,* used what amounts to a slur on the woodpecker by interjecting a quote from a local Arkansas resident who referred to all three "creatures" at once in that film.[51] In reviewing *The Grail Bird,* the *New York Times* flippantly dishes out the insult that: "it could as easily be a book about the hunt for Bigfoot."[52] Hoose even refers to Tanner's almost singular familiarity with the ivory-billed woodpecker as "a little like being a world expert on UFOs or the Loch Ness Monster."[53]

In referring to David Kulivan's 1999 report of two ivory-bills in the Pearl River, Weidensaul falls for the same drivel: "If you're going to tell a sceptical world that you've seen the avian equivalent of a unicorn, it's probably best not to break the news on April Fool's Day."[54] Scientists seem to be no better at avoiding dissemination of the same tomfoolery. In sorting through incoming sightings about possible ivory-bills in Alabama, Hill likens the process of appraisal to "getting a report of Bigfoot or the Loch Ness Monster."[55] No, actually, it is not. Drawing such equivalency is ludicrous.

Likening ivory-bill to some beast not substantiated by

science is riddled with logical blunder and cognitive spin, each over-seasoned with insults to our intelligence. Drawing a cryptozoological parallel is intended by the originator to divert listeners from the central issue of ivory-bill extinction or survival. The tactic instead compels defense of another matter entirely, one that *is* untenable. Drawing a cryptozoological parallel exemplifies in textbook fashion the ***availability cascade*** (through constant repetition), the ***Dunning-Kruger effect*** (idiotic but never realized as such by the originator), ***equivalency bias*** and ***false equivalence*** (searches for the woodpecker and Bigfoot are the same), ***framing effect*** (forcing the listener to accept equivalency as a legitimate option in the first place), and ***neglect of probability*** (chances of existence for the woodpecker and some undocumented entity like unicorn are about equally likely).

Box 9.3. ***Dunning-Kruger effect*** is a dire cognitive error wherein those who are incompetent in a particular subject area not only wildly inflate their own skill level but *then also lack the metacognition ability to ever realize their error*. It is a level of incompetence so boundless that the cognitive deficiency is unable ever to come to terms with the depth of its own witlessness.

No other extinct or nearly extinct animal in North America has ever been obliged to lug around so much abstract cargo as has the ivory-billed woodpecker. Obsolete folklore that we aimed at this bird began inflicting its mockery during the youngest decades of the twentieth century. Within a hundred years, our unease over handling doubt led us to uncritically adopt both ancient and modern metaphor as the primary cognitive adjustment to enormous ambiguity over the woodpecker's fate.

Surely this much encumbrance from symbol, fable, and allegory is enough to crush any resemblance to truth.

Endnotes

1 When evaluating truth of a claim, like extinction, juxtaposing a tangential graphical object, like Eckelberry's painting of the "last ivory-bill," is more likely to lead people to believe in the claim of extinction – a ***truthiness*** effect is the result. ***Truthiness*** then goes on to have persistent, distorting effects on human judgment. See: Fenn, E., E.J. Newman, K. Pezdek, and M. Garry. 2013. The effect of non-probative photographs on truthiness persists over time. *Acta Psychologica* 144: 207–211.

2 Between the end of Tanner's study in the late 1930s and Eckelberry's portrait in 1944, ivory-bills were still seen irregularly in the Singer Tract by a number of ornithologists, including Roger Tory Peterson. However, a combination of random levels of credibility assigned to and by various "experts," lack of tangible evidence, and arbitrary skepticism aimed at future observers led to 1944 being touted widely as one of the woodpecker's putative extinction dates.

3 Jackson, J. A. 2006. *In Search of the Ivory-billed Woodpecker,* 2nd ed. HarperCollins, New York, p. 152.

4 Zickefoose, J. 2012. *The Bluebird Effect: Uncommon Bonds with Common Birds*. Houghton Mifflin Harcourt, Boston, MA, p. 251.

5 Remark by Gene Laird, son of Jessie Laird, who took Don Eckelberry out to find the "last" ivory-bill; see Severson, F.J. 2007. Memories from the Singer Tract. *Birding* 39: 42–45.

6 Despite the oft-used 1944 date for the woodpecker's final disappearance, correspondence from visitors to the Singer Tract still gave second-hand accounts of ivory-bills persisting at this site until at least November 1948. Michaels, M. 2015. Last letter on the Singer Tract ivorybills: January 1949. https://projectcoyoteibwo.com/2016/04/14/last-letter-on-the-singer-tract-ivorybills-january-1949 (accessed 30 April 2016).

7 And so became solidified a capricious treatment of sight records and observer credibility. For example, Herbert L. Stoddard, Jr., a prominent ornithologist and wildlife biologist, saw three ivory-bills on two different occasions during the 1950s, yet these were never afforded the same official or literary "credibility" as the earlier, purely sight report by Eckelberry.

8 Hoose, P. 2004. *The Race to Save the Lord God Bird.* Farrar, Straus, and Giroux, New York, p. 132.

9 Hitt, J. 2012. Science and truth: we're all in it together. *New York Times.* http://www.nytimes.com/2012/05/06/opinion/sunday/science-and-truth-were-all-in-it-together.html?_r=0 (accessed 25 February 2015).

10 Gallagher, T. *The Grail Bird: The Rediscovery of the Ivory-billed Woodpecker.* Houghton Mifflin Harcourt, Boston, MA, p. 16.

11 A cognitive social bias wherein we believe information after being exposed to it repeatedly such that the perception increases in plausibility through its greater prevalence in our social discourse.

12 Bale, S.L. 2010. *Ghost Birds: Jim Tanner and the Quest for the Ivory-billed Woodpecker, 1935–1941.* The University of Tennessee Press, Knoxville, p. 3.

13 Hill, G.E. 2007. *Ivorybill Hunters: The Search for Proof in a Flooded Wilderness.* Oxford University Press, London, p. vii.

14 Thompson, M. 1885. A red-headed family. *The Elzevir Library* 4: 11.

15 Audubon, J.J. 1842. *The Birds of America,* Volume 4, p. 214.

16 Keen, S. 2007. *Sightings: Extraordinary Encounters with Ordinary Birds.* Chronicle Books, San Francisco, CA, p. 107.

17 Core, J.R. Undated. *The Lord God Bird: A Season of Rain and Death.* Unpublished manuscript of a novel of the 1927 flood in Arkansas; Cremeens, C. 1968. The Lord God bird. *The Georgia Review* 22: 400–415; Gallant, T. 2012. *The Lord God Bird.* Quantuck Lane Press, New York; Hill, R. 2009. *The Lord God Bird.* Caravel Books, New York; Hoose 2004.

18 Cremeens, p. 400.

19 Sibley, D.A. 2005. Return of the King: In the Arkansas swamps,

Tim Gallagher tracks the majestic ivory-billed woodpecker, until recently believed lost forever. *Boston Globe*, June 26, 2005.

20 See Artforum review at: http://artforum.com/film/id=25411 (accessed 27 April 2015).

21 Steinberg, M.K. 2008. *Stalking the Ghost Bird: The Elusive Ivory-billed Woodpecker in Louisiana.* Louisiana State University Press, Baton Rouge.

22 Vance, E. 2008. Extinct, maybe, but digitally in flight. *The Chronicle of Higher Education* 11 July 2008: A5.

23 See: https://news.google.com/newspapers?nid=1683&dat=20060430&id=u7IaAAAAIBAJ&sjid=JEUEAAAAIBAJ&pg=6461,8066406&hl=en (accessed 28 April 2015).

24 Fara, E. 2001. What are Lazarus taxa? *Geological Journal* 36: 291–303.

25 Bale, p. 35.

26 See: http://animals.nationalgeographic.com/animals/birds/ivory-billed-woodpecker/ (accessed 27 April 2015).

27 See: http://news.nationalgeographic.com/news/2005/04/0428_050428_extinctwoodpecker.html (accessed 27 April 2015).

28 Jackson, J.A. 2010. *Ghost Bird* – The Ivory-billed Woodpecker: Hopes, Dreams, and Reality. *PLoS Biology* 8: e1000459.

29 Narechania, A., and W. Yandik. 2005. Hearing is believing: ivory-billed sightings leave field biologists wanting to hear more. *The American Scholar* 74: 84–97.

30 King Arthur: "On second thought, let's not go to Camelot. It is a silly place." Monty Python and the Holy Grail, 1975 film.

31 A critics consensus for the film as given at Rotten Tomatoes: http://www.rottentomatoes.com/m/monty_python_and_the_holy_grail/ (accessed 27 April 2015).

32 See: http://www.npr.org/templates/story/story.php?storyId=4632991 (accessed 27 April 2015).

33 Gibson, J.W. *A Reenchanted World: The Quest for a New Kinship with Nature*. 2009. Henry Holt and Company, New York, p. 61.

34 "I've always been the kind of person who gets caught up in

obsessive quests.... If I weren't hooked on quixotic quests before that moment, I certainly was after it." Gallagher, pp. xii–xiv. Gallagher's buddy "Harrison had intermittently searched for the ivory-bill for thirty years." Steinberg, p. 9.

35 Gabbard, G.O. 2008. Review of: Rosen, J. 2008. *The Life of the Skies: Birding at the End of Nature*. Farrar, Straus, and Giroux, New York. *American Journal of Psychiatry* 165: 1,616–1,617.

36 Murillo, L.A. 1990. *A Critical Introduction to Don Quixote*. Peter Lang International Publishers, New York.

37 Steinberg, *ibid*.

38 Gallagher, p. 238.

39 Hitt.

40 Steinauer-Scudder, C. 2020. The lord god bird: apocalyptic prophecy and the vanishing of avifauna. *Emergence Magazine*. https://emergencemagazine.org/essay/the-lord-god-bird/ (accessed 15 July 2021).

41 Hitt, J. 2012. *Bunch of Amateurs: A Search for the American Character*. Crown, pp. 93–95. Hitt describes expert woodsman J.J. Kuhn, for example, as a "folksy local yokel." Steinberg, p. 81, provides a more balanced portrait of ivory-bill sightings that originate from lay observers. "Many credible reports of ivory-bills ... have come from individuals outside of the scientific establishment."

42 Jackson, p. 250.

43 Klinkenberg, J. 2004. His hope has wings. St. Petersburg Times Online. http://www.sptimes.com/2004/10/07/Weekend/His_hope_has_wings.shtml (accessed 17 April 2015).

44 Peterson, M.C.E. 2007. The truth is out there. Pp. 17–36 in *The Philosophy of The X-Files* (Kowalski, D.A., ed.). The University Press of Kentucky, Lexington, p. 17.

45 Dawkins, R. 1996. Science, delusion and the appetite for wonder. 1996 Dimbleby Lecture. Available at: http://hermiene.net/essays-trans/science_delusion_wonder.html (accessed 29 April 2015).

46 Kowalski, D.A. 2007. Introduction: Mulder, Scully, Plato, Aristotle, and Dawkins. Pp. 1–16 in *The Philosophy of The X-Files* (Kowalski,

D.A., ed.). The University Press of Kentucky, Lexington, p. 8.

47 Flannery, R., and D. Louzecky. 2007. Postdemocratic society and the truth is out there. Pp. 55–76 in *The Philosophy of The X-Files* (Kowalski, D.A., ed.). The University Press of Kentucky, Lexington, p. 55.

48 Under conditions of uncertainty, we tend to lose entirely our grasp of probability, ultimately leading us to seriously misjudge how likely something is to really happen.

49 Undue faith that one knows the truth; an unwarranted precision in one's own belief.

50 Searches on Google Web were used with each of the exact phrases inside quotation marks and separated by a "+" sign, 29 April 2015.

51 Genzlinger, N. 2010. In Arkansas, they brake for the woodpeckers. *New York Times*. See: http://www.nytimes.com/2010/04/28/movies/28ghost.html?_r=0 (accessed 29 April 2015).

52 Burdick, A. 2005. 'The Grail Bird': new hope from Arkansas. *New York Times*. See: http://www.nytimes.com/2005/07/03/books/review/03BURDICK.html (accessed 29 April 2015).

53 Hoose, p. 150.

54 Weidensaul, S. 2002. *The Ghost with Trembling Wings: Science, Wishful Thinking and the Search for Lost Species*. Farrar, Straus and Giroux, New York, p. 57.

55 Hill, p. 4.

Chapter 10

Desperate Guessing

Don't keep a man guessing too long – he's sure to find the answer somewhere else.
—Mae West

Our own species has such pitiful yearning to escape uncertainty.[1] Between 1900 and the 1930s, the woodpecker's fate was so opaque that we tried to find a sense of confidence by entombing the bird alive. When that tactic for assurance didn't work, we researched the ivory-bill hoping that science could pin down causes for its scarcity. If we could isolate threats to its survival, surely we might reverse the death spiral and save the woodpecker before it was too late.[2] Instead, a caprice of history launched thousands of miles away in Europe intervened in isolated, rural Louisiana. Conservation politics failed in the end to save the ivory-bill in its Singer Tract home.

Botched politics joined the persuasive but deceptive finales handed us already by science and art. A sham closure lent the extinction myth enough traction that it became virtually impregnable to reasoned criticism. Sham closure set up a resilient system overflowing with ***focalism***, or ***anchoring bias***,[3] an obstinate fallacy that served as the primary grounds for contesting future evidence put forth for the ivory-bill's survival. Our haste to accommodate uncertainty and establish closure[4] with the woodpecker made future social and cognitive discords inevitable. Stories about the ivory-bill became cross-contaminated both by *conscious ignorance* (knowing what we did not know) and by *meta-ignorance* (not knowing what we did not know).[5]

What so fortified the bogus closures by art (Eckelberry

painting) and science (Tanner study) was conservation's last-ditch failure to protect the ivory-bill in the Singer Tract.[6] The sole purpose of Tanner's study was to identify protections that were vital to rescue the woodpecker. His work was financially sponsored by a multi-year Audubon Fellowship granted to Cornell University's Graduate School. No mere academic venture, the Audubon Society made no secret of the real intention behind Tanner's field study:

> *The National Audubon Society considers facts obtainable through scientific research the essential basis for wise policies governing the conservation of wildlife resources.... The present world-wide war, with its pressure for maximum lumber production, underscores the predicament of the Ivory-bills. That the Ivory-bill shall not, as a part of America's natural heritage, go the way of the Passenger Pigeon and the Great Auk, is an objective of the National Audubon Society.*[7]

Expectations were high that Tanner's study would produce applied findings from which imperative conservation directions would be taken. This was rather a lot to expect from a 23-year-old student. Nevertheless, Tanner seemed to have delivered on this demand.

Tanner's conservation proposal[8] in many ways was stunningly farsighted in its specificity on how to best safeguard the ivory-bill. His recommendations were both logically and ecologically plausible. His first proposal stressed how relatively large areas of forest were needed to protect just a single pair, up to six square miles, representing a substantial investment in reserve size. Given the relatively undisturbed, old-growth forest he thought necessary, Tanner recognized that high costs posed a big hurdle. But he thought costs could be minimized if some logging was allowed such that a matrix of strict reserves, partial cutting areas, and heavy logging were combined inside these reserves. Although entirely speculative, Tanner likely drew such parallels from the Singer Tract where he had seen juxtaposed disturbed areas, clearings, and forest stands of mixed age.

Management direction from Tanner for these ivory-bill reserves was detailed. In addition to a sustained yield management, he recommended retention of standing and fallen dead wood, leaving live sweet gum showing signs of crown die-back, spare cutting of the Nuttall oak, and retention of all hackberry, each a tree favored by the foraging woodpeckers. In perhaps the most unusual advice, Tanner gave extensive instruction on how to artificially increase the woodpecker's food supply by the intentional girdling of prime trees. That proposal was based on previous observation of ivory-bills elsewhere working trees after they had been killed in preparation for agricultural clearing.[9]

Woodpecker protections also included wardens to patrol the reserves. Due to risks from curiosity shooting, Tanner advised that hunters be barred entirely, and even legitimate visitors first obtain entry permits. In cases where suitable habitat was lost, he gave instructions for translocation of ivory-bills to new areas, including optimal seasons to accomplish trapping, and how to localize captures around non-breeding roost holes. Tanner's recommendations are so thorough that they likely would not differ substantially from what conservation biology might propose today. His restrictions still probably could not be enforced completely, however, on public-use lands like national wildlife refuges.

Despite the proposal's thoroughness, conservation actions of the day fell way short. What neither Tanner nor anyone else could pledge was implementing measures to save the ivory-bill's presumed last stronghold during the midst of World War II. That some of the Singer Tract was very nearly saved anyway is testament to the enduring will of those arrayed to rescue the big woodpecker. Oddly, it was neither regional nor national politics that in the end stopped the preservation, but rather global factors related to the war effort.

Support to save the Singer Tract for ivory-bill was widespread, diverse, and persistent,[10] though not without its theoretical and applied detractors.[11] Once Singer Manufacturing Company exercised its options to log, a sense of time running out accelerated mutual efforts to save the parcel. A resolution from the American Ornithologists' Union earmarked the Tract as essential. Louisiana senator Allen Ellender introduced national legislation to designate Tensas Swamp National Park. In a show of regional support unimaginable today, the governors of Tennessee, Arkansas, Mississippi, and Louisiana all urged Singer to sell its rights to the land. To expand public support, John Baker published Tanner's study as the first ever National Audubon Society Research Report. Baker also tried, unsuccessfully, to negotiate directly with logging contractor, Chicago Mill and Lumber Company, to set aside some land for the bird.

By the early 1940s, Baker had even appealed directly to President Franklin D. Roosevelt. After consults among Harold L. Ickes, Interior Secretary, Ira Gabrielson, director of the U.S. Fish & Wildlife Service, Newton Drury, director of the National Park Service, and Earle Clapp, chief of the Forest Service, the latter lent his support for retaining some of the Singer Tract's old-growth for the bird. Those efforts seemed to show promise, even as the nation was mobilizing for an intense war effort. Indeed, armed forces recruitment had so depleted the South's regional labor supply that for a little while it seemed as if an accidental shortfall in manpower might rescue the ivory-bill's home.

All efforts came to naught though. Wood demand was so high for shipping containers and other material requested by the War Department that Chicago Mill and Lumber refused to negotiate out of its contract with Singer. Instead of local labor to extract every last bit of the timber, Chicago Mill and Lumber found an unforeseen work-around: German prisoners of war were trucked in from nearby POW camps to do the cutting of the Tract's timber.

Our battle to save the ivory-bill's home was lost. National conservation was handed a callous defeat, one that looked to be everlasting. "With the Singer Tract, it seemed, went the ivory-bill."[12] The Singer Tract then became the ivory-bill's figurative burial ground. When it was obvious that all trees in the Tract would be felled eventually, ornithological dignitaries Ludlow Griscom, Roger Tory Peterson, John Baker, along with other National Audubon Society staff and various notables, made their respective pilgrimages to the ragged fragments of the Tract. As if paying last respects, supported by background science (Chapter 8) and an artistic capstone (Chapter 9), the story of the living ivory-bill came to an end.

Except that it didn't. Almost nothing of what seemed to point so clearly to the bird's then extinction was ever truly persuasive or even sensibly coherent. Riddled with illogical guesswork and plagued by thinking fallacies, by the middle 1940s the ivory-bill's fate moved into a cognitive limbo where, for all intents and purposes, the species remains tightly fastened today. The chain of events that led us into this limbo began with much uncertainty. That ambiguity demanded a closure, one induced by a ***primacy effect***[13] that delimited the particular closure that was ultimately selected. This in turn cemented a deep ***anchoring bias***, with yet more secondary fallacies arising from the deviations in cognitive logic that invariably followed.

Contradictions about the woodpecker's habits (Chapter 7), and meager study of its life history before Tanner, had already facilitated much uncertainty about this bird. The human mind, however, is stunningly hostile towards holding for very long any lack of clarity. In both our individual and evolutionary development, humans react to vagueness by forming feasible but impulsive rationalizations. In order to eliminate the stress of hesitating over the unknown, we seek cognitive closure with a mental model that reflects our "desire for a firm answer to a question and an aversion to ambiguity."[14]

Although not all individuals or situations demand this instant closure to the same degree, those nearer the closure end of the spectrum routinely jump to impetuous thinking.[15] Leaping to judgment with scarce evidence and rigid thought patterns creates strong resistance to further evaluation. High need for closure leads individuals to generate fewer hypotheses, and to be quicker to achieve an "unfounded confidence" in their rash belief.[16] Ever since the first half of the twentieth century, a collective need for closure followed by rigidity in human perception about the ivory-bill manifested itself most primordially (but not exclusively) in a hasty fixation on the woodpecker's extinction.

Flights to closure typically expose two cognitive tendencies in us: one is of *urgency* and the other is of *permanence.*[17] The first inclines us to "seize" on closure quickly, as a delay is felt as awkward, burdensome, or irritating. We simply cannot wait any longer for a resolve. The second inclines us to "freeze"

Box 10.1. When handed choices, we are more likely to avoid options in which the outcome feels unknown. This bias of probability and belief is termed the ***ambiguity effect***. In matters of ivory-billed woodpecker, this bias manifested when individuals struck a position of utter certainty in the bird's extinction, as this is far less ambiguous to hold than a position of "I don't know." ***Ambiguity effect*** may also explain an undisputed belief in the bird's survival out of similar cognitive motives that seek to dodge uncertainty.

knowledge in place so as to preserve both its current and future forms. We try to end the irksome uncertainty as soon as possible, then keep it from ever recurring again. This drive prejudices us towards a closure having a known over an unknown outcome, thereby manipulating us through the ***ambiguity effect***.[18]

What made seizing and freezing so compelling with ivory-bill during the 1940s was the deep crisis mentality that roused conservation to act without delay, and for what was then thought sincerely to be the very last remnant of a species in dire peril. Because closure superficially affords us some predictability and solid footing for more action,[19] the intense burden to save the woodpecker must have created enormous impetus for the bird's then-protectors to achieve an outcome that was both quick and enduring.

High need for closure functions to preserve stability and conformity.[20] Closure reflects our penchant for a consensual platform that is less likely to be challenged by socially prominent others, and that also expresses our penchant for consistent knowledge that can be extended to any new, related situations that might arise. In-group consensus caused by this permanence tendency may inspire in-group favoritism and resistance to any external inputs, a status quo bias that can have disastrous consequences. In the Yom Kippur War, for example, Israeli intelligence ignored the most recent warnings about an imminent attack on the country because analysts had already "frozen" on an earlier conventional wisdom that regarded such an assault as unlikely.[21]

Seeking cognitive closure led to our adopting unbending curbs to the diffusion of alternative knowledge about the ivory-billed woodpecker. Our inclinations to seize and freeze on early judgments about the bird are especially conspicuous in views about the species' putative extinction. "In his dissertation, Tanner all but declared the species extinct.... Tanner's assessment indicated that the ivory-bills were doomed."[22]

Seizing and freezing then cascaded to the most itemized beliefs about the bird's habitat use, foraging habits, and behavior. Over time these increasingly stubborn judgments operated a closed thought system that hampered our cognitive perspectives about the woodpecker.

One conspicuous effect of ***anchoring bias*** was Tanner's own practice for surveying habitat and occupancy of ivory-bill populations beyond northeast Louisiana. Because in the Singer Tract he had found that "ivory-bills there lived almost entirely in the parts of the forest where the sweet gum-oak association predominated,"[23] he inhibited his evaluation of suitability in other areas unless they were "most resembling places that ivory-bills did inhabit, such as the Singer Tract where big sweet gums and oaks are abundant…."[24] If an area was not large, virgin timber dominated by certain key trees, or was missing the feeding sign with which he was familiar, the location was snubbed. This ***primacy effect***,[25] caused by some initial, random information that is disproportionately imprinted for ensuing purposes, commandeered a cognitive filter that Tanner (and later, others) would use for assessing habitat suitability in this rapidly vanishing species.

Even locations having recent reports of ivory-bill were discounted. An ivory-bill was reported in the Savannah River Waterfowl Refuge, South Carolina, in 1936, but Tanner considered the site "unsuitable" because the "only swamp nearby is of small trees."[26] In Indian House Hammock, Sumter County, Florida, an ivory-bill was seen not long before his 1937 visit, yet Tanner concluded "suitable territory is so small that there is little or no chance of ivory-bills being there."[27] Although the last large tract of cypress-gum swamp in Louisiana was in Tangipahoa Parish, he concluded that the "habitat is poor for ivory-bills and I found no possibility of the birds being there."[28]

Tanner's search bias was perhaps most evident in the Big Thicket of east Texas. He described it as "low country of mixed pine and swampy forest, thoroughly logged over, and greatly over-rated as a wilderness area," then went on to depreciate recent ivory-bill sightings as cases of mistaken identity.[29] In later years, he repeated his bias even more stridently. John Dennis, who was familiar with the ivory-bill already from earlier field work in Cuba, reported the woodpecker from the Big Thicket in 1966.[30] But Tanner dismissed utterly that sighting in a public interview.

This predisposition in habitat appraisals was not the only anchor set by Tanner's study. ***Anchoring bias*** was linked to feeding sign he considered diagnostic of ivory-bill presence – prominent scaling or chiseling of tight bark away from recently dead trees such as sweet gum. For the Stewards Neck-Wadmacaun section of Santee Bottoms, South Carolina, ivory-bills had occurred repeatedly in prior years, but Tanner concluded "there are no signs of any of the birds residing in that area."[31]

Box 10.2. ***Confirmation bias*** is used to describe a large category of thinking and decision errors that reflect our inclination to look for, translate, prefer, remember, or report information in ways that confirm pre-existing beliefs or hypotheses. A tendency to only testify about evidence that supports a theory, for example, constitutes ***outcome reporting bias***. "Seeing what one wants to see" reflects a kind of anticipatory priming from ***observer expectancy effect***. Never expecting the ivory-bill to occur anywhere but mature bottomland hardwoods is an ***observer expectancy effect***, whereas not reporting this woodpecker from pine forest because it does not confirm to expectations even if seen would trigger an ***outcome reporting bias***.

Evaluating further the Santee Swamp, Tanner's penchant for direct comparison with the John's Bayou pair of ivory-bills was plainly influenced by a ***confirmation bias***.[32] His reasoning was that he "did not find in any part of the swamp enough ivory-bill sign to indicate that birds were residing steadily in one place,"[33] as had been the case with the only pair he was ever able to monitor in Louisiana. ***Confirmation bias*** led Tanner to overlook entirely the Choctawhatchee River in Florida, a region that lacked either specimens or reports. Yet this locale would reveal promising signs consistent with the ivory-bill 70 years later.[34]

If a locale seemed too small, lacked proper old-growth condition, or was absent the right tree species composition, Tanner dismissed it. Of 45 localities he inspected, 10 were disregarded over an unproven confusion by observers with the pileated woodpecker, but Tanner's screening criteria for this exclusion were never disclosed. Similarly, 10 more localities were eliminated due to lack of virgin timber.[35] Because some geographic sites were visited for as few as two or three days, finding no ivory-bills during such limited search effort is not an adequate basis for reaching conclusions of absence for a rare and/or elusive species (Chapter 14).[36] And yet, these very findings ordained Tanner's stark conclusion that by 1939 a mere 24 individual ivory-bills remained across the entire country![37]

Focalism entrenched by the Tanner study bred additional bias and fallacy. Because of its singular nature, the Singer Tract study justified appeals to conventional wisdom or special pleadings dictated by ***argument from authority***. For example, in reference to the species' foraging tactics "according to Jim Tanner and others, bark peeling is far more prevalent with ivory-bills than with other species."[38] Despite a contrary experience of his own Cornell colleagues in Florida where the woodpecker fed in an open, upland pine forest in the same manner as hairy and pileated woodpeckers, "Tanner believed that by 1915 [ivory-bills] only remained hidden in the largest of swamps."[39]

Decades later, as Cornell surveyed Arkansas' Big Woods, an ***observer expectancy bias*** was again activated for the single ivory-bill putatively present and assumed to have the same habitat relationships *and* sedentary habits claimed for the species by Tanner.[40] In 2006, Cornell surmised: "If it was an established bird then, it is surprising the bird left the site considering that [Bayou de View] includes the largest, oldest trees and highest dead wood volumes of all forest stands sampled in the Big Woods."[41] In not considering that a single Arkansas bird might have dispersed from a source population elsewhere, even from a considerable distance, reliance on the "established bird" premise narrowed expectations through a ***congruence bias***.[42] During wider explorations in South Carolina, Florida, Mississippi, Louisiana, and Texas, Cornell again borrowed expectations from Tanner to constrain their focus mainly on "mature stands of bottomland hardwood forests or cypress

> **Box 10.3.** ***Belief bias*** occurs when an argument's validity and believability don't match up. Essentially, the logical dimensions of an association get over-ridden by the intuitive appeal of a belief. ***Belief perseverance*** is our tendency to hold onto an idea even after it is contradicted. Believability of a link between the ivory-bill woodpecker and its feeding on large beetle larvae found on big trees in mature hardwood forest, plus the bird's scarce detections, led to conclusions that the bird was doomed to extinction by the 1930s. In this case, the inherent appeal of a tidy storyline trumped evidence that showed the woodpecker used other foods and habitats, or that its rarity was difficult to interpret entirely because of the species' extreme wariness.

tupelo ... and ... concentrations of dead and dying trees."[43]

After more than 70 years of widespread, nearly exclusive obedience to the Tanner doctrine for what constituted valid habitat and behavior in the ivory-bill (the ***bandwagon fallacy***),[44] this ***anchoring bias*** about the species became pervasive and deep. Neither Tanner nor later searchers ever truly "got away from using the Singer Tract as the benchmark for the area needed to support a population of ivory-bills."[45] Tanner's conclusions for the woodpecker were entirely plausible in an ecological sense, but without trusty statistical backing (Chapter 8) this very conceivability is what fosters our getting so duped by ***belief bias***.[46]

Plausible claims are not truth. Because of the murky correlations among ivory-bill scarcity/extinction, behavior, and habitat, our views about the bird were also unduly narrowed by the ***illusion of validity***,[47] the notion that whatever more information we may be able to secure will invariably prove useful. Since the woodpecker was often not searched for or thoroughly studied in conditions beyond those framed by Tanner, we simply do not know how truly specialized or potentially adaptable this species actually was (is).

For a time, legitimate doubts about ivory-bill were admitted candidly in one venue: from the 1940s nearly through the end of the twentieth century, the woodpecker's status was still generally listed as uncertain in our field guides and reference books.[48] In the 1946 Audubon Guide, authored by Richard Pough and illustrated by Don Eckelberry, the ivory-bill was described guardedly as "apparently doomed ... so rare that any record of one is noteworthy and should be passed on at once to the National Audubon Society, which is trying to save the bird from extinction."[49] Roger Tory Peterson's field guide listed ivory-bill in early editions (1934–1947) as "close to extinction" yet still "to be looked for in Florida and South Carolina."[50] His 1980 edition gave a somewhat more pessimistic reading, "very close to extinction, if, indeed, it still exists," but nevertheless noted that recent reports from Florida, Louisiana, South Carolina, and Texas warranted "further verification."[51]

In 1983 Chandler Robbins and his co-authors of the Golden Guide illustrated the woodpecker on both a tree trunk and in flight, listing it as "on verge of extinction."[52] Four years later, *National Geographic* gave a similar take on the ivory-bill's status: "probably extirpated in North America" due to loss of habitat believed to have brought this "never-common species to the brink of extinction."[53] The next decade, the *National Geographic* guide continued to depict ivory-bill, but edged closer to writing the species off entirely. By 1999, that guide listed it as "thought now to be extinct in North America," then walked back the declaration somewhat by closing with a "probable extinction of this never common species."[54] During this era, then, a small nod to cognitive integrity still depicted honestly the unsettled status for the ivory-bill.

After a half-century of prudence, however, courtesy and caution once given to the woodpecker's ambivalent standing were ditched in newer birding guides. An opening salvo was fired in 1997 when the National Audubon Society's photographic guide, no longer bothering to portray the woodpecker, openly proclaimed it "recently became extinct."[55] Other field guide authors soon followed suit.[56] Kenn Kaufman baldly avowed that the ivory-bill was "now extinct in North America."[57] David Sibley's first-edition field guide also ignored the woodpecker because the species did not, in his view, "meet ... criteria for inclusion."[58] Indeed, subsequent scholars who examined the cultural impact of the woodpecker's 2004 "rediscovery" noted that the ivory-bill was "conspicuously absent" from Sibley's guide.[59]

Respect for the ivory-bill's murky station had vaporized. With absolutely no evidence, sport birding decreed its own conjectured verdict. Bird watching first adopted wholeheartedly the same ***anchoring bias*** about the species' extinction, then promoted it openly to a next generation of field observers and would-be ivory-bill hunters.[60] An amateur-driven hobby sought to ground the ivory-bill's non-existence on nothing other than its own ***argument from authority*** plucked out of thin air.

Endnotes

1 Uncertainty can be experienced as distressful, anxiety-provoking, and even paralyzing (see Berenbaum, H., K. Bredemeier, and R.J. Thompson. 2008. Intolerance of uncertainty: exploring its dimensionality and associations with need for cognitive closure, psychopathology, and personality. *Journal of Anxiety Disorders* 22: 117–125). Consequently, our coping strategies for uncertainty are often defective. Fear of hesitation and its attendant unease cause us instinctively to impose cognitive stability on our circumstances in order to regain control, whether or not those impositions are rational. This profound need affects both individuals and groups. At times and when unchecked it can lead to social pathology and extremism (see Hogg, M.A., and D.L. Blaylock, eds. 2012. *Extremism and the Psychology of Uncertainty*. Wiley-Blackwell, Hoboken, NJ). Uncertainty may originate from three constructs that arise routinely from everyday life: *randomness, delay* (in consequences or outcomes), or *lack of clarity* (see Smithson, M. 2008. Pp. 205–217 *in* Bammer, G., and M. Smithson (eds.) *Uncertainty and Risk: Multidisciplinary Perspectives*. Earthscan, London). Uncertainty about the ivory-bill is due in part to the *lack of clarity* over unknown aspects of its life history, but a very long *delay* in resolution over its fate also contributes substantially to the illogical discourse arising out of this frantic search for closure.

2 "Is it too late?" is an ill-advised framing device for a species that might be headed towards imminent extinction. As Kim Stanley Robinson points out astutely "either answer promotes inaction. If it's too late, you don't need to act; if it's not too late, you don't need to act." See: Johnson, R. 2010. Kim Stanley Robinson maps

the future's gray areas. Los Angeles Times, February 21 edition. http://articles.latimes.com/2010/feb/21/entertainment/la-ca-kim-stanley-robinson21-2010feb21 (accessed 16 May 2015).

3 ***Anchoring*** represents a kind of semantic priming characterized as strong, robust, and reliable; it may consist of rather inflexible fixation on a procedure, a result, or a psychological process. See: Strack, F., and T. Mussweiler. 1997. Explaining the enigmatic anchoring effect. *Journal of Personality and Social Psychology* 73: 437–446. ***Anchoring bias*** may originate from the ***primacy effect***, i.e., the first information we just happen to receive, the ***recency effect***, i.e., the last information we receive, or from an ***availability cascade*** due to its constant repetition, all of which are very prevalent in matters related to the ivory-bill. ***Anchoring bias*** is very difficult to avoid even when we are forewarned of its likelihood and asked explicitly to avoid falling victim to it. See: Wilson, T.D., C.E. Houston, K.M. Etling, and N. Brekke. 1996. A new look at anchoring effects: basic anchoring and its antecedents. *Journal of Experimental Psychology: General* 125: 387–402.

4 Closure refers to our acute desire for securing definite knowledge on some issue, driven by our psychological leanings of urgency and permanence. Urgency represents an individual's inclination to attain closure as rapidly as possible, whereas permanence represents our inclination to maintain closure for as long as possible. See: Kruglanski, A.W., and D.W. Webster. 1996. Motivated closing of the mind: "seizing" and "freezing." *Psychological Review* 103: 263–283.

5 Bammer and Smithson, p. 208. In more formal statistical terms, we have seriously misjudged various aspects of the ivory-bill's fate because we have not accounted properly for the likelihood of *risk* (known probabilities) or *uncertainty* (unknown probabilities). *Sensu lato* Shaw, W.D., and M. Wlodarz. 2013. Ecosystems, ecological restoration, and economics: does habitat or resource equivalency analysis mean other economic valuation methods are not needed? *Ambio* 42: 628–643.

6 Louisiana's lease on the Singer Tract expired in November 1936, after which Singer Manufacturing Company contracted first with Tendall Lumber and then later with Chicago Mill and Lumber to cut the tract's timber. See: Jackson, J.A. 2006. *In Search of the Ivory-billed Woodpecker*, 2nd ed. HarperCollins, New York, pp. 141–142.

7 Foreword to Tanner, J.T. 1942. *The Ivory-billed Woodpecker*. Dover Publications, Inc., Mineola, NY, by John H. Baker, Executive Director, National Audubon Society, September 15, 1942, p. V.

8 Tanner, pp. 94–97.

9 Attraction of ivory-bills to girdled trees dates back to the time of Audubon who watched them feed on "belted trees of newly cleared plantations." W.E.D. Scott observed *11* ivory-bills feeding in such girdled timber. Tanner interviewed locals near the Singer Tract who had seen the same behavior. Tanner, p. 96.

10 Struggles to save the ivory-bill in northeast Louisiana lasted almost a full decade. For a more complete history of the nation's attempts to save the Singer Tract for the woodpecker, see: Hoose, P. 2004. *The Race to Save the Lord God Bird*. Farrar, Straus, and Giroux, New York, pp. 116–135; Bale, S.L. 2010. *Ghost Birds: Jim Tanner and the Quest for the Ivory-billed Woodpecker, 1935–1941*. The University of Tennessee Press, Knoxville, pp. 231–245; and Jackson, pp. 138–154.

11 In a foretaste of future conservation battles, not everyone agreed on tactics for saving the Singer Tract ivory-bills. Aldo Leopold weighed in on behalf of the bird, chastising impotency by the Park Service for its abstract notion of inviolate sanctuary and its tendency instead to protect scenic, charismatic landscapes with little regard for wildlife and other unique natural resources. He also pointed out the undue restrictions that would be caused by using "duck money" for a national wildlife refuge, or the Forest Service's management directive to cut timber, as bureaucratic constraints that further thwarted efforts to save ivory-bills. Prominent conservation activist Rosalie Edge lambasted the National Audubon Society for funding Tanner's research, believing that instead this money should have been spent directly on land

protection for the ivory-bill. See: Jackson, pp. 141–146. Edge even referred to Tanner's study as "a sad commentary on 'research' in lieu of protection." *Ibid.*, p. 146.

12 Hill, G.E. 2007. *Ivorybill Hunters: The Search for Proof in a Flooded Wilderness*. Oxford University Press, London, p. 10.

13 A ***primacy effect*** occurs when we become overly attached to and fixated (or anchored) on the first bits of information that we receive. We do not update our position sufficiently when new information arrives.

14 Kruglanski and Webster, p. 264.

15 Need for closure can be manifest by the individual preference for order, preference for predictability, discomfort with ambiguity, close-mindedness, or decisiveness. See: Calogero, R.M., A. Bardi, and R.M. Sutton. 2009. A need basis for values: associations between the need for cognitive closure and value priorities. *Personality and Individual Differences* 46: 154–159.

16 Kruglanski and Webster, p. 269.

17 Webster, D.M., and A.W. Kruglanski. 1997. Cognitive and social consequences of the need for cognitive closure. *European Review of Social Psychology* 8: 133–173.

18 Frisch, D., and J. Baron, J. 1988. Ambiguity and rationality. *Journal of Behavioral Decision Making* 1: 149–157.

19 Webster, D.M., and A.W. Kruglanski. 1994. Individual differences in need for cognitive closure. *Journal of Personality and Social Psychology* 67: 1049–1062.

20 Calogero et al., p. 157.

21 Bar-Joseph, U., and A.W. Kruglanski. 2003. Intelligence failure and need for cognitive closure: on the psychology of the Yom Kippur surprise. *Political Psychology* 24: 75–99.

22 Hill, p. 10.

23 Tanner, p. 15.

24 *Ibid.*, p. 21.

25 ***Primacy effect*** stems from a mental tendency to rely on first impressions. For example, individuals having otherwise exact but

reversed patterns of failure and success are assigned [irrationally] different intellectual ability, with those recording early success granted a higher score via this ***primacy effect*** (Jones, E.E., L. Rock, K.G. Shaver, G.R. Goethals, and W.M. Ward. 1968. Pattern of performance and ability attribution: an unexpected primacy effect. *Journal of Personality and Social Psychology* 10: 317–340). Presentation order for stimuli (objects, persons, or events) also significantly modifies our later judgments and inferences. See: Dennis, M.J., and W-K. Ahn. 2001. Primacy in causal strength judgments: the effect of initial evidence for generative versus inhibitory relationships. *Memory & Cognition* 29: 152–164. Absent a suitable plan for vigilance (Tetlock, P.E. 1983. Accountability and perseverance of first impressions. *Social Psychology Quarterly* 46: 285–292), this ***primacy effect*** can also influence fairness judgments and distort some judicial outcomes. See: Stone, V.A. 1969. A primacy effect in decision-making by jurors. *Journal of Communication* 19: 239–247; Lind, E.A., L. Kray, and L. Thompson. 2001. Primacy effects in justice judgments: testing predictions from fairness heuristic theory. *Organizational Behavior and Human Decision Processes* 85: 189–201.

26 Tanner, p. 24.

27 *Ibid.*

28 *Ibid.*, p. 25.

29 *Ibid.*, pp. 25–26.

30 Dennis, J.V. 1967. The ivory-bill flies still. *Audubon Magazine* 69: 38–44.

31 Tanner, p. 23.

32 Seeking or interpreting evidence in ways that are partial to pre-existing beliefs, expectations, or a hypothesis in hand. See: Nickerson, R.S. 1998. Confirmation bias: a ubiquitous phenomenon in many guises. *Review of General Psychology* 2: 175–220.

33 Tanner, p. 27.

34 Hill, p. 66.

35 *Ibid.*, p. 26.

36 Exceedingly low (if not wholly inadequate) effort also characterized later searches for the ivory-bill. Despite 40 years of intermittent looking as described in one account, close scrutiny reveals that most attempts consisted of one or a very few days, in almost no cases did searches even last weeks or longer for any single location. Sykes, P.A. 2016. A personal perspective on searching for the ivory-billed woodpecker: a 41-year quest. Pp. 171–182 *in* The History of Patuxent – America's Wildlife Research Story (Perry, M.C., ed.), U.S. Geological Survey Circular 1422.

37 *Ibid.*, p. 99.

38 Gallagher, p. 156.

39 Bale, p. 4.

40 "When the food supply is sufficient, the woodpecker is probably resident or sedentary." Tanner, p. 99.

41 Rorhbaugh, R., K. Rosenberg, M. Lammertink, E. Swarthout, R. Charif, S. Barker, and M. Powers. 2006. *Summary and conclusions of the 2005-6 ivory-billed woodpecker search in Arkansas*. Final Report, Grant Agreement #401816G060, submitted by Cornell Laboratory of Ornithology to U.S. Fish & Wildlife Service. http://www.birds.cornell.edu/ivory/pastsearches/0607season/0607stories/FinalReportIBWOtext.pdf (accessed 16 May 2015).

42 See for another example: Helgeson, V.S., and K.G. Shaver. 1990. Presumption of innocence: congruence bias induced and overcome. *Journal of Applied Social Psychology* 20: 276–302.

43 Rorhbaugh, R., M. Lammertink, K. Rosenberg, M. Piorkowski, S. Barker, and K. Levenstein. 2007. 2006-7 ivory-billed woodpecker surveys and equipment loan program. Final Report, Grant Agreement #401816G060, submitted by Cornell Laboratory of Ornithology to U.S. Fish & Wildlife Service, U.S. Geological Survey, National Fish and Wildlife Foundation, and National Ivory-billed Woodpecker Recovery Team. http://www.birds.cornell.edu/ivory/pdf/FinalReportIBWO_071121_TEXT.pdf (accessed 16 May 2015).

44 Also known as the come-on-in-the-water's-fine fallacy (Briggs, W.M. 2014. Common statistical fallacies. *Journal of American*

Physicians and Surgeons 19: 58–60), the ***bandwagon fallacy*** is group think in which conduct or beliefs spread as fads or trends that lead to their adoption by ever more individuals. It may arise from our preference to conform or because we lazily rely upon information from other people as a mental shortcut. See: Maxwell, A. 2014. Bandwagon effect and network externalities in market demand. *Asian Journal of Management Research* 4: 527–532.

45 Hill, p. 68.

46 A tendency to rate an argument as legitimate if we think that the conclusion is empirically true, and vice versa, regardless of the actual, formal validity of the argument. Although sensitivity to logic-belief conflict may exist to some degree in everyone, it may be more dramatic in high-logic problem solvers, thereby posing a particular risk to scientists and researchers. See: Stupple, E.J.N., L.J. Ball, J.St.B.T. Evans, and E. Kamal-Smith. 2011. When logic and belief collide: individual differences in reasoning times support a selective processing model. *Journal of Cognitive Psychology* 23: 931–941.

47 Although we generally place great confidence in our fallible reasoning, both experts and non-experts display flawed judgment unless the influences from prior experience are subject to constant modification via frequent, meaningful feedback, both positive and negative. See: Einhorn, H.J., and R.M. Hogarth. 1978. Confidence in judgment: persistence of the illusion of validity. *Psychological Review* 85: 395–416.

48 Birding field guides are startlingly ideological, serving to both reflect and shape the political world of bird watching, e.g., "field-guide authors represent birds and the environments they live in as strangely detached from and unaffected by a wide array of currently pressing environmental challenges." See: Schaffner, S. 2011. *Binocular Vision: The Politics of Representation in Birdwatching Field Guides*. University of Massachusetts Press, Amherst, p. 3.

49 Pough, R.H. 1946. *Audubon Guides: All the Birds of Eastern and Central North America*. Doubleday & Company, Garden City, NY,

p. 51.

50 Peterson, R.T. 1947. *A Field Guide to the Birds*. Houghton Mifflin Company, Boston, MA, p. 145.

51 Peterson, R.T. 1980. *A Field Guide to the Birds*, fourth edition, Houghton Mifflin Company, Boston, MA, p. 188.

52 Robbins, C.S., B. Bruun, and H.S. Zim. 1983. *A Guide to Field Identification: Birds of North America*. Golden Press, New York, p. 194.

53 National Geographic Society. 1987. *Field Guide to the Birds of North America*, second edition. National Geographic Society, Washington, DC, p. 274.

54 National Geographic Society. 1999. *Field Guide to the Birds of North America*, third edition. National Geographic Society, Washington, DC, p. 274. When the Society updated its field guide in 2008, it neither listed the ivory-bill as extinct or not-extinct, instead expanding the account to include recent reports and leave open the bird's status. National Geographic Society. 2008. *Field Guide to the Birds of Eastern North America*. National Geographic Society, Washington, DC, p. 252. But three years later, it again ditched all pretext at balance on the woodpecker's prospects, giving its authors free rein even to proclaim that would-be observers were filled with "wishful thinking." Dunn, J.L., and J.K. Alderfer. 2011. *National Geographic Field Guide to the Birds of North America*. National Geographic Books, Washington, DC, p. 324.

55 Bull, J., and J. Farrand, Jr. 1997. *National Audubon Society Field Guide to North American Birds*, revised edition. Alfred A. Knopf, New York, p. 567.

56 One field guide that refused to follow suit and eliminate the ivory-bill was the Stokes guide, which furnished both color and black-and-white photos of the woodpecker. See: Clark, H.O. 2011. Review of The Stokes Field Guide to the Birds of North America. *Canadian Field-Naturalist* 125: 167–168.

57 Kaufman, K. 2000. *Birds of North America*. Houghton Mifflin Company, New York, p. 218.

58 Sibley, D.A. 2000. *National Audubon Society: the Sibley Guide to Birds.* Alfred A. Knopf, New York. See also: http://www.sibleyguides.com/bird-info/ivory-billed-woodpecker/ (accessed 26 May 2015). Ironically, then, the very conservation organization that had devoted its national influence to saving the ivory-bill during the 1930s and 1940s was among the first, if not the first, to abandon belief in the species' survival inside its own sponsored field guides.

59 Lynch, M. 2011. Credibility, evidence, and discovery: the case of the ivory-billed woodpecker. *Ethnographic Studies* 12: 82.

60 Thereby also creating a self-fulfilling prophecy wherein bird watchers cannot readily identify the ivory-billed woodpecker because they lack the necessary identification tools and resources to do so.

Chapter 11

Scorned Witness

There is no fate that cannot be surmounted by scorn.
—Albert Camus

After 1944, observer credibility toppled material evidence as the measure for ivory-bill existence. Credibility, both the personal and professional, would be variously ascribed, withheld, questioned, disparaged, and mocked for persons who might report the living woodpecker in years after destruction of the Singer Tract. "A lot of good people have been ruined because they claimed they saw an ivory-bill."[1] Seeing a living ivory-bill became an act of social defiance against a tidy narrative and arbitrary thought system that anchored wild speculation about the bird. Seeing a living version of the woodpecker carried mounting jeopardy over time: reputations of observers reporting an ivory-bill in the 1950s were less clouded than for those individuals who did so in later decades.[2]

With little else to go on, we refereed the motives and guessed at the field skills of these witnesses to stabilize our cognitive strain over the ivory-bill's unknown providence. More spectators than players took part in this shaky contest. Like any sport, a few players achieved superior acclaim, mostly for showing tenacity in refusing to be talked out of what they saw or heard.

Rendering verdicts on our peers is a pervasive trait of human nature. "Not that I think she's lying, but she has such a rich imagination."[3] We may be compelled to make these social contrasts, but it is a toxic proclivity that also stresses, depresses, and divides us. Collateral damage from such comparison typically leads us either to convey our envy upwards, or direct our scorn downwards.[4]

For fifty years after Eckelberry's painting, a monolithic ***system justification***,[5] cemented by ***anchoring bias*** and ***argument from authority***, enforced a shared "truth" that we had invented for the ivory-billed woodpecker. What began with cognitive distortions of a particular nature, a plague of one fallacy or the odd bias here and there, morphed into a caricature of truth maintained for the purposes of social cohesion. Prohibitions on deviating from this group identity hijacked the frames for processing our beliefs and knowledge about the ivory-bill.

Box 11.1. ***System justification*** organizes a very resilient, socially-mediated partiality that compels us to maintain the status quo, even if that current situation is no longer coherent or useful to us any longer. Maintaining a sense of certainty and stability, a need to feel safety and reassurance, and the relational needs to affiliate with others who share the same group identity are all critical elements that help foster and maintain this ***system justification***.

Each of us was still left to wonder exactly what had happened to the woodpecker. The nature of our dilemma was revealed tellingly by Roger Tory Peterson:

> *Unlike the passenger pigeon, which officially expired at the Cincinnati Zoo at 1:00 P.M. Central Standard Time, September 1, 1914, no one knows the exact time of the ivory-billed woodpecker's passing.*[6]

Or, if the ivory-bill had passed at all. But our sparse insight had adjudicated a final doom on the bird. "In his dissertation, Tanner all but declared the species extinct."[7] Extinction then became the definitive paradigm, the starting point for all else. Henceforth we felt compelled to prove the woodpecker's existence. Death was certain, now go verify life. A foolish test of legitimacy was ordered for authenticating this ostensible phoenix.

Lacking alternatives, and given highly rigid suppositions about the woodpecker's habitat, foraging, and behavior, the Tanner beliefs acted as both a filter and an obstacle course for all who came along thereafter. If this pseudo-science were not a sufficient deterrent for anyone who might flaunt the status quo, Tanner's crystal-clear ***argument from authority*** was there to keep order: "I am the world expert on an extinct bird."[8] Any would-be observer of a living ivory-bill was forced to contend with both premises in that loaded statement.

Into this distrustful milieu, Whitney H. Eastman pioneered a ten-year foray to look for the ivory-bill starting in 1949. On a dare from John Baker, president of National Audubon Society, and guided by a two colorful snake hunters, Eastman trekked across Florida looking for the big bird, ranging from Big Cypress Swamp in the south to the Chipola and Apalachicola River basins in the northwest. He diagnosed astutely "that so many of our southern swamp areas are so inaccessible and so extensive, I felt from the very beginning that there must be a small remnant of this rare species still in existence."[9] Eastman didn't assume the bird was gone, just *rare*.

Although Big Cypress came up empty, Eastman had better luck elsewhere. Prompted by an earlier sighting of ivory-bills in the Chipola River swamp, on March 2, 1950, Eastman and his search party heard ivory-bill calls and drumming. The next day as they got too close, they spooked a male woodpecker that flew deep into the swamp. The day after they saw a female winging her way across the river. The following year, tracking down other "reliable reports" near Silver Springs, they were disappointed to not see or hear any ivory-bills, but they found "extensive ivory-bill stripping of pine trees in the area and large numbers of wood borers" in a "vast, heavily timbered and swampy area."[10]

Some curious behaviors were noted in the birds by Eastman's team. "We were not sure from reports received whether the birds were nesting in the area or were nomadic in their habits and constantly on the move in search of food ... they appear to roam the country like gypsies, and are seldom successful in breeding due to insufficient dead timber in any one locality to make it possible for them to lead a sedentary life."[11] As the team struggled to keep up with the birds through the Chipola swamp's boggy terrain, the "ivory-bills appeared to be constantly on the move."[12]

Eastman's team saw more than one ivory-bill, extensive feeding sign, roost and nesting cavities, and heard the species' distinctive drumming and calling. He was sufficiently optimistic by 1958 to conclude: "with the large land concentrations which have been assembled in recent years by purchase as private game preserves, by ranchers, and by large timber and paper mill interests, the bird now has a better chance for survival ... the ivory-bill will make a slow come-back."[13] None of this could be trusted, apparently, because the bird had been ruled dead, or maybe because Eastman, by now a prominent businessman from Minneapolis, was just an amateur bird watcher.[14]

After a few days of exploring in 1950, James Tanner would declare later to Jerome A. Jackson that there was no evidence of ivory-bills along the Chipola. A pattern of depreciating the sightings of others set a persistent standard. In the eventual retelling, and like so many other ivory-bill incidents, the Eastman search was after a time gradually diminished through constant repetition that he had merely "claimed to have spotted a pair of ivory-bills...along the Chipola River."[15]

Even greater drama branded ivory-bill hunts during the 1960s. John Dennis, like Whitney Eastman, had been a graduate student at the University of Florida. Dennis was part of Eastman's team initially, but he had abandoned that search too early, only to learn a few days later that Eastman had seen a pair of ivory-bills. Dennis first reacted as most did: "In spite of a strong wish to believe this report, I found myself in the same camp as the skeptics."[16] This is peerless demonstration of the ***myside bias***,[17] a brand of ***confirmation bias*** in which Dennis's reaction was skewed by his preconceived inclination to depreciate the credibility of others.[18] "If they can't see an ivory-bill they do not want to believe that anyone else has done so."[19]

> **Box 11.2.** A subset of ***confirmation bias***, the ***myside bias*** is a tendency to see and hear things through the lens of our own peculiar values and personal experience. This thinking blunder leads us to construe the meaning of what we sense in accordance with our own idiosyncratic tendencies. When one observer fails to find ivory-bill despite persistent searching under auspicious conditions, a reaction to depreciate that anyone else could be more successful and actually find the woodpecker reveals a ***myside bias***.

Later on, Dennis faced much the same kind of incredulous opposition to his own ivory-bill reports, only stronger, including an oft-quoted put-down delivered by Tanner himself. Dennis and Davis Crompton saw and photographed the Cuban ivory-bill in the Oriente region of that country in 1948. The birds occurred in habitat nothing at all like that described by Tanner, an area so logged-over that Dennis described it as "scrub growth."[20] He published the last widely accepted photographs ever taken of the Cuban ivory-bill, an adult male near its nest cavity.[21]

But searching later in east Texas, Dennis ran into resolute

disbelief over his ivory-bill sightings. The Trinity and Neches rivers in the Big Thicket had produced fresh reports by Whitney Eastman of two pairs and a lone female in the early 1960s. Dennis reported hearing and seeing ivory-bill in the Thicket between 1966 and 1968.[22] After days of searching, following a bird he first heard and then tracked by swimming naked through deep water, he finally stumbled on the woodpecker with her wings outstretched, perched in the open on a stump.[23]

This vibrant story was shot down summarily "as a mistake or, worse, a fabrication."[24] Ornithologist George Miksch Sutton just labelled Dennis as "overly optimistic." But when Paul Sykes and James Tanner spent a week in 1968 with Dennis in the Big Thicket, searching by foot, car, boat, and plane, the first two concluded there were no ivory-bills. Jerome A. Jackson still refers to the second-hand reports circulating about ivory-bills from the Thicket as "poetic license."[25] Tanner directed an unmistakable jibe at the searcher: "Dennis wants to believe he saw something, but he didn't."[26]

Promising reports of ivory-bill also came from Florida during the same decade. H. Norton Agey and George M. Heinzmann spent 41 days on a private ranch in Highlands County where they (and as many as five witnesses total) saw or heard up to two ivory-bills on 11 occasions between 1967 and 1969. From a toppled roost or cavity tree they retrieved a secondary flight feather, and possibly some body contour feathers, the former identified conclusively by Alexander Wetmore as belonging to the woodpecker.

Their account, too, became diminished with ensuing suspicion: "some shadow of doubt is cast over these records because Agey and Heinzmann also tape-recorded what they said were ivory-billed woodpeckers, and ... Cornell Laboratory of Ornithology identified the birds on those tapes as pileated woodpeckers."[27] Actually what happened is that Agey and Heinzmann had, in their own words, managed to tape record a bird call that was "faint and distant, several times" but "it proved too faint, with too much external noise, for Cornell University to positively identify."[28]

Early in the next decade, one of the most contentious ever reports of the ivory-bill arose after two Louisiana photographs of a male woodpecker on two different tree trunks were presented by Dr. George Lowery at the 1971 annual meeting of the AOU. The incident is enlightening for revealing how a rigid conviction in the bird's extinction had by now fueled a collective trend in run-away cognitive distortion:

> *The very fact that the photos have remained so controversial demonstrates the intensity of emotions associated with the search for the ivory-billed woodpecker. The photos became a lightning rod for debate not only because they were of a supposedly extinct bird, but also because the name of the photographer and the exact location in which the photos were shot were kept secret – at least from the birding public – until fairly recently.*[29]

Presence of strong emotion is revealing.[30] Lowery directed the Museum of Natural Science at Louisiana State University. He was a prominent ornithologist of national reputation. Wanting to prevent hordes of birders from over-running the woodpeckers (a pair was seen), Lowery refused to disclose either the location or the photographer, Fielding Lewis. Lewis was an avid outdoorsman from Franklin who had seen ivory-bills multiple times, then was finally put into contact with Lowery and eventually secured the photographs. Lewis wished to remain anonymous out of concern that federal or state authorities might restrict access to the private lands belonging to a hunting club then frequented by the birds.

Lowery apparently anticipated some reaction to his disclosure, but "criticisms and accusations of fraud remained a sore point … until his death."[31] The report's vague details irritated Lowery's accusers. "Most declared the pictures a fraud. They insisted that the man [Lewis] had climbed forty feet up a tree, set a stuffed ivory-bill in place, and snapped a picture of it – twice."[32] Thirty years later, Jackson still cast aspersion on the incident, probing "whether or not the photos are of a live bird. The pose is certainly similar on the two trees."[33] With yet more re-telling, the story achieved a level of mendacious nonsense. Hitt claimed, ignorantly and contrary to all fact, that Lowery had "died a disgraced birder" after his "ignominious exile from ornithology."[34]

What is striking about several of these incidents is the energy with which credibility has so often been attacked. Neither Lowery nor Lewis had clear motive for a hoax. Indeed, both men had transparent motives to not disclose the photos at all, albeit for different reasons. These otherwise inexplicably harsh cross-examinations when others report a living ivory-bill may reveal ***motivational bias***[35] in the interrogators themselves. In describing Tanner's stinging rebuke of Dennis, Steinberg pondered this possibility:

> *perhaps there was even a bit of academic egotism in [Tanner's] claim that he had studied the last ivory-billed woodpeckers. Tanner was surprisingly hostile to individuals who claimed to have seen the ivory-bill. For example, he openly questioned the credibility and honesty of John Dennis, who claimed to have seen the ivory-bill in the Big Thicket of east Texas. It remains unclear why Tanner was so unbending in his belief that the ivory went extinct in the 1940s.*[36]

Whereas cognitive biases tend to be "cold" and widespread in human nature, motivational biases are expressively "hot" and peculiar to a given situation. Motivational bias is a plausible reason for the caustic reactions of some individuals to continued reporting of a living ivory-bill, but this alone it is not sufficient to explain the collective uproar. Within a social context, ***confirmation biases*** arising from ***anchoring bias*** and ***argument from authority*** better account for the widespread rigid beliefs held about this species.

Turkey hunter David Kulivan was dressed in camouflage waiting quietly for a gobbler early on April 1, 1999, in the Pearl River Wildlife Management Area when a pair of ivory-bills landed in a nearby water oak. Watching them for 10 minutes, he later described to his professors key features of the species not depicted in field guides. Dr. Van Remsen warned Kulivan that he'd be treated as if he had seen a UFO and then "sicced my best attack dogs on him."[37] A widely-syndicated story by the *Associated Press* eventually labeled him a fool, delusional, pathological.[38] After enough doubters had assaulted Kulivan's credibility, he finally grew "weary of the ordeal and clammed up."[39]

Skeptics and deniers[40] routinely invoke cognitive priming to malign the credibility of observers who see ivory-bill. Hypothetically triggered by a previous sighting, or buoyed by mere hope, this reasoning postulates that the anticipatory prospect of wanting to see the bird creates an ***observer expectancy bias***[41] that leads observers to believe they saw the woodpecker when in fact they did not. All that an observer needs, supposedly, is an object in a suitable location that fits the approximate shape or pattern of ivory-bill, and a sighting will be generated. "If you want to see an ivory-bill bad enough, a crow flying past with sunlight flashing on its wings can look pretty good."[42]

To be sure, ***observer expectancy bias*** is a well-documented cognitive defect in observational ornithology.[43] It arises regularly in behavioral studies that rely on variables that are measured subjectively.[44] Experienced birders also are more likely to report false-positives of rare as opposed to common species, apparently too overconfident in their superior ability.[45]

But this side of observer expectancy bias does not fit tidily into the narrative for most ivory-bill reports. The dominant presumption for ivory-bill is that it is *not* there: it is extinct. Indeed, five discrete, persuasive forces all strongly predispose observers to *under-report*, not over-report, sightings of the ivory-billed woodpecker.

One factor causing this ***outcome reporting bias***[46] stems from observers not reporting ivory-bill out of concern for its welfare. Given the species' long history of persecution, perceived risks of disturbance or injury to the rare woodpecker served as legitimate reason to keep quiet.[47] "We are obligated to keep secret all information that might lead to the discovery of any remaining ivory-bills,"[48] said Dennis about the Cuban birds. Later he also remarked: "Although he was very close-mouthed, it was well known in ornithological circles that Stoddard knew a great deal about the recent whereabouts of the ivory-bill."[49] Agey and Heinzmann stated "we are not at liberty to make public the exact location because of its possible return" and "only harm could result from … publicity resulting in a flood of curiosity seekers."[50]

John Terres, editor of *Audubon Magazine* for 12 years and recipient of the John Burroughs Medal, kept secret for 30 years a 1955 Florida sighting of a pair of ivory-bills. His reasoning was: "I didn't want them to be captured and used as specimens. It was better that they be untouched in the wild where they belonged."[51] In a letter to James Tanner, George Lowery explained why he kept quiet the circumstances surrounding a pair of ivory-bills found in Louisiana during the early 1970s:

> *You know what would happen if the information became general knowledge. There would be two hundred amateur bird watchers on planes from all corners of the United States descending on the area tomorrow. And I think that would be the worst possible development so far as the birds themselves are concerned.*[52]

More than three decades later, John Fitzpatrick had identical justification for maintaining an early discretion about the Arkansas search for ivory-bill in that region. He "feared the place would become Coney Island with birders piling in all over the place."[53]

Attacks on one's reputation constitute a second predisposition for ***outcome reporting bias***. Interviewed about the ivory-bill in 2001, Van Remsen was candid: "I don't want to be written off as a lunatic."[54] Later, Hill seconded this outlook: "We avoid anything that smacks of the lunatic fringe."[55] Vernon Wright pointed to culprits behind this fear of ridicule: "professionals in the birding community are being disrespectful. They should not be attacking people for reporting what they see."[56] Fielding Lewis, no shrinking violet, advised his duck hunting buddy Jay Boe: "Keep your mouth shut because people will attack you like they did me and Lowery."[57] Nancy Higginbotham, a witness to two separate ivory-bill sightings in Louisiana, became "just tired of being hassled about her experience."[58] David Kulivan early wondered if "people would think he was crazy, or making a story up to attract attention."[59] Gene Sparling too had trepidations about reporting his sighting in Arkansas: "I wouldn't want to be lumped in with the alien abduction hunters and Sasquatch stories."[60]

A third reason behind ***outcome reporting bias*** comes from the reduced capability of hunters, fisherman, outdoor recreationists, and other lay or rural people to identify the ivory-bill readily, to understand the significance of their sighting, or to decide how and to whom it might be reported.[61] Hill emphasizes that "it would be rare for anyone living near a swamp in rural northern Florida to recognize an ivory-billed woodpecker as something noteworthy," and even if they did, "the gulf between the world of rural southerner and ... professional ornithologist is too wide to allow for exchange of information."[62] Worse still, after being rebuffed harshly after reaching out to a professional, one Hammond, Louisiana resident "did not want to experience another critical rejection by an academic."[63] Many people who believe they have seen the ivory-bill thus cannot or will not campaign actively for the legitimacy of their own sightings.[64]

Xenophobic distrust of government and "outside" environmental organizations constitutes a fourth source of ***outcome reporting bias*** about the ivory-bill. In a telephone interview with one veteran outdoorsman in Louisiana, Steinberg suspected that the person knew more about ivory-bills than he was willing to disclose. The interviewee, in turn suspecting Steinberg was a fed, did not want "the same sort of attention to the Atchafalaya Basin that the 2002 Pearl River search had attracted."[65] Another "who saw an ivory-bill on state-owned land even told [Steinberg] that he would consider shooting the bird if it showed up on his property."[66]

In the Big Woods region of Arkansas "several residents ... now claim to know that [ivory-billed woodpeckers] have persisted on private lands in Arkansas throughout the 20th century."[67] One resident reported watching ivory-bills nesting annually in the same old oak tree from 1960 until the mid-1990s. Tellingly, however, "none of the local residents" wished to report the woodpecker to "any outsiders ... for fear that it could interfere with the duck-hunting season."[68]

Box 11.3. For 70 years, ***outcome reporting bias*** has been an insidious thinking error linked to our reactions to the ivory-billed woodpecker's fate. No fewer than five distinct, powerful motivations (see above) predispose us to under-report or fail to report the possibility of a living ivory-bill. For example: knowing or upon learning that the woodpecker is already declared extinct, observers are strongly impelled to discount their own senses about what they see or hear.

Strong cognitive priming anchored by the ivory-bill's putative extinction, however, is the most imposing underpinning to ***outcome reporting bias***. "Only crackpots and kooks, after all, see extinct birds."[69] The mental readjustment exacted after viewing a species that supposedly no longer exists obligates an observer to undergo profound internal reorientation. One must talk oneself out of the sighting entirely, aiming the doubt backward and inward, essentially attacking the self.[70] Or, alternatively, the arduous cognitive resistance to what one has just seen must be vanquished, so that the observer moves into a new mental posture that can then isolate them well outside the comfort zone of peer acceptance and group cohesion.

Cognitive readjustments after seeing an ivory-bill are not easily masked. Steinberg and Tommy Michot describe the latter's perpetual doubts over his 1981 sighting at Duck Lake, Louisiana, as a virtual tug-of-war, a deliberation that ultimately arrived at only a 50% expressed certainty.[71] In her contemporaneous field notes for a sighting made on April 11, 2004, in Arkansas, Melanie Driscoll penned: "I am shaking and feel like I could cry."[72] While being interviewed by the Cornell team about a bird she identified as an ivory-bill on April 10, 2004, reactions of once deeply skeptical team member, Mindy LaBranche, are described sagely by Gallagher:

> *[I]t was interesting to watch the expressions on her face while conversations went on all around her. People were laughing and joking nearby, and she would occasionally laugh along with them, throwing out a comment or two. But every time she stopped talking, she would withdraw into herself. Her face at times looked almost horror-struck, almost like that of a person in shock….*[73]

These reactions are symptoms of keen stress. Gallagher was perplexed over why his fellow team members were ascribing levels of certainty to their Arkansas sightings using probabilistic terms.[74] "What is it about your sighting that gives you that one percent of doubt?" And LaBranche fired back: "Because the bird is freaking extinct! For years I've been convinced of that. And that's why I can't be a hundred percent sure."[75]

Forced by new evidence to change core beliefs can be experienced somatically as threat. When a frame that anchors our perceptions falls away, shock is the instinctual reaction. Cognitive threat forms a powerful incentive to maintain silence, especially as this can sometimes be a less stressful means to deal with the prospect of a living ivory-bill. Fixation only on the anticipatory side of ***observer expectancy bias*** for seeing the woodpecker gives us a splendid display of ***bias blind spot*** in the human species. The cognitive priming that is truly approved socially favors our being *sightless* to the big woodpecker, not conjuring it up imaginarily. The former is rewarded, the latter vilified.

About the ivory-bill, we suffer from impulsive thinking that we impose on ourselves. Our fundamental psychological need to reach closure is so ardent that we retained opinions about the woodpecker that are held true to the extent that our peers endorse them. Wildlife conservation is not immune from this fear of "upsetting important others."[76] Those who believe that the ivory-bill's fate is only about biological science are conning themselves. When subjectively held beliefs become adopted communally to a passable extent, they achieve *de facto* objectivity.[77] With this bird, we came to depend on socially authoritative opinion to appraise even our direct knowledge as accurate, reliable, and valid.[78]

When a putative ivory-bill emerged in the Big Woods region of Arkansas during the first few years of the new twenty-first century, however, these social constructs would get provoked like never before. A grand collision between professional ornithology and amateur bird watching marked the next stage of cognitive scandal waged over the ivory-bill's existence.

After yet another "re-discovery" of the extinct woodpecker gripped national attention, the elaborate pretenses that we had fabricated out of mythical farce finally started to fall apart.

Endnotes

1 Gallagher, T. *The Grail Bird: The Rediscovery of the Ivory-billed Woodpecker*. Houghton Mifflin Harcourt, Boston, MA, p. 158.

2 "The more these inconclusive sightings a searcher has, the more his or her credibility diminishes." *Ibid*. p. 57.

3 *Ibid*., p. 68.

4 Fiske, S.T. 2010. Envy up, scorn down: how comparison divides us. *American Psychologist* 65: 698–706.

5 ***System justification*** refers to motivational tendencies that exist for rationalizing or defending a given social, economic, or political arrangement. Because it is thought to represent members' group need to view the system as stable, fair, and legitimate, alternatives to the status quo can be deprecated or avoided so as to preserve "the way things are," even at the expense of objective social interests. See: Jost, J.T., M.R. Banaji, and B.A. Nosek. 2004. A decade of system justification theory: accumulated evidence of conscious and unconscious bolstering of the status quo. *Political Psychology* 25: 881–919.

6 Peterson, R.T. 2007. *All Things Reconsidered: My Birding Adventures*. Houghton Mifflin Harcourt, Boston, MA, p. 123. "For more than one hundred years, the grave has been dug, but no one can confidently fill it in. Dressed in black, we stand around the gravesite with no final corpse to bury." Bale, S.L. 2010. *Ghost Birds: Jim Tanner and the Quest for the Ivory-billed Woodpecker, 1935–1941*. The University of Tennessee Press, Knoxville, p. 2.

7 Hill, G.E. 2007. *Ivorybill Hunters: The Search for Proof in a Flooded Wilderness*. Oxford University Press, London, p. 10.

8 Hoose, P. 2004. *The Race to Save the Lord God Bird*. Farrar, Straus, and Giroux, New York, p. 150.

9 Eastman, W. 1958. Ten-year search for the ivory-billed woodpecker.

Atlantic Naturalist 13: 216–228.

10 *Ibid.*, p. 224.

11 *Ibid.*, p. 218.

12 *Ibid.*, p. 221.

13 *Ibid.*, p. 227.

14 Ivory-billed woodpecker seen in Florida sanctuary. The Times-News, Hendersonville, NC, January 12, 1951.

15 Alderson, D. 2007. *The Ghost Orchid and Other Tales from the Swamp.* Pineapple Press, Sarasota, FL, p. 114.

16 Gallagher, p. 18.

17 Exhibiting little relation to our innate intelligence, the ***myside bias*** arises when individuals appraise bits of evidence, generate support, or test hypotheses in a manner slanted by their own prior opinions and attitudes as offset against those of others. See: Stanovich, K.E., and R.F. West. 2007. Natural myside bias is independent of cognitive ability. *Thinking & Reasoning* 13: 225–247.

18 Dennis doubted the Chipola ivory-bill reports only until he returned to the locale and heard the woodpecker give *kent* calls himself. Gallagher, p. 18. An inclination to use the "I-don't-believe-ivory-bill-sightings-unless-I-can-confirm-it-personally" became a standard reaction to ivory-bill reports thereafter. This fueled some of the cognitive distortions witnessed over the Arkansas report of ivory-bill in 2004 (Chapter 12).

19 Gallagher, p. 102.

20 Dennis, J.V. 1948. A last remnant of ivory-billed woodpeckers in Cuba. *The Auk* 65: 497–507. Cuban ivory-bills were still living in a region that had been logged over seven years before. *Ibid.*, p. 504. Dennis emphasized that the birds were finding "such a food supply more advantageously in cut-over pine forests where millions of trees were already dead or were in the process of being killed by fire and, presumably, by the attacks of insects." *Ibid.*, p. 505. Dennis also noted that the foraging pine trees were small and defective, generally less than 5 inches in diameter, even a nesting snag reached a mere 12 inches. Given such stark deviation in the Cuban

birds' habitat associations from that observed in Tanner's study, one cannot help but ponder the scorn that might have been heaped on this 1948 report had no photograph been secured.

21 *Ibid.*, plate 13.

22 Shackelford, C.E. 1998. A compilation of published records of ivory-billed woodpecker in Texas: voucher specimens versus sight records. *Bulletin of the Texas Ornithological Society* 31: 35–41.

23 Gallagher, p. 20.

24 *Ibid.*

25 Jackson, J. A. 2006. *In Search of the Ivory-billed Woodpecker*, 2nd ed. HarperCollins, New York, p. 190.

26 Gallagher, p. 21.

27 Jackson, p. 161.

28 Agey, H.N., and G.M. Heinzmann. 1971. The ivory-billed woodpecker found in central Florida. *The Florida Naturalist* 44: 46–47, 64.

29 Steinberg, M.K. 2008a. *Stalking the Ghost Bird: The Elusive Ivory-billed Woodpecker in Louisiana.* Louisiana State University Press, Baton Rouge, p. 85.

30 Van Remsen singled out the disproportionate passion behind reactions to the ivory-bill's ongoing persistence: "It's amazing how strongly some people came out against those sightings." Gallagher, p. 122.

31 Steinberg 2008a, p. 89.

32 Gallagher, p. 22.

33 Jackson, p. 173. Jackson's contention that the photographs were "greeted with scientific skepticism" (p. 172) at the 1971 AOU meeting in Baton Rouge is a logically-dissonant juxtaposition. Although being greeted with skepticism was undoubtedly true, there was (and is) nothing whatsoever scientific about it. Whenever skepticism strays into matters of another's motive or intent, our cognitive processing, not the scientific method, is being engaged. The fact that scientists were involved is irrelevant.

34 Hitt, J. 2012. *Bunch of Amateurs: A Search for the American Character.*

Crown Publishing, New York, NY, p. 95.

35 ***Motivational biases*** consist of errors in judgment that we make when trying to evaluate the rationale behind either our own behavior or the behavior of others, i.e., a faulty perception of causation about actions of others and/or explanations for self-behavior. See, for example: Burger, J.M. 1981. Motivational biases in the attribution of responsibility for an accident: a meta-analysis of the defensive-attribution hypothesis. *Psychological Bulletin* 90: 496–512.

36 Steinberg 2008a, p. 119.

37 Martel, B. 2000. Reported sighting of "extinct" woodpecker drives bird-watchers batty. Los Angeles Times, November 19 http://articles.latimes.com/2000/nov/19/news/mn-54363 (accessed 27 May 2015).

38 *Ibid.*

39 Tower, W. 2006. The thing with feathers. Outside Online, March 1, 2006 http://www.outsideonline.com/1824091/thing-feathers (accessed 27 May 2015).

40 In disputes over the existence of the ivory-bill, Deniers (also known as Atheists) possess an unbending belief in the species' extinction. Although Deniers may like to refer to themselves as Skeptics (also known as Agnostics), this semantic disguise is a ***red herring*** wherein obstinate belief poses as scientific caution. Very limited research suggests that most of us are content to remain conventionally skeptical about this woodpecker. Although the representation of the sample is constrained by self-reporting, many birders and ornithologists reside in a middle ground of uncertainty, with the vast majority of those queried (>75%) believing that the ivory-bill "possibly" exists or "probably" does or does not still exist. See: Hayes, F.E., and W.K. Hayes. 2007. The great ivory-billed woodpecker debate: perceptions of the evidence. *Birding* 39: 36–41.

41 "Seeing what one is looking for," or the ***observer expectancy bias,*** is one of several different manifestations of a larger, broader category of ***confirmation bias.*** See: Nickerson, R.S. 1998. Confirmation bias: a ubiquitous phenomenon in many guises. *Review of General*

Psychology 2: 175–220.

42 Gallagher, p. 164.

43 Balph, D.F., and M.H. Balph. 1983. On the psychology of watching birds: the problem of observer expectancy bias. *The Auk* 100: 755–757.

44 Balph, D.F., and H.C. Romesburg. 1986. The possible impact of observer bias on some avian research. *The Auk* 103: 831–832.

45 Farmer, R.G., M.L. Leonard, and A.G. Horn. 2012. Observer effects and avian-call-count survey quality: rare-species biases and overconfidence. *The Auk* 129: 76–86.

46 ***Outcome reporting bias*** occurs when the availability of the information conveyed through publication, research, or other means is skewed via the nature or direction of the results. For example, it can occur often in the field of epidemiology because individual subjects do not wish to truthfully or fully disclose their actual health risk(s). See: McGauran, N., B. Wieseler, J. Kreis, Y-B. Schüler, H. Kölsch, and T. Kaiser. 2010. Reporting bias in medical research – a narrative review. *Trials* 11: 39. See: http://www.trialsjournal.com/content/11/1/37 (accessed 26 May 2015).

47 Carelessly promoting the rediscovery of a once-thought extinct species can back-fire if the publicity leads to conservation threats that were previously absent. "As conservationists we need to learn that some information is better kept secret than made public." See: Meijaard, E., and V. Nijman. 2014. Secrecy considerations for conserving Lazarus species. *Biological Conservation* 175: 21–24.

48 Dennis, p. 498.

49 Gallagher, p. 18.

50 Agey and Heinzmann, pp. 47, 64.

51 Interview of John K. Terres, August 1, 2005, by Mark D. Kutner, The Southern Nature Project http://www.southernnature.org/interview-profile.php?Interview_ID=19 (accessed 25 May 2015).

52 Gallagher, p. 101.

53 Stokstad, E. 2007. Gambling on a ghost bird. *Science* 317: 888–892.

54 Williams, J.J. 2001. Ivory-billed dreams, ivory-billed reality. *Birding*

33: 515–522.

55 Klinkenberg, J. 2007. A bird worthy of Melville. Tampa Bay Times, March 3, 2007 http://www.sptimes.com/2007/03/03/Floridian/A_bird_worthy_of_Melv.shtml (accessed 3 June 2015).

56 Williams, p. 518.

57 Steinberg 2008a, p. 98.

58 *Ibid.*, p. 76.

59 Hoose, P. 2004. *The Race to Save the Lord God Bird*. Farrar, Straus, and Giroux, New York, p. 153.

60 Niskanen, C. 2005. Eyewitness accounts of ivory-billed woodpeckers from the Cache River, Arkansas. *BirdWatching* http://www.birdwatchingdaily.com/featured-stories/eyewitness-accounts-of-ivory-billed-woodpeckers/ (accessed 31 August 2015).

61 "There also exists a distinct hierarchy in the birding world, and reports of rare birds made by hunters and fishers rank low in the birding caste system." Steinberg, M.K. 2008b. Bottomland ghost: southern encounters and obsessions with the ivory-billed woodpecker. *Southern Cultures* 14: 6–21.

62 Hill, pp. 55–56.

63 Steinberg 2008a, p. 104.

64 *Ibid.*, p. 68.

65 *Ibid.*, p. 118.

66 *Ibid.*, p. 83.

67 Bivings, A.E. 2006. Rediscovery and recovery of the ivory-billed woodpecker. *Journal of Wildlife Managemen*t 70: 1,495–1,496.

68 *Ibid.*

69 Hill, p. 66.

70 "Emotion, skepticism, and downright ill-feelings … even led some very credible individuals who have reported ivory-bill sightings over the past thirty years to change their stories and to question their own sightings." Steinberg 2008a, p. 22.

71 *Ibid.*, pp. 67–70.

72 Gallagher, p. 217.

73 *Ibid.*, p. 216.

74 Such explicit incorporation of probability in this setting indicates that not all ivory-bill observers ignored the possible bias that can arise out of ***neglect of probability*** or the ***base rate fallacy***. The former misjudges entirely how likely or unlikely any event can be, whereas the latter depreciates the likelihood of events that have a probability not equal to zero or one.

75 *Ibid.*, p. 215.

76 Meek, M.H., et al. 2015. Fear of failure in conservation: the problem and potential solutions to aid conservation of extremely small populations. *Biological Conservation* 184: 209–217.

77 Kruglanski, A.W., E. Orehek. 2012. The need for certainty as psychological nexus for individuals and society. Pp. 1–18 in Hogg, M.A., and D.L. Blaylock (eds.), *Extremism and the Psychology of Uncertainty*. Wiley-Blackwell, Hoboken, NJ.

78 Hardin, C., and E. Higgins. 1996. Shared reality: how social verification makes the subjective objective. Pp. 28–84 in Sorrentino, R.M., and E. Higgins (eds.), *Handbook of Motivation and Cognition*, Volume 3. Guilford Press, New York, NY. Revealing leanings toward both ***fundamental attribution error*** and ***myside bias***, some who failed to ever find ivory-bill were quite impulsive in their opinion of other's perceived credibility: "Therefore, many searchers may have been subconsciously biased and, as a result, not sufficiently cautious in their identifications under field conditions. In other words, their perception was in error – and they did not actually see what they believed they saw." Sykes, P.A. 2016. A personal perspective on searching for the ivory-billed woodpecker: a 41-year quest. Pp. 171–182 *in* The History of Patuxent – America's Wildlife Research Story (Perry, M.C., ed.), U.S. Geological Survey Circular 1422, p. 181.

Chapter 12

Rorschach Test

The trouble with the world is that the stupid are cocksure and the intelligent are full of doubt.
—Bertrand Russell

Rediscovery[1] of the ivory-bill might have played out differently had it not taken place under the divisive presidency of George W. Bush. Given suspicions rife over faith-based internationalism, imaginary weapons-of-mass destruction, and other fabricated certainties, it was hardly the most propitious era to herald a lost woodpecker's return. Yet at a hastily-arranged press conference[2] on April 28, 2005, various cabinet officials, the Nature Conservancy, and Cornell Laboratory of Ornithology jointly disclosed to a flabbergasted world that at least one ivory-bill had been found in the Big Woods of central Arkansas. The explosive story went on to headline 459 U.S. newspapers, 174 television, and 43 radio shows.[3]

Only a few years later, though, this once-optimistic news had been vilified as "Peckergate"[4] and pilloried by *Fleecing of America*. The Arkansas incident launched more than 14,000 comments, many exceedingly hostile and bullying, from aggrieved birders on a *single* discussion thread.[5] Initial hopes withered for finding a conspicuous source population for this apparently solo woodpecker. Credibility in all parties was questioned. "In my 35-year career, no other ornithological issue that I can think of has created such intense and sometimes acrimonious debate among my ornithological colleagues."[6] Many puzzled over whom to blame amidst polarized shouting matches, and how to explain what seemed like so much incoherent behavior. How and why a mere bird came to galvanize that much social passion and cultural heat pleads for introspection.

And yet, despite the huge row, the Arkansas saga is not insoluble. Alleged detection of one ivory-billed woodpecker spurred other hunts, some of these ostensibly encountering the species too. Contrasting these searches, and probing the conventional premises and ***framing*** devices relied upon by the involved parties, the saga can be distilled as a massive collapse in social cognition.[7] Main parties to this breakdown were: science, especially professional ornithology; amateur bird watching, or birding; conservation organizations, both public and private; mainstream media; and bloggers.

Two obstinate liabilities set up the colossal breakdown. First, after 60 years of near-limbo over the bird's true status, no ornithologist or birding expert could truthfully lend any broad, first-hand expertise to identifying a live ivory-bill.[8] Some claims of what constituted decisive criteria for the Arkansas bird's true identity turned out to be dead wrong or woefully incomplete.[9] This confusion led to a skewed media coverage that muddled a reciprocally balanced dialogue about the woodpecker's identity and its fate.

A second contribution to social breakdown concerned evidentiary standards adopted to substantiate the species' ongoing presence, and whether such evidence ought necessarily to conform either to "scientific" or "birding" criteria at all. Evidentiary standards are normative,[10] hence intrinsically subjective, disputable in their own right. Different conventions about evidentiary standards set off discordant ***framing***. Verifying an extant species in its original range is a dubious purpose for which to deploy authentic science.[11] Even firm evidence of one or a few individual ivory-bills is not *scientifically* gripping. Given ample historical proof already, one more photograph, video, or other shred of incontestable physical evidence on a species' mere existence represents a trivial contribution to genuine hypothesis testing.

Recourse to a bird records committee[12] is not a compelling evidentiary standard, either. Finding an endemic bird inside its original range, no matter how scarce, does not equate to verifying a first-time extralimital species far outside prior known geographic limits.[13] An American robin *Turdus migratorius* may be notable in Mongolia: not in Alabama. Observing rare (i.e., putatively extant) ivory-bill inside its native range and archetypical habitat falls between these cases. If birders play their competitive sport such that only sightings of rare[14] species judged to be "good" or "accepted" get tallied in a score-keeping pastime, so be it. But that longing does not obligate other consortia, including science or conservation practice, to follow suit. Birding is a leisure activity.[15] Hobbies are not customarily granted the deciding vote on what forms socially sanctioned "truth."[16]

So it was that the 2005 Arkansas declaration of ivory-bill became blighted instantly by two arguable conventions: diagnostic identification criteria and evidentiary standards for extant status in the woodpecker.[17] Conflict ensued because unstated and rival premises were thrust uninvited into opposing social orders. Once avian science and amateur birding were triggered, they battled each other over the rediscovery's narrative construct, its central meaning. What constitutes suitable documentation for bird occurrence has been disputed between these interests for over a century (Chapters 3–6).[18] Media loved a circus with these back-and-forth broadsides, but its reportage failed to ever criticize deeply the fundamental reasons for this strife. Indeed, ultimately it would fall to bloggers to track down a few of the more obscure reasons for breakdown, then hold to account those responsible for some of the worst shortcomings.

An opening salvo in a very prolonged, contentious dispute

arose from the precise way in which announcement of the ivory-bill's putative rediscovery was couched (i.e., a ***framing effect***). Cornell, led by spokesperson Dr. John Fitzpatrick, published in *Science*[19] a set of multiple visual encounters, audio recordings, plus the 4-second Luneau video clip from April 2004 that were taken to "confirm the existence of at least one male [ivory-bill]." In later interviews, Fitzpatrick amplified this evidence as possessing "all the necessary ingredients for a definitive identification."[20] Cornell's own media release also quoted Fitzpatrick as saying: "We have *conclusive* proof that the ivory-billed woodpecker has survived into the 21st century" (emphases supplied).[21]

Words like "conclusive" and "definitive" had binding repercussions. Those adjectives were not essential to depict what was already intriguing evidence, certainly by norms of science reporting. Dangling them in front of birding, though, was akin to waving a bright red banner in front of a hostile bull. When it comes to criteria used to identify *live* rare birds, "conclusive" and "definitive" arise directly from the ***framing*** language used to communicate social identity in that hobby. "In the birding community nothing brings scorn as does lack of definitive proof."[22] Indeed, because evidence presented by Cornell need not have been portrayed absent some reference to scientific uncertainty (thus constituting also a ***neglect of probability***),[23] that ill-fated selection of adjectives could be perceived by birding as little other than a direct challenge, if not an outright taunt.[24]

Cornell thus fell victim to a ***false consensus effect***.[25] Authors of the Arkansas ivory-bill report(s) over-appreciated how much the evidence that they deemed convincing would remain uncontested by others.[26] Our beliefs, values, and habits distort perceptions of how widely our positions are truly shared.[27] Although such reasoning seems judicious at first glance, this ***projection bias*** is not truly rational.[28] By definition we are unconscious of our subjective readings, so conflict arises when divergent narrative frames are imposed upon, and then compete over, the same set of facts.

> **Box 12.1.** ***False consensus effect*** is an attributional category of egocentric bias that refers to our tendency to overvalue the extent to which other people actually share our beliefs, attitudes, and behaviors. ***False-consensus effect*** is closely allied with projection (attributing one's own characteristics to others), making this error readily aggravated across different social identities, each of which typically rely upon highly discordant framing in the first place.

Partly to avoid the "tactical blunder in claiming proof for the species too fast in Arkansas,"[29] another search team working later in north Florida wisely switched course. Hill's team used more tempered language to describe the evidence for ivory-bill that they collected along the Choctawhatchee River from 2005 to 2008. That team was "content to call [their] evidence 'highly suggestive' rather than 'definitive.'"[30] Related terms like "consistent with ivory-bill" when referring to the evidence kept that search from drawing nearly the same level of scathing criticism leveled at the Arkansas efforts.[31]

Cornell also depicted the Arkansas rediscovery so that species identity was framed *as* science. Some bystanders caught on to the fact that identification analytics had essentially confiscated the formal hypothesis-testing of science. Frank Gill, retired chief scientist of National Audubon Society, puzzled: "It was cast as a scientific analysis of these [video] pixels. It had all the pizzazz of technology. That was brilliant on Fitz' part, but it was weird to go to this length."[32] In the Florida search, Hill seemed to catch on, too, and, "determined not to make the same errors," flatly avowed that "we are not doing science,"[33] and instead "would ask a rare bird committee to review [their] evidence."[34]

Professional ornithology thus interjected itself squarely into what for the six previous decades had been mostly a layperson's story about seeking, finding, and identifying the lost woodpecker. "The professionals started to tell an amateur story."[35] Strong push-back from birding was inevitable, its forcefulness driven by ***Semmelweis reflex,***[36] wherein birding rebuffed evidence for extant ivory-bill because it contradicted paradigm that already prejudiced belief in extinction.[37] Contemporary birding celebrities Kaufman and Sibley had expunged ivory-bill from then-modern field guides,[38] broadcasting their ***argument from authority*** to promote a widespread conjecture in extinction by the wholly unqualified (see ***Dunning-Kruger effect***).[39]

Box 12.2. ***Semmelweis reflex*** is an impulse for us to reject new testimony or updated information because it contradicts our established rules, beliefs or paradigms. The ***Semmelweis reflex*** often results from our "freezing" on some prior cognitive closure, giving this error special intensity behind its adamant defense. An extinction ***meme*** for the ivory-bill had been so widely broadcast and reinforced through social discourse that any likelihood of the bird's survival provoked immediate challenge, if not obstinate defiance.

Strong push-back was also compelled by ***reactance,***[40] the inclination to resist staunchly having one's choices overly curtailed by others. Because Cornell had already declared identity of the Arkansas woodpecker as "definitive," birding had been excluded utterly from what it considered to be its own turf for making final verdicts about species' identity in field settings. Birding thus saw its social order and freedom of choice as under raid by outside interests.[41] This attitude polarization pushed the identities of birding and professional ornithology further apart, reinforcing a strong ***backfire effect***.[42]

Primed by ***reactance*** and ***backfire effect***, an avalanche of criticism would be mustered against first the identity and eventually the very existence of ivory-bill in Arkansas. In due time the "voices of skeptics were beginning to rival the cheers of supporters."[43] Much of this criticism was unmerited, however; practically none of it matched the original evidence or sophistication of analyses provided by the Cornell team. Still, the criticisms reveal how far cognitive bias and fallacious reasoning had appropriated the ivory-bill's story. And they show just how much our irrational natures, in both individual and collective expressions, could drown out the pivotal biological, statistical, and conservation questions that needed to be asked (even if not fully answered) about the woodpecker (see Chapters 1, 13, 14).

One especially harsh critic was Jerome A. Jackson, a long-time ivory-bill searcher and an acknowledged woodpecker authority. Not invited to join the Arkansas search,[44] Jackson published a long opinion in *The Auk* that ascribed various motivational intentions to the Cornell search team.[45] Claimed to have been neither peer-reviewed nor fact-checked by the journal's editor,[46] the piece excoriated Cornell about its supposed lapses over the quality of evidence, hidden agenda for timing of the announcement, intentional leaks to the media, a compromised peer-review, misdirection of conservation funds, and more.

Absent any true data analysis, Jackson's piece was sheer rebuke (***ad hominem*** and ***appeal to motive***). Scientists faulted Jackson's censure of Cornell's findings over sound recordings as "unfair" on technical grounds.[47] Others were "bothered by the implication in Jackson's article that the Cornell group played to the media and [was] motivated by the prospect of soliciting funds."[48] Jackson's wild guessing at others' motivation was thus riddled with ***fundamental attribution error***.[49] His essay over-estimated personal disposition inside Cornell, and under-valued the rediscovery's situational nature.

To whit, Jackson claimed that the *Science* publication and Gallagher's book[50] were "arranged to coincide with the [public] announcement."[51] But the press conference entailed other actors (e.g., federal government, Nature Conservancy), its proximate timing was induced by a media leak outside Cornell's control, and researchers also sought to present their evidence only after peer review. Jackson's misattributions began with a ***naïve cynicism***[52] that led to untamed generalization. Miscues in truly having understood our inner states by others expose blunder from ***illusion of transparency***.[53]

Box 12.3. In seeking to explain others' behavior, and rather than consider external characteristics that govern a situation, we place undue emphasis on internal (and hidden) characteristics like personality or intention. This ***fundamental attribution error*** reflects deep miscalculation in our ability to really see inside others' motives, or even if motive is involved at all. ***Fundamental attribution error*** was rampant during and after the 2005 media revelation of a putative ivory-billed woodpecker seen in Arkansas' Big Woods.

As others jumped into the fray, yet another socially-transmitted error, now from ***myside bias***,[54] commandeered debate. At first glance this error seems to emanate from old-fashioned envy: "If they can't see an ivory-bill they do not want to believe that anyone else has done so."[55] ***Myside bias*** arises principally, however, from a faulty mindset that leads us again to perceive the world egocentrically. We steer our external readings towards our own interior opinions and beliefs. This often manifested during various ivory-bill sagas as one party believing that other parties could not possibly find, see, or record what that same party did not or could not manage to do themselves.[56]

Myside bias ran virtually unchecked. Cornell amassed repeated ivory-bill sightings (including by two observers at once), sound recordings, and a video. In contrast to this feat, unprecedented since the 1930s, Jackson had scoured decades for the big woodpecker but never collected any physical evidence at all.[57] Still, Jackson wrote "Malcolm [Hodges] and I both believe that the bird that responded to our tape was an ivory-billed woodpecker. The bases ... are ... that the bird flew in from a great distance; it was extremely wary...; it called incessantly ... with invariable cadence and pitch; we never heard a blue jay...; and the habitat seemed excellent."[58] Jackson thus applied a much flimsier standard to himself than the one he imposed so pitilessly on Cornell.

Florida searchers were more transparent about their penchant for ***myside bias***. "It was going to be a struggle to remain objective ... it was all too easy to point a finger of criticism at one ivorybill searcher for overstating evidence and then turn around and be just as emphatic about one's own."[59] Hill admitted he "had to face my own biases," and thus question why he exalted his team's "Florida sightings ahead of the ivorybill sighting by Tim Gallagher and Bobby Harrison in Arkansas."[60] ***Myside bias*** led the Florida team to accept "things that we saw, heard, and recorded as definitive while at the same time dismissing evidence of the same quality coming from other sources."[61]

Myside bias also irrupted throughout amateur bird watching. The pastime was irked mightily over having what had long been its uncontested domain for certifying the identity of rare birds infiltrated by "science." Factors behind the pushback against those who gave credence to extant status for the ivory-bill were easy to find.[62] According to a host of critics representing a cross-section of birding interests,[63] any factual interpretation for an ivory-bill rediscovery:

> *overstates the credibility of recent sight reports and does not explain how their merit was evaluated ... recent sight reports are not credible or convincing since all involve brief encounters by observers steeped in expectations, all lack reconfirmation of characters thought to have been seen, all include descriptions flawed in significant ways, and not one was repeated by independent, let alone experienced or expert, observers immediately or soon after the original report ... for decades bird records committees have provided critical judgments on the acceptability of such reports for the historical record. We recommend ... [adopting] a mechanism that follows the bird records committee model for evaluating and archiving such reports.*[64]

In this sermon to the choir, an indictment by Sibley and others used ***argument from authority*** to demand allegiance to validation standards of bird records committees. No rationale was furnished for that standard, and indeed if adopted it would just bolster more ***system justification*** as dictated entirely by the social identity of birding. This demand falsely stated that all re-sightings of putative ivory-bill in Arkansas lacked reconfirmation of field characters seen.[65] The indictment also incorrectly claimed that independent or expert observers never repeated sightings.[66]

By insisting on "unequivocal" and "irrefutable"[67] support for an extant ivory-bill, birding tried to impose criteria that were not only *expressly* non-scientific (due to the total ***neglect of probability***), birding went further to insist that evidence could be disqualified by sheer refutation alone. Birding's indictment selectively (but conveniently) stressed a single cognitive fault, ***observer expectancy effect*** (assumed largely through

fundamental attribution error), but then neglected entirely all of the other personal fallacies and social biases (thus exemplifying ***bias blind spot***) that have long shaped the cultural narratives we have constructed about the ivory-billed woodpecker over the last century.

Given their intrinsic reliance on ***argument from authority,*** findings from bird record committees are not amenable to replication or to measurement. Outcomes depend solely on subjective trust.[68] To be seen as competent, birders must balance whether their sightings are reported (highly confident) or instead withheld (if not confident). That dynamic pits communal trust against individual self-interest and secrecy.[69] Hierarchies, including eBird editors and the American Birding Association (ABA), dictate how and when a specific bird sighting is accepted into the official record. With a few standard field guides, plus an array of self-designated experts,[70] the entire enterprise of birding is thus molded uniformly.

Social networks like birding are notoriously disposed to homophily, i.e., its group members adopt a set of identical traits. This tight cohesion limits worldview, curtails information flow, restricts attitudes formed, and constrains interactions experienced.[71] On the basis of social approval, then, birding differs quite radically from science. The former group prizes strict conformity in its practitioners. The latter group well-nigh detests it given that research novelty typically garners more acclaim from one's peers.

Birding was already vested in denouncing a still-living ivory-bill. Even with so much contrary evidence,[72] the hobby preserves a smug arrogance in its staunch belief that if the big woodpecker were present, it ought to be easily and repeatedly found. Critic Sibley opined that "If the ivory-billed woodpecker still existed, the odds that it could consistently elude this army of skilled searchers is vanishingly small."[73] Oblivious about human liability to ***illusory superiority***[74] and ***omniscience bias,***[75] birders were left incredulous that such a large bird could have ever escaped their flawless detection skills.

In contrast to the plodding ways of science, birding's homophily enabled its criticisms to be mobilized quickly,[76] then honed to a consistent, crowd-sourced message grounded in ***argument from incredulity***. Abandoning any pretense of ***Bayesian caution***,[77] the fitting response to appraise *all* of Cornell's evidence, birding fell to using ***subadditivity effect***[78] to divide material in an attempt to more easily refute it.[79] Tossing aside anecdotal sightings as wishful thinking by the unfit,[80] deploying snark to patch over its own naiveté about ivory-bill identification,[81] mostly ignoring audio recordings altogether (unless haphazardly attributing each matching sound to duck wings, gunshots, other bird species, and so on),[82] birding honed in on the fuzzy Luneau video. Emboldened over finding such a susceptible target, birding unleashed an unvarnished contempt:

- "Why in the world didn't they pass this around to a wide circle of people, mainly expert birders, before they went public with it? I can't be sure what it is, but I can be sure it's not an ivory-billed woodpecker."—Brent Whitney, co-founder of Field Guides, a leading bird tour company[83]
- "The bird cannot be an ivory-billed woodpecker. Why? Because the bird exhibits black on the trailing edge of the wing."—Louis Bevier, Colby College[84]
- "The bird in the video is a normal pileated woodpecker, and to date there is no irrefutable data to indicate the presence of ivory-billed woodpecker in Arkansas."—Mark Robbins, University of Kansas[85]
- "I've never seen such awful documentation on any record. I just look at the video and say, 'God it's hopeless.'"—Jon Dunn, field guide consultant and leader for Wings Birding Tours[86]
- "The Luneau video by itself as presenting anything other

than an unidentified woodpecker falls below the standards of proof normally required for scientific publication." —J. Martin Collinson, geneticist, British birder, and self-described "universal data skeptic"[87]

- "[I]f thousands of people can be shown a few seconds of blurry video of a pileated woodpecker and be convinced that it's an ivory-billed woodpecker, then the sky's the limit." —Steven N.G. Howell, field guide author and tour leader for Wings Birding Tours[88]
- "[T]he thylacine, ivory-billed woodpecker, the Loch Ness Monster and Bigfoot: although sightings of the aforementioned are largely or entirely discredited by the scientific community, reports will continue to dribble in indefinitely." —Robert L. Pitman, NOAA Fisheries Ecosystem Studies Program[89]
- "[searchers] have yet to get any kind of real documentation ... it becomes extremely implausible that even one individual could be sneaking around in the woods of Arkansas and evading the cameras of the entire Cornell search team for the last two winters." —Kenn Kaufman, field guide author[90]

During the Arkansas incident, professional ornithology had snatched the license on field identity away from bird watching. Fully in keeping with expectations of ***Semmelweis reflex***, amateurs then responded by trying to gain an upper hand with "science," leading to outcomes that were by turns comical and calamitous. In lambasting the U.S. Fish & Wildlife Service for ignoring "detailed, well-founded critiques from some of the world's most accomplished ornithologists," the real substance behind this peevishness was revealed to be because birders "agreed that, were we members of a rare bird records committee that would be evaluating this record, we would not be inclined to accept it."[91]

After science's fox broke into birding's chicken coop, a few hens thought they could take on that toothy carnivore. Clueless about biological traits he got exactly wrong (see Chapter 7), Kaufman opined that: "Many people are clinging to the myth that ivory-bills were extraordinarily elusive birds.... This runs counter to everything we know about ... the past."[92] Leading the charge against Cornell, illustrator and author David Sibley faultily claimed ivory-billed flapped slower than pileated woodpecker,[93] erroneously depicted pileated underwings as having too much white *and* an overly narrow black border in his own field guide,[94] made the staggering claim (with no support) that a flapping bird in caudal view could conceal dorsal wing surfaces during both up- *and* down-strokes, and repeatedly anchored an imprudent criticism on image artifacts[95] interpreted from visual distortions arising from non de-interlaced video.[96]

Birding marshalled another line of attack, claiming the bird in the Luneau video fit the identity of pileated because its wing-beat frequency closely matched the smaller woodpecker. But those advancing that claim used statistical sleights of hand to make their case.[97] One claimant used faster wing beats on immediate takeoff from birds that had been tossed into the air,[98] so that slower beats later in the flight sequence still overstated the wing-beat rate. Not providing the original data so that the allegedly supportive images could be verified, he was finally shut down by astute bloggers who caught on to the trick.[99] A second claimant, Martin Collinson, provided material openly,[100] but critics pointed out his data in fact refuted the notion that pileated woodpeckers could sustain for as long the high wing-beat frequencies that were detected in the peculiar bird depicted in the Luneau video.[101]

Far from debunking the ivory-bill hypothesis, in due time new information showed that the Arkansas woodpecker's wing-beats fit closely to the ivory-bill's near cousin, the imperial woodpecker *Campephilus imperialis*. Based on rediscovery of a 16-mm film from Mexico taken in 1956, analysis of flight mechanics in those images refuted two key arguments tendered by Sibley and other critics: 1) in normal takeoff, a woodpecker holds its tail against the trunk until after its wings are extended and primed for the first down-stroke (*it does not*), and 2) woodpeckers larger than pileated necessarily beat their wings more slowly than that smaller species (*they do not*).[102] Of course, given the snail's pace of science reporting, media did not cover these key updates with anything resembling the interest shown at the original rediscovery and its immediate controversy.

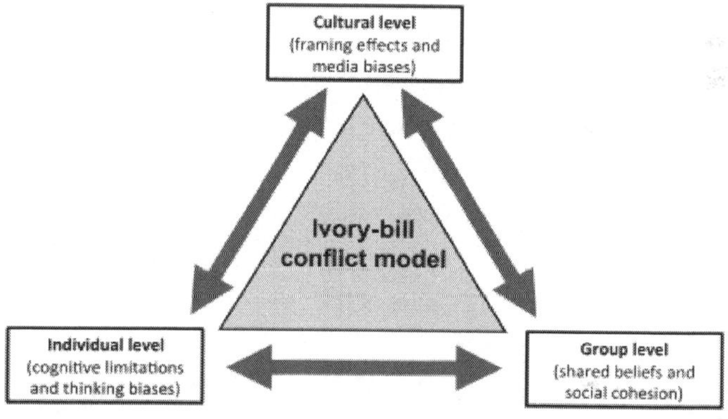

Image 12.1 Conceptual model for understanding the reasons that drive and then also reinforce conflicts over the identity and persistence of the ivory-billed woodpecker. Conflict is amplified because of the synergistic discord within and among the individual, social, and cultural dimensions to human cognition.

Putative rediscovery of the ivory-bill in Arkansas and Florida set off an arresting if entirely impromptu trial in the cognitive perils brought on by our rash individual, social, and cultural psychology (***Image 12.1***). Disputes over finding the ivory-bill have long served as Rorschach tests that disclose fundamental differences in our belief systems and social identities. The Arkansas saga revealed weaknesses but it also validated certain strengths from the colossal power of ***framing*** among each of five parties to that saga.

Professional ornithology abandoned its historical course as gate-keeping enforcer,[103] instead lending a forum where some discussion might occur without total censure to those who might dare report a living ivory-bill.[104] As a venue to argue merits for and against the species' existence, to pilot new forensic techniques for detecting rare, cryptic species, and to formulate statistical tools to gauge probability of extinction or survival, science had relinquished (at least temporarily) its complete suppression of new information.[105] Avian science could have bettered its reactions had it avoided the primal snags brought on by ***false consensus effect*** and ***myside bias***.

Conservation entities – the U.S. Fish and Wildlife Service, Nature Conservancy, and others with missions to protect endangered species or habitats – engaged in a protracted synthesis of multiple lines of evidence before ever arriving at much of a real decision at all. Once an ivory-bill endangered species recovery plan was adopted in 2010, conservation practice merely reaffirmed an indeterminate (and rather status quo) standing for the woodpecker as uncertain rather than declare it gone prematurely. Conservation practice dodged more cognitive hazards by waiting, not rushing to reach closure(s) so impulsively, and for explicitly conceding the lingering ambiguity about the ivory-bill. Conservation also shunned some earlier missteps: it did not default to extinction without proper evidentiary backing, and it largely avoided advocating specific land set-asides as the only protective measure for ivory-bill.

Media coverage of the Arkansas incident was erratic, blemished,[106] and intolerably lazy, substituting for investigative rigor fanciful references to ghosts, Elvis, and Bigfoot (Chapter 9). Media concocted even more myth about ivory-bill, framing the Arkansas controversy as "a tale of professionals erecting a citadel of expert opinion around a new truth, with a sequel about a messy band of amateurs assaulting that fortress and tearing it brick by brick to the ground." Crude distinctions between professionals and amateurs miss entirely the point of this episode[107] in large part because amateur bird-watching went on to fabricate its own house of cards out of cognitive mirage. Media's feeble response stemmed from a failure to see how stove-piped are conservation, science, and birding, and how the frames relied upon by each endeavor lead to mostly discrete narratives for the ivory-bill.

In covering the dispute between Jackson and Cornell, media neglected ever to separate raw opinion from data-driven analysis. In granting the crowd-sourced ***framing*** by competitive birding so much egocentric clout, the media imposed an ***equivalency bias*** on readers. Despite a large competency gap between competitive birding and science over ivory-bill biology, individuals' opinions were portrayed as having essentially the same merit,[108] thwarting a measured perspective from ever reaching public view because of this allegiance to ***equality bias***.[109] Media succumbed to wholesale pursuit of a false balance, and thereby rendered lopsided coverage under ***Okrent's Law***.[110]

Blogging's contributions were decidedly mixed. Initial skepticism fostered by Tom Nelson at the *Ivory-bill Skeptic* helped brake gullible extrapolation that overran social discourse right after the Arkansas announcement. But giving credence to doubt later devolved into "besmirching the integrity, character and professionalism of the individuals who presented the evidence."[111] Entirely heedless of their own ***fundamental attribution error***, Nelson's posters shouted that "faceless TNC and [Cornell Laboratory of Ornithology] bureaucrats were thinking [only] of funding," portraying even this kind of core mission fidelity[112] as "a war crime if your parent institution was the Third Reich."[113] Once Nelson had tired of repudiating the woodpecker, he turned instead to denying climate change.

Out of this self-immolation, a quieter corner of the blogosphere took notice. New sites were launched where searchers, genuine skeptics, and the open-minded might dialogue with civility, "precisely because of the high level of personal vitriol" found elsewhere.[114] New sites aggregated historical documents, shared info about the ivory-bill,[115] and began a laborious task of deciphering many lingering mysteries about the woodpecker. Some bloggers were so thorough that after years-long appraisal, they pointed out their own earlier misconceptions, and used facts to craft novel analyses. At least one came at last to accept the Arkansas bird as having been identified correctly by the Cornell team.[116]

Despite its striking conformity over the Arkansas incident, birding's group-think came at a steep price. The hobby's fundamental illiteracy about the historical biology of ivory-bill ascended from ***argument from authority*** cemented by this identity's tight homophily.[117] Critics, unfamiliar with past literature and the severe shortcomings to Tanner's study (Chapters 8, 10), never stopped insisting that "experience with other rare birds, especially resident species, suggests that any valid sighting should very quickly lead to more sightings."[118] That overly rigid ***framing effect*** meant that "sight records, both of [ivory-bill] and generally, have been not always treated with an open mind."[119] Birding showed a particular liability for socially-facilitated error and group-entrenched cognitive bias.[120]

Professional ornithology once steered a communal disbelief about the ivory-bill, but after the Arkansas incident, bird-watching took over as the institutional bastion of unwavering certitude. Birding did not just grill field marks that we used to identify a living ivory-bill, it fronted a rigid denial of the species' very existence. "The reason the bird cannot be confirmed is because it is not there."[121] Buoyed by this unflinching conceit from ***Dunning-Kruger effect***, birding patronized everyone else to constrain social truth about the ivory-bill. "We need to accept the tragic loss of the ivory-billed woodpecker, and move beyond it."[122]

Yet despite this dramatic change in reigning critic, the highly polarized thought system that we had long cultivated around this species still remained fully intact. The Arkansas incident made very obvious something else: fissures in ivory-bill myth had by now become too glaring and numerous to be concealed any longer. None of our standard plotlines for the woodpecker could account for so many tenacious discrepancies, glaring conflicts, and stubborn contradictions.

Endnotes

1 "There is a crucial distinction between thinking that you have made a discovery (or even announcing that you have made one) and actually making and getting credit for one ... a 'discovery' can turn out to have been an inadvertent 'replication' of an earlier achievement made by others" (Lynch, M. 2011. Credibility, evidence, and discovery: the case of the ivory-billed woodpecker. *Ethnographic Studies* 12: 78–105). Indeed, it is reasonable to ask why the Arkansas incident was ever hailed as a "rediscovery" at all given that few if any of the other more than 100 incidents of putative ivory-bill occurrence since 1944 (see Chapter 15) were ever granted this same acclaim. One explanation for the need to label these "rediscoveries" as such is an attempt or even need to overcome the potent narrative power of "extinction" from the earlier claims that the bird is already "dead."

2 Scheduling for this press conference had to be pushed up once the *Science* report of the rediscovery had been leaked prematurely. A newspaper in Cornell's hometown (Ithaca, New York) had broken an embargo request. See: Hackler, T. 2006. Flashback: the secret of the ivory-billed woodpecker expedition. http://www.prsa.org/SearchResults/view/205/105/Flashback_The_secret_of_the_ivory_billed_woodpecke

3 Stokstad, E. 2007. Gambling on a ghost bird. *Science* 317: 888–892.

4 Wall, J. 2012. Peckergate: the ivory-billed woodpecker hoax,

Campephilus principalis still extinct – pileated woodpecker misidentified in Arkansas as elsewhere. http://www.worldtwitch.com/ivorybill.htm (accessed 5 June 2015).

5 http://www.birdforum.net/showthread.php?t=33968 (accessed 5 June 2015). Use of derision, sarcasm, and harassment exemplified large swaths of this birding commentary. "Well, the game's up minus the spurious straw-clutching. The only people clinging to the video as proof are Cornell and a rag-tag bunch of believers. Do any serious birders out there still not get it?" Comment #10031 by "Edward woodwood" at: http://www.birdforum.net/showthread.php?t=33968&pages=402 (accessed 24 August 2015). "Good luck with conserving a species nobody (other than you) seems to be able to see." Comment #26 by 'lewis20126' at: http://www.birdforum.net/showthread.php?t=297876&page=2 (accessed 25 August 2015). "in truth, some of them left here because they could ... no longer stand the scorn, ridicule and personal abuse...." Comment #11553 by 'salar53' at: http://www.birdforum.net/showthread.php?t=33968&page=463 (accessed 25 August 2015). Referring to all the "'armchair' bickering going on," ironically much of it by birders outside the U.S., one poster resigned to take a "much-needed year-long break from this senseless arguing" because it "turns out the same few 'dominants' continue to pound their il-informed [sic] opinions down the throats of a few good people." Comment #11619 by 'Snowy1' at: http://www.birdforum.net/showthread.php?t=33968&page=465 (accessed 25 August 2015).

6 This observation about his peers was offered by David Bird, McGill University, available at: http://www.canada.com/montrealgazette/news/arts/story.html?id=b7b907a3-2285-46cd-9aff-2d305b210660 (accessed 30 March 2015).

7 Social cognition refers broadly to how humans interpret, archive, and apply the bits of information that they receive about other people, groups, and social situations, with an emphasis on the cognitive processes that influence those interpretations.

8 In one of few candid admissions from the birding perspective,

Ned Brinkley wondered aloud: "who among us would qualify as an 'I.D.' expert on ivory-billed woodpecker?" See: http://listserv.arizona.edu/cgi-bin/wa?A2=ind0603c&L=birdwg01&T=0&O=D&P=2444 (accessed 18 March 2006). Throughout this work, and unless indicated otherwise, the term "birding" and "birders" refers to the particular kinds of ***framing*** tactics and strategies customarily adopted by this social identity to enforce its internal cohesion, not to stereotype each and every member of the group (a form of ***ultimate attribution error***). Individual persons may, at any particular moment, also adopt the ***framing effects*** from more than one social identity.

9 A more thorough synthesis of technical merits used to debate the Arkansas bird's identity, particularly evaluations levied at the Luneau video, is presented in Appendix C.

10 Evidentiary standards used in human society vary widely. In judicial systems, probative (but also, in most cases, still disputable) standards for weighing evidence range from a laxer "preponderance of evidence" (>50% probability) in certain civil matters to a far more stringent "beyond a shadow of a doubt" (close to 100%) for criminal trials. A third standard, "clear and convincing," is intermediate with respect to its quantitative probability. Similarly, in science it is acceptable practice to report findings that either support or refute a particular hypothesis at virtually any value of probability. Readers are left to decide for themselves whether the uncertainty is adequate or not for the claim(s) made. In this vein, there was nothing about the evidence *per se* as reported in the 2004–2005 Arkansas ivory-bill incident that fell outside science norms. Lynch (p. 90) underscored what others mostly chose to ignore: "Taken alone, the evidence provided by the [Arkansas] eyewitness accounts, field notes, and sketches, might be viewed as powerful, and even conclusive. If this were a court of law, the eyewitness testimony would be considered quite strong, if not dispositive."

11 Evidentiary rules for gauging ivory-bill persistence are irrational. Such evidence must first be indisputable and irrefutable, but then it

must also be instantly replicated to the same or even better degree. "If no additional evidence of the ivory-bill is found this field season, we may well join the ranks of ornithological pariahs." Gallagher, T. *The Grail Bird: The Rediscovery of the Ivory-billed Woodpecker*. Houghton Mifflin Harcourt, Boston, MA, p. 264. Finally, albeit based entirely on ***argument from authority***, such evidence must be both acoustic *and* visual: "In most of the reports made in the last several decades, the bird was either heard but not seen, or seen but not heard. These reports are contrary to accepted knowledge about how to locate the species." Sykes, P.A. 2016. A personal perspective on searching for the ivory-billed woodpecker: a 41-year quest. Pp. 171–182 *in* The History of Patuxent – America's Wildlife Research Story (Perry, M.C., ed.), U.S. Geological Survey Circular 1422, p. 180.

12 In a later ivory-bill search conducted in north Florida, team leader Hill devoted generous ink to discounting any view that such searches were truly scientific, that the activity constituted more than mere bird watching, and that their team instead would try to get social acceptance for their findings. Thus, this team would "seek outside opinion from experts like Jerry Jackson to ascertain whether our evidence was definitive. We would ask a rare bird committee to review our evidence." See: Hill, G.E. 2007. *Ivorybill Hunters: The Search for Proof in a Flooded Wilderne*ss. Oxford University Press, London, p. 147. Conceding that a likely intention was to try and avoid the blowback that happened to Cornell, this plea just clutched at a different error, ***argument from authority***, this time from a birding ***frame*** instead of one from professional ornithology (e.g., constrained by Tanner's strong views on the woodpecker's habits or fate; see Chapter 8).

13 Based on its being unprecedented, the latter might be argued reasonably as constituting an "extraordinary" event; the former cannot be.

14 Species rarity strongly influences the numbers and behaviors of bird watchers as they engage in their pastime (Booth, J.E., K.J.

Gaston, K.L. Evans, and P.R. Armsworth. 2011. The value of rarity in biodiversity recreation: a bird watching example. *Biological Conservation* 144: 2,728–2,732). During the Arkansas incident, birding, the media, and other parties failed to ever distinguish between alternative (and conflicting) meanings of "rarity." Rarity from low population size stems from a different biological process (e.g., demography) than does rarity caused by vagrancy (e.g., dispersal). For purposes of enumeration and bragging rights in the competitive branches of this sport, it may matter little why a bird is rare in the first place. However, by imposing its own evidentiary standards and ***framing*** values on the Arkansas incident, then insisting that all others adopt and abide by those standards too, birding fell sway to its own profound error from ***false consensus effect***.

15 Sheard, K. 1999. A twitch in time saves nine: bird-watching, sport, and civilizing processes. *Sociology of Sport Journal* 16: 181–205.

16 So-called "committed" birdwatchers, that is, those who are both highly active and amply knowledgeable, make up only about 3% of the American public. See: Kellert, S.R. 1985. Birdwatching in American society. *Leisure Sciences* 7: 343–360.

17 "We may need to rethink our idea of proving the existence of ivory-bills 'beyond a shadow of a doubt' before we accept their presence." Steinberg, M.K. 2008. *Stalking the Ghost Bird: The Elusive Ivory-billed Woodpecker in Louisiana.* Louisiana State University Press, Baton Rouge, p. 66.

18 "There is to-day, most unhappily, too often a gap between the amateur naturalists and the pure field-workers on the one side, and the trained biologists on the other. The blame, as usual, cannot be laid to the account of either, for both are guilty." Huxley, J.S. 1916. Bird-watching and biological science. *The Auk* 23: 142–161. Hence, a tension between the ***frames*** of birding and professional ornithology have had an entire century to become even more entrenched.

19 Fitzpatrick, J.W., M. Lammertink, M.D. Luneau, Jr., T.W. Gallagher, B.R. Harrison, G.M. Sparling, K.V. Rosenberg, et al. 2005. Ivory-

billed woodpecker (*Campephilus principalis*) persists in continental North America. *Science* 308: 1,460–1,462.

20 Owen, J. 2005. "Extinct" woodpecker found in Arkansas, experts say. *National Geographic News*, April 28, 2005. http://news.nationalgeographic.com/news/2005/04/0428_050428_extinctwoodpecker.html (accessed 31 July 2015).

21 Stokstad, p. 888.

22 http://www.sdakotabirds.com/blog/?p=260 (accessed 4 September 2015).

23 Science ***framing*** that avoids entirely any reference to uncertainty constitutes a major lapse from ***neglect of probability***. "Science is never certain" (Sheil, D., and E. Meijaard. 2012. Purity and prejudice: deluding ourselves about biodiversity conservation. *Biotropica* 45: 566–568). Journal editors who oversaw a later paper that reported on evidence gathered for ivory-bill persistence in Florida felt that they needed to stress that putting "good science first implies embracing uncertainty.... To deny publication of a controversial paper simply because it did not present a definitive conclusion to an ongoing debate ... would only mean that we abrogated our responsibility." See: Nudds, T.D., J.R. Walters, and M.-A. Villard. 2006. On evidence of absence. *Avian Conservation and Ecology* 1: 3 http://www.ace-eco.org/vol1/iss3/art3/. In contrast, birding (and other actors, including some in the media) often conflated this ordinary realm of scientific uncertainty with raw "faith," thereby resorting to ***argument from incredulity*** and other breakdowns in logic in attempting to equate less-than-conclusive evidence with mere belief (***false equivalency***).

24 Lynch, p. 88, points out that certain language used by Cornell may have been quite intentional, and due to "strenuous efforts to head-off or rebut skeptical dismissal of the sightings..." entirely. The Cornell "team evidently anticipated disbelief, and even mistrust." *Ibid.*, p. 91. Similarly, Ned Brinkley stressed that "every member of the Cornell team I've known or met is very aware of the thousands of consequences of their claims, as well as being aware of the ultimate possible damage to reputations – dozens who have reported this

species been [sic] discounted as quacks, hoaxers, or delusional in decades past." See: http://listserv.arizona.edu/cgi-bin/wa?A2=ind0603c&L=birdwg01&T=0&O=D&P=3216 (accessed 19 March 2006). Fitzpatrick also grasped the wider implications of their announcement: "our lives will never be the same from this day on." Dickinson, R. 2005. The best-kept secret. *Audubon Magazine* http://magazine.audubon.org/features0507/rediscovery.html (accessed 4 August 2007). Attempts to address the issue of credibility, then, but also not be constrained to that issue alone, are entirely consistent with the title of Cornell's original *Science* paper, one that stressed the woodpecker's *persistence* rather than its *identity*.

25 Also termed ***egocentric attribution bias***, this socially discernible projection occurs when an individual or a group tends to perceive that their position is more widely accepted than it truly is. "Cornell is showing far too much certainty in relation to the quality of its evidence." Moore, S. 2006. Cornell's unwarranted level of certainty in the ivory-billed woodpecker debate is disturbing. https://www.mail-archive.com/ecolog-l@listserv.umd.edu/msg01310.html (accessed 29 September 2015). A bias from ***false consensus effect*** arises from assuming that other individuals or social identities think like we do, or that they hold the same habits, values, and behavioral responses. ***Projection bias***, a tendency to assume that either others or our future self will also share our present-time thoughts, values, emotional states, or beliefs, can go on to contribute to even wider social manifestation of this ***egocentric attribution bias***.

26 Although many of its critics remained unmoved, 10 years later Cornell was still "convinced that this [the Luneau] video provides documentary evidence confirming the numerous sight records of ivory-billed woodpecker in the Cache River National Wildlife Refuge between February 2004 and February 2005." http://www.birds.cornell.edu/ivory/evidence/segments/conclusions (accessed 25 August 2015).

27 See: Gilovich, T. 1990. Differential construal and the false consensus effect. *Journal of Personality and Social Psychology* 59: 623–634. We

perceive ourselves asymmetrically from how we typically see others. Whereas we can access our own particular sensations, emotions, and cognitions, our view of others generally relies only upon what can be most easily observed externally. Pronin, E. 2008. How we see ourselves and how we see others. *Science* 320: 1,177–1,180.

28 Krueger, J., and J.S. Zeiger. 1993. Social categorization and the truly false consensus effect. *Journal of Personality and Social Psychology* 65: 670–680.

29 Hill, p. 209.

30 *Ibid.*, p. 201. Both the quantity and quality (e.g., very large cavities, concurrent acoustic and visual detection) of evidence collected for ivory-bill in Florida were actually superior in some ways (absent, perhaps, the Luneau video) than the body of material otherwise assembled from Arkansas. Hill and his associates saw or heard more than one putative woodpecker at once, and they also saw the birds dispersed across distances that they considered as representing potentially different individuals or pairs.

31 "Hill and colleagues [were] much more circumspect than Fitzpatrick and colleagues (2005) in their conclusions." McKelvey, K.S., K.B. Aubry, and M.K. Schwartz. 2008. Using anecdotal occurrence data for rare or elusive species: the illusion of reality and a call for evidentiary standards. *BioScience* 58: 549–555.

32 Stokstad, p. 891.

33 Hill, p. 125.

34 *Ibid.*, p. 147.

35 Wright, R. 2013. Who you callin' amateur? [a review of *Bunch of Amateurs: A Search for the American Character*. J. Hitt]. *Birding* 45: 67.

36 The ***Semmelweis reflex*** arises from automatic propensity to reject new facts (usually those that conflict with the status quo) without proper reflection or other due consideration.

37 This paradigm was so deeply ingrained in birding that logic was turned on its head by heedless resort to ***argument from ignorance***, e.g., "Believers should shut up until they have some proper evidence

to support their case for the continued existence of [ivory-bill]." Commentary by Rob Stoff, available at: http://www.birdforum.net/showthread.php?t=33968&pp=25&page=507 (accessed 10 August 2015).

38 Media never earnestly questioned leanings of critics who had already staked out earlier positions on ivory-bill status, and who might allow their pre-existing opinions (***belief perseverance***) to influence their widely-disseminated views. See: Tetlock, P.E. 1983. Accountability and perseverance of first impressions. *Social Psychology Quarterly* 46: 285–292. "David [Sibley's] mind was already largely made up (years before the Arkansas story came along) that the Ivory-bill was extinct," blogger cyberthrush reporting firsthand his interaction with Sibley even before the Arkansas incident; http://ivorybills.blogspot.com/2014/04/memory-lane.html (accessed 24 August 2015). Being heavily vested in a prior decision leads readily to ***sunk cost fallacy***. Rigid clutching at earlier decisions despite new evidence can have tragic, deadly consequences (see Roberto, M.A. 2002. Lessons from Everest: the interaction of cognitive bias, psychological safety, and system complexity. *California Management Review* 45: 136–158). Also, individuals already anchored to an inflexible position were unreliable representatives for the "other side," unable to serve as legitimate counterparts for a supposedly balanced debate on the woodpecker (see ***equivalency bias***).

39 A conspicuous lack of neuropsychological humility, a type of blind ignorance that is wholly unable to properly evaluate or even comprehend the true level of its own witlessness. "In a competitive birding world full of large egos, rediscovery is the ultimate achievement for a member of the birding tribe, and the ivory-bill is the Holy Grail among birders." Steinberg, M.K. 2008b. Bottomland ghost: southern encounters and obsessions with the ivory-billed woodpecker. *Southern Cultures* 14: 6–21.

40 Theoretical underpinnings for ***reactance*** are believed to arise from a motivated arousal that is triggered by attempts to restore a threatened or lost freedom. See: Brehm, S.S., and J.W. Brehm. 1981.

Psychological Reactance: A Theory of Freedom and Control. Academic Press, New York.

41 Another illustration of ***reactance*** arises from hunters' staunch resistance to the conservation of large predators that they perceive as curtailing human access to the same prey animals customarily sought by people (e.g., deer). See: Lüchtrath, A. and U. Schraml. 2015. The missing lynx – understanding hunters' opposition to large carnivores. *Wildlife Biology* 21: 110–119.

42 Several revealing examples of the unintended consequences that can arise by attempts to censor alternative views, i.e., the ***backfire effect***, can be found in Jansen, S.C., and B. Martin. 2003. Making censorship backfire. *Counterpoise* 7: 5–15.

43 Hill, p. 45.

44 Crewdson, J. 2006. Woodpecker sighting a flight of fancy? *Chicago Tribune*, May 19, 2006 http://articles.chicagotribune.com/2006-05-19/news/0605190139_1_ivory-billed-woodpecker-cornell-university-bird/3 (accessed 19 May 2005).

45 Jackson, J.A. 2006. Ivory-billed woodpecker (*Campephilus principalis*): hope and the interfaces of science, conservation, and politics. *The Auk* 123: 1–15.

46 Fitzpatrick, J.W., et al. 2006. Clarifications about current research on the status of ivory-billed woodpecker (*Campephilus principalis*) in Arkansas. *The Auk* 123: 587–593.

47 Hill, p. 146.

48 Comment by John T. Rotenberry, University of California-Riverside. See: Mendenhall, M. 2006. Other ornithologists gave Jackson's article mixed reviews; the reactions ranged from unease over his fundraising allegations to expressing support for the technical arguments that lay behind some of his skepticism. *Birders World Magazine* http://www.birdersworld.com/brd/default.aspx?c=a&id=632 (accessed 1 April 2006). As Fitzpatrick admitted during interviews, Jackson's allegation that the Arkansas rediscovery had been exploited for financial gain "hurts the most." *Ibid.*

Accusations of skewing the search purpose and/or its outcome for an agenda of fundraising were common (***fundamental attribution error***). They were leveled at Cornell and at the Nature Conservancy, among others, e.g., "hysterical public officials redirected $14 million of precious conservation funds into developing the ivory-billed woodpecker recovery plan – never mind the fact that this species has not been seen in more than 50 years and is extinct in the USA" (this criticism manages to combine two errors of fact along with a wholly unsubstantiated assertion [extinction]; blogger GrrlScientist at http://scienceblogs.com/grrlscientist/2010/02/12/faith-based-birding-201-fraudu/ [accessed 29 August 2015]).

49 ***Fundamental attribution error*** arises from giving too much weight to internal, unseen factors, such as personality, when attempting to explain others' behavior, and also not enough weight to external factors that arise out of the key circumstances found in a particular situation. Consequently, we tend to give the worst interpretation to others' behavioral expression based on their guessed-at hidden motivations, but reserve the best intentions to ourselves, i.e., we fail to put ourselves adequately in others' shoes.

Attacks on motive based on ***fundamental attribution error*** were quite prominent in the criticisms aimed at Cornell, e.g., "It is also useful to examine motive, including subconscious motives in issues of controversy dependent on disputed evidence. The Cornell team and its individual members have much to gain from a position of certainty regarding the ivory-billed woodpecker's rediscovery." Moore.

50 Gallagher, T. *The Grail Bird: The Rediscovery of the Ivory-billed Woodpecker*. Houghton Mifflin Harcourt, Boston, MA.

51 Jackson, p. 2.

52 An expectation that one will find more ***egocentric bias*** in other people than in oneself creates a demonstration of ***naïve cynicism***. See Kruger, J., and T. Gilovich. 1999. "Naive cynicism" in everyday theories of responsibility assessment: on biased assumptions of bias. *Journal of Personality and Social Psychology* 76: 743–753.

53 ***Illusion of transparency*** refers to miscalculations in our ability to perceive (or have perceived) our real inner states. See, for example: Gilovich, T., K. Savitsky, and V.H. Medvec. 1998. The illusion of transparency: biased assessments of others' ability to read one's emotional states. *Journal of Personality and Social Psychology* 75: 332–346. Such miscalculations were prevalent when others attributed false motivations to nondisclosure agreements and other attempts at secrecy practiced by the Arkansas search team. Rather than some darkened agenda, in truth that confidentiality was because "none of us wanted to be responsible for putting this bird in jeopardy before protection could be put in place." See Dickinson.

54 Natural ***myside bias*** is the inclination to evaluate propositions from within one's own perspective when given no instructions or other strong prompting to do otherwise. Stanovich, K.E., and R.F. West. 2007. Natural myside bias is independent of cognitive ability. *Thinking & Reasoning* 13: 225–247. Despite considerable training in critical thinking, the science disciplines have no special immunity to this fallacy, in part because ***myside bias*** is divorced from our thinking disposition, cognitive ability, and basic intelligence. Stanovich, K.E., R.F. West, and M.E. Toplak. 2013. Myside bias, rational thinking, and intelligence. *Current Directions in Psychological Science* 22: 259–264. Moreover, it is particular *beliefs* rather than particular *people* that are more likely to foster the ***myside bias***. Toplak, M., and K.E. Stanovich. 2003. Associations between myside bias on an informal reasoning task and amount of post-secondary education. *Applied Cognitive Psychology* 17: 851–860.

55 Gallagher, p. 102.

56 David Sibley's criticisms began not long after he tried but failed to find the Arkansas woodpecker during a field search of a few days in the Big Woods. Sibley was hardly alone, however, as others trained in field biology also succumbed to ***myside bias***. "At this point, not having seen or heard an [ivory-bill], I and others began to doubt the reliability of the reported sightings on the Bayou de View and in the surrounding areas." Sykes, p. 178.

57 Jackson did, however, have one auditory (pp. 182–184) and one visual encounter (pp. 7–8).

58 Jackson, p. 184.

59 Hill, p. 95.

60 *Ibid.*, p. 94.

61 *Ibid.*, p. 141.

62 Some critics, like Louis Bevier, readily divulged what actually lay behind their dispute with the Cornell findings: "I became interested with the claim as an identification problem. Having spent 21 years on records committees...." http://www.birdforum.net/showthread.php?t=33968&pp=25&page=507 (accessed 21 August 2015).

63 Social identity inside birding is not monolithic (e.g., Cole, J.S., and D. Scott. 1999. Segmenting participation in wildlife watching: a comparison of casual wildlife watchers and serious birders. *Human Dimensions of Wildlife* 4: 44–61). Thus, generalizing a set of stereotypical traits to each and every individual member of this group would be susceptible to ***ultimate attribution*** and ***group attribution errors***.

Nonetheless, in the U.S. this hobby is composed of two primary constituencies: female-biased, leisure-oriented recreational bird-watchers, and heavily male-biased (among adults) sport birders who are focused on displaying their level of knowledge to each other, taking more bird watching trips, and practicing an ability to identify species by sight and sound, all related to a more competitive style of list-keeping [Lee, S., K. McMahan, and D. Scott. 2015. The gendered nature of serious bird watching. *Human Dimensions of Wildlife* 20: 47–64; Cooper, C.B., and J.A. Smith. 2010. Gender patterns in bird-related recreation in the USA and UK. *Ecology and Society* 15: 4. http://www.ecologyandsociety.org/vol15/iss4/art4/ (accessed 14 May 2015)]. "Competitive birding is a highly masculine endeavor, and involves ... essentially a form of hunting where the game is 'bagged' in non-lethal ways." See: Moore, R.L., D. Scott, and A. Moore. 2008. Gender-based differences in birdwatcher's participation and commitment. *Human Dimensions of Wildlife* 13:

89–101. Harshest critics of ivory-bill persistence and identity came substantially from the latter segment, a group that commands most of the hobby's positions of authority, consider themselves more advanced in knowledge and ability, and show a strong preference for competitive engagement with each other (see Kellert, pp. 355–356). Those who chase rare birds "comprise a particular subset of society. Most strikingly, all surveyed individuals were male and white." Booth et al.

Confirming the pastime's adherence to a rigid authority structure, and given the rules under the American Birding Association that forbade ivory-bill from being "counted" in U.S. sport birding, in hindsight it cannot be viewed as a "big mystery [as to] why so few birders have come to Arkansas up to now." http://tomnelson.blogspot.com/2006/02/john-acorns-grail-bird-review.html (accessed 25 August 2015). The fact that fewer birders than expected rushed to see a putative ivory-bill caught the U.S. Fish & Wildlife Service by surprise. The agency had moved extra law enforcement agents from elsewhere to Arkansas in order to manage an anticipated rush of ivory-bill seekers that never materialized (Jon Andrew, *pers. comm.*). However, the lack of widespread interest by birders can be explained as fully in keeping with the sport's hierarchical and commanding hegemony over most of its members.

64 Sibley, D.A., et al. 2007. Ivory-billed woodpecker draft recovery plan comment: letter commenting on Draft Recovery Plan. 6 pp. Interests, affiliations, and occupations of the 72 co-signees to Sibley's letter expose a strong birding (rather than scientific) identity and ***framing*** outlook behind their perspective. Although a few were employed as professional ornithologists, most of these and the vast majority (more than 80%) of all other co-signees were entirely or mostly concerned with bird status, i.e., avian distribution, abundance, and identification. At least 60 co-signees were tour guides, "top birders," members of rare bird committees, natural history curators and collection managers, field guide authors, bird artists, or breeding bird atlas coordinators and others routinely

engaged primarily with geographic status of birds. There were few or no endangered species professionals, wildlife demographers, biostatisticians, conservation biologists, or forest and ecosystem ecologists represented among this set of critics.

65 Sibley repeatedly disparaged the kind and number of field marks actually noted by the observers of putative ivory-bill in Arkansas and Florida. He claimed that nobody ever saw the light bill of an ivory-bill, or the species perched, when in truth Taylor Hicks in Florida observed both field marks on a bird from 40 feet away. One commentator finally took issue with Sibley's blatant inaccuracies at the latter's own website, stating: "it is patently untrue as you ought to be already aware" that "'all sightings emphasize a single field mark – the white trailing edge of the wing.'" For example, seven of Cornell's sightings included descriptions of "a narrow red crescent on the bird's folded crest" (signifying a male bird), as well as other diagnostic field marks. See: Sibley Guides Notebook, p. 15 (of 17) http://sibleyguides.blogspot.com/2007/10/ivory-billed-woodpecker-status-review.html (accessed 27 May 2015).

66 Not only had Cornell observer Mindy LaBranche obtained her Ph.D. in woodpecker biology, prior to the search she was an avowed skeptic of the bird's existence. Speaking later about her own sighting on April 10, 2005, she declared: "I was convinced the bird was extinct.... I'm not a lucky birder, and I certainly knew I wasn't going to see it.... I've doubted myself on a lot of other bird sightings, but not this one. There is nothing else it could have been. Nothing." http://www.birdwatchingdaily.com/featured-stories/eyewitness-accounts-of-ivory-billed-woodpeckers/ (accessed 31 August 2015). LaBranche's predisposition and explanation do not conform to the cognitive priming theory so often speculated about by critics (***fundamental attribution error***), a conjecture wherein searchers were putatively driven by their run-away ***observer expectancy bias***, i.e., seeing what they wanted or expected to find. Instead, and due to the exceedingly strong bias towards the ***meme*** of extinction over survival, LaBranche's comments here better align

with a predisposition towards ***outcome reporting bias***.

67 Sibley 2007, p. 1.

68 Donnelly, P. 1994. Take my word for it: trust in the context of birding and mountaineering. *Qualitative Sociology* 17: 215–241. Trust between professional ornithology and amateur birding can be characterized as faltering and conditional. Except along a zone of contact wherein birders' efforts on Christmas Bird Counts, Breeding Bird Surveys, banding records, and so forth might be used by science for true hypothesis-testing, each of these two social groups focused on birds have maintained largely separate identities since they diverged in the early twentieth century (see Chapters 3–6). Individual persons, however, may well straddle each of these social identities. Unless individuals are consciously aware of such dual identity, however, ***framing*** effects from one identity can easily cross-contaminate the outlooks under another identity.

69 Fine, G.A., and L. Holyfield. 1996. Secrecy, trust, and dangerous leisure: generating group cohesion in voluntary organizations. *Social Psychology Quarterly* 59: 22–38.

70 No professional certification process, standardized guidelines, or transparent set of credentials distinguish those birders who are considered to be "expert" or who come to occupy the more rarified echelons of decision making in birding leadership.

71 Consequently, such groups become cut off from alternative viewpoints, localized in social space, with the upshot that individual members learn to "obey" and conform with the group's inside culture and stereotypical behaviors. See: McPherson, M., L. Smith-Lovin, and J.M. Cook. 2001. Birds of a feather: homophily in social networks. *Annual Reviews of Sociology* 27: 415–444. Within such isolated networks, members are prone to ***in-group bias***, a tendency to grant preferential treatment and over-consideration to those perceived to be members of their own group. See Wilke, A., and R. Mata. 2012. Cognitive bias. *Encyclopedia of Human Behavior, Second Edition* 1: 531–535.

72 "Birders typically watch from the road, a bridge, or a canoe on open

water. They rarely penetrate more than a hundred yards into a forest unless there is a road or trail, making chances of an ivory-bill sighting slim." Jackson, p. 241. Also: "most birders, even experts who should know better, vastly overestimate the efficiency of the transcontinental birding community as a bird-finding machine." Bill Pulliam, http://www.bbill.blogspot.com/2009/09/why-rush.html (accessed 10 August 2015).

73 Sibley Guides Notebook, p. 3 (of 17).

74 ***Illusory superiority*** is generated from an unrealistic optimism coupled with self-deception. See: Hoorens, V. 1995. Self-favoring biases, self-presentation, and the self-other asymmetry in social comparison. *Journal of Personality* 63: 793–817.

75 The ***illusion of omniscience*** arises from an inbuilt psychological bug, greatly exacerbated by the internet, which leads us to believe that we cannot possibly learn anything new. See: Leslie, I. 2014. Omniscience bias: how the internet makes us think we already know everything. *New Statesman* http://www.newstatesman.com/internet/2014/06/omniscience-bias-how-internet-makes-us-think-we-already-know-everything (accessed 18 August 2015).

76 Compared to the individual autonomy that is so cultivated in science, the tight unity inside birding enabled more consistent ***framing*** by the group's elite to shape public messaging about ivory-bill. This social-psychological involvement and strong level of commitment are tied very tenaciously to each other inside the hobby of birding. See Kim, S-S., D. Scott, and J.L. Crompton. 1997. An exploration of the relationships among social psychological involvement, behavioral involvement, commitment, and future intentions in the context of bird-watching. *Journal of Leisure Research* 29: 320–341.

77 ***Bayesian caution*** is the logical avoidance of assigning an extreme (0% or 100%) probability to a fallible proposition. See: Weintraub, R. 1993. Fallibilism and rational belief. *British Journal for the Philosophy of Science* 44: 252–261.

78 ***Subadditivity effect*** is an inclination to appraise the probability of

the whole as being less than the probabilities of its constituent parts. Shanteau, J. 1974. Component processes in risky decision making. *Journal of Experimental Psychology* 103: 680–691. Sibley, apparently not realizing his analogy furnished a textbook demonstration of fallacy from ***straw man*** and ***subadditivity effect***, remarked: "the body of evidence is only as strong as the single strongest piece – ten cups of weak coffee do not make a pot of strong coffee." In fact, not only can the amount of caffeine in ten weak cups be the same as (or stronger than) that found in a pot of coffee, it only takes one bit of evidence to make true either the identity or the persistence of ivory-bill. Regardless of the strengths in each bit of evidence, they necessarily sum to a probability greater than zero (after due diligence being made for adherence to ***Bayesian caution***). Quote taken from: http://www.sibleyguides.com/2007/10/ivory-billed-woodpecker-status-review/ (accessed 1 September 2015).

79 One anonymous (as usual) poster disclosed the objective in this way over at Tom Nelson's site on ivory-bill skepticism: "Focus on the misidentified video, the crappy cluster of sight records, and the driving force of RemFitzenPatrick." http://tomnelson.blogspot.com/2006/07/some-humor.html (accessed 17 August 2015). In this fashion, various bits of the Arkansas evidence were isolated and attacked more readily using a wider (albeit quite arbitrary) range of tactics, including, as the last part of the quote underscores, the use of ***ad hominem***.

80 Not all of birding was so ready to dismiss outright the anecdotal sight records of a putative Arkansas ivory-bill. Rick Wright, editor of the American Birding Association's *Winging It*, bemoaned birding's apparent obsession with replacing the "shotgun with digital camera," and wondered why "there has been precious little attention paid to the sight records made by normally credible observers." His essay also noted "birding today runs the risk of becoming so heavily technologized that we abandon the artifact our hobby was found upon nearly 100 years ago." See: *Winging It*, July/August 2006, p. 2. The word "artifact" here may have revealed far

more than what was intended.

81 E.g., Kaufman's mocking barb: "I do feel qualified to comment on some things in the Luneau video, even without the mystical enlightenment that would come with actually seeing the species in life." See http://listserv.arizona.edu/cgi-bin/wa?A2=ind0603c&L=birdwg01&T=0&O=D&P=2992 (accessed 19 March 2006).

82 Jones, C.D., J.R. Troy, and L.Y. Pomara. 2007. Similarities between *Campephilus* woodpecker double raps and mechanical sounds produced by duck flocks. *The Wilson Journal of Ornithology* 119: 259–262.

83 See Crewdson. The first sentence is a complaint for more allegiance to ***argument from authority***; the second is an inherent contradiction, not to mention disquieting from an expert who ostensibly makes their living by correctly identifying birds for paying clientele.

84 *Ibid.* The speaker was not technically adroit to comprehend that the black color was not from the trailing edge of the bird's wing, but rather a video artifact caused by lighting contrast in adjacent pixels in very low-resolution imagery before it has been digitally de-interlaced. See ***Dunning-Kruger effect***, and compare to Fitzpatrick, J.W., M. Lammertink, M.D. Luneau, T.W. Gallagher, and K.V. Rosenberg. 2006. Response to comment on "Ivory-billed woodpecker (*Campephilus principalis*) persists in continental North America." *Science* 311: 1555b.

85 *Ibid.* Insistence on "irrefutable data" here as the appropriate evidentiary standard reveals cognitive bias. Robbins, a collections manager at a natural history museum, ignores that uncertainty is embedded inherently in *all* genuine science-based reporting. Instead, he borrows a ***meme*** and/or the ***framing*** language from birding to insist on the unassailable. Moreover, throughout the prolonged disputes over ivory-bill, birding routinely conflated issues of the bird's *identity* with the bird's *persistence*.

86 *Ibid.*

87 Hagner, C., and M. Mendenhall. 2007. A new challenge to the Luneau video. *Birders World Magazine* http://bwfov.typepad.com/birders_

world_field_of_vi/2007/03/a_member_of_the.html (accessed 13 August 2005). Presumably trained in research practice and protocol, Collinson should have known that a standard of 'proof' is not what qualifies as acceptable content in a scientific publication (***neglect of probability***). Collinson was likely influenced by ***framing*** language of birding, however, as words like "proof" and "conclusive" are ubiquitous in the group communications of that pastime.

88 As cited in 2006 for a book review appearing in *Western Birds* 37: 118.

89 Stringing together fictional beasts along with real species in order to discredit the possibility of an extant ivory-bill exemplifies the logical fallacy of ***false equivalence***.

90 http://listserv.arizona.edu/cgi-bin/wa?A2=ind0603c&L=birdwg01&T=0&O=D&P=4587 (accessed 21 March 2006). Kaufman used ***fallacy of division*** here, i.e., he assumed that something that is true of one thing must also be true of all or some of its related parts. He also falls to ***neglect of probability***, as research showed that in fact the likelihood of detecting a single ivory-bill during the Arkansas search given the survey effort actually expended was rather low, only about 12% (see: Scott, J.M., F.L. Ramsey, M. Lamertink, K.V. Rosenberg, R. Rohrbaugh, J.A. Wiens, and J.M. Reed. 2008. When is an "extinct" species really extinct? Gauging the search efforts for Hawaiian forest birds and the ivory-billed woodpecker. *Avian Conservation and Ecology* 3: http://www.ace-eco.org/vol3/iss2/art3/).

91 September 17, 2007 letter "Comments on the Draft Recovery Plan for the ivory-billed woodpecker (*Campephilus principalis*)" from Robert A. Hamilton to Deborah Fuller, U.S. Fish & Wildlife Service, Lafayette, LA. After first admitting that he had no particular expertise in woodpecker ecology and could not comment on technical aspects of the Plan, the author tellingly references his "considerable experience in critically evaluating the validity of sight records."

92 http://listserv.arizona.edu/cgi-bin/wa?A2=ind0603c&L=birdwg01&T=0&O=D&P=4587 (accessed 21 March 2006). In one set of

statements, Kaufman tumbled to the ***availability heuristic, illusory correlation, insensitivity to sample size*** (Chapter 8), ***argument from authority*** (Chapter 10), plus ***primacy effect, anchoring bias,*** and ***bandwagon fallacy*** (Chapter 11). His illiteracy of 30 or more sources that for over two centuries pointedly drew attention to the exceptionally wary nature of the ivory-bill (Chapter 7) exemplify a classic manifestation of the ***Dunning-Kruger effect***. Birding served as a vast echo chamber for spreading false information about the bird's supposedly tame nature. For example: "I see no reason why the ivorybill would stay half-tame through over 200 years of hunting pressure, then make a quantum leap to ultra-wariness during 60+ years of no hunting pressure." Comment by Tom Nelson, available here: http://tomnelson.blogspot.com/2005/08/just-how-wary-was-ibwo.html (accessed 21 May 2015).

Even conservation professionals fell to ill-informed supposition from ***Dunning-Kruger***. Keith Ouchley of the Nature Conservancy "disputed the notion that the birds are extremely shy," then doubled down with: "There is no indication given by past writers or naturalists that these birds were exceedingly shy." Steinberg, p. 42.

93 http://listserv.arizona.edu/cgi-bin/wa?A2=ind0603c&L=birdwg01&T=0&O=D&P=3757 (accessed 21 March 2006).

94 Fitzpatrick, et al. 2006. *Science* 311: 1,555b. Others also pointed out error in Sibley's depiction: "the wings of the bird in this painting are held too far down … the wingstroke is too deep." Gallagher, p. 167.

95 Sibley, D.A., L.R. Bevier, MA. Patten, and C.S. Elphick. 2006. Comment on "Ivory-billed woodpecker (*Campephilus principalis*) persists in continental North America." *Science* 311: 1,555a.

96 Fitzpatrick, et al. 2006. *Science* 311: 1,555b.

97 See commentary at: http://bbill.blogspot.com/2011/11/imperial-film.html.

98 Birds were apparently tossed into the air by hand, then filmed, hardly a comparable set of experimental conditions to an

unrestrained woodpecker launching itself into flight from a vertical tree trunk.

99 Those never-substantiated claims about pileated woodpecker beating its wings just as fast as the putative Arkansas ivory-bill were originally put forth on the web by Louis Bevier, here: http://web.mac.com/lrbevier/ivorybilled/Wingbeat.html. However, that site was taken down and disappeared from the internet after blogger dave_in_michigan thoroughly denounced Bevier's reasoning and the latter's refusal to back up such claims by providing the original data so that it could be independently verified by others (see comments #13561 through #13569 at http://www.birdforum.net/showthread.php?t=22968&page=543 (accessed 18 August 2015).

100 Collinson, J.M. 2007. Video analysis of the escape flight of pileated woodpecker *Dryocopus pileatus*: does the ivory-billed woodpecker *Campephilus principalis* persist in North America? *BMC Biology* 5: 8. Collinson's claim that "pileated woodpeckers can display a wingbeat frequency equivalent to that of the Arkansas bird during escape flight" was demonstrably false. See, e.g., footnotes 98 and 101.

101 See extensive comments by David Martin and Bill Pulliam available at: http://www.biomedcentral.com/1741-7007/5/8/comments (access ed 7 March 2009). To this very day, no pileated woodpecker has ever been documented as sustaining its wing-beats as rapidly and for as long as those measured in the bird that is depicted in the Luneau video.

102 Lammertink, M., T.W. Gallagher, K.V. Rosenberg, J.W. Fitzpatrick, E. Liner, J. Rojas-Tomé, and P. Escalante. 2011. Film documentation of the probably extinct imperial woodpecker (*Campephilus imperialis*). *The Auk* 128: 671–677. Birders went seriously astray about the heuristic value of woodpecker body size and flapping rates because they failed to factor in wing loading and aspect ratios when taxa like the ivory-bill have proportionately narrower wings (see Chapter 1). See also: Collins, M.D. 2017. Periodic and transient motions of large woodpeckers. *Scientific Reports* 7: 12551.

103 One cannot help but ponder whether professional ornithology would have been quite as accommodating if Tanner had been still alive at the time of the Arkansas incident.

104 Despite fractious debate over the Luneau video, dealings inside professional ornithology generally stayed cordial at a superficial level in the public sphere (Jackson *contra* Cornell being a notable exception: allusions to "idiots," "garbage," "clowns," and "smoke and mirrors" were apparently communicated in the background for some media interviews. See Crewdson). During and after Cornell's presentation at the Santa Barbara American Ornithologists Union conference in 2005 (where this author was present), the well-attended session was almost anti-climatically quiet, with the large audience respectful and attentive.

105 Nevertheless, science just barely gets a "passing" grade for how it handled the 2000's Arkansas and Florida ivory-bill incidents. Some editors at the time felt obliged to make extraordinary justifications as to why work on ivory-bill submitted to them had merit, going so far as a need to explain why science deals with uncertainty. Still later, several key and novel analyses lending support for a living ivory-bill were delayed for reasons that had nothing to do with scientific merit. In some cases, it took 40 submissions and 10 years for these key contributions to reach print. Collins, M.D. 2019. Statistics, probability, and a failed conservation Policy. *Statistics and Public Policy* 6: 67–79, including the accompanying supplement.

106 Stokstad's "Gambling on a ghost bird" was perforated with inaccuracies, several of which came lock, stock, and barrel from elite birding's attempts to discredit the Arkansas sight records. Stokstad misstated that: 1) all sightings were by lone, amateur observers; 2) only one additional sighting occurred after team members were paired (there were actually 10 more sightings); 3) volunteers were obliged to sign legal confidentiality agreements; and 4) secrecy by a few in the Cornell teams controlled the decision-making process (another prime example of ***fundamental attribution error***). See: Luneau, D. 2007. Errors in *Science* article "Gambling on a ghost

bird" by Erik Stokstad (17 August 2007), http://www.ibwo.org/science_errors.html (accessed 26 October 2007).

107 Wright.

108 In particular, *New York Times* contributor Jack Hitt tried to portray the crowd-sourced, -reviewed, -corrected and "-improved" statements by amateur birders and bloggers as having the same technical merit for adjudicating truth as the scholarly, peer-reviewed reports about the Arkansas woodpecker. Hitt, J. 2012. Science and truth: we're all in it together. *New York Times*, May 5, 2012 http://www.nytimes.com/2012/05/06/opinion/sunday/science-and-truth-were-all-in-it-together.html?_r=0 (accessed 25 February 2015).

109 Unless individuals' opinions are scaled according to some measure of their empirical reliability, however, costlier or faultier decisions are usually the result. See Mahmoodi A., et al. 2015. Equality bias impairs collective decision-making across cultures. *Proceedings of the National Academy of Sciences* 112: 3,835–3,840. ***Equality bias*** comes from the misguided notion that "we're all right, nobody's wrong, and nobody gets hurt feelings." Mooney, C. 2015. The science of protecting people's feelings: why we pretend all opinions are equal. Washington Post, March 10. http://www.washingtonpost.com/news/energy-environment/wp/2015/03/10/the-science-of-protecting-peoples-feelings-why-we-pretend-all-opinions-are-equal/ (accessed 27 August 2015). ***Equality bias*** thus refers to false value granted across *individuals'* positions or opinions whereas ***equivalency bias*** refers to a false worth assumed and then given to different socially-mediated frames, hypotheses, or other *collective* memes.

110 Some media bias was so one-sided that it betrayed itself through transparent hubris: "weekend bird enthusiasts, field guides and others produced reams of counter-evidence and arguments, and so completely dismantled each piece of ivory-bill evidence that few outside the thin-lipped professionals at Cornell still believed in the bird." See: Hitt, J. 2012. Science and truth: we're all in it together. *New York Times*, May 5, 2012 http://www.nytimes.com/2012/05/06/

opinion/sunday/science-and-truth-were-all-in-it-together.html?_r=0 (accessed 22 February 2015).

111 Comment by anonymous poster, July 17, 2006, at: http://tomnelson.blogspot.com/2006/07/so-what-wereare-these-people-thinking.html (accessed 19 August 2015).

112 Land set-asides are fully in keeping with a fundamental mission orientation of most conservation organizations. This particular tactic for protecting the ivory-bill essentially had been the only response available historically to address the perceived conservation threats to the woodpecker; see Chapters 5, 10. Consequently, those who ascribed some different, unseen motive to such a routine orientation fell to ***fundamental attribution error***. For example: "Cornell … and its individual members have much to gain from a position of certainty regarding the ivory-billed woodpecker's rediscovery." Moore.

113 *Ibid*. Resorting to a Nazi analogy often signifies the lost side of an argument.

114 *Ibid*. Also: "Just as importantly, our forum is to be a place of warmth and respect for all, quite independent of the debate issue. I like to think that our signal/noise ratio here is pretty good, and we appreciate everyone's restraint." Comment by fangsheath, March 31, 2010, at: http://www.ibwo.net/forum/showthread.php?t=67&page=4

115 A partial list of these web sources include: *Ivory-bill Researchers Forum* http://www.ibwo.net/; *Ivory-bills Live???!* http://ivorybills.blogspot.com/; *Project Coyote* http://projectcoyoteibwo.com/; and *Kints* http://www.ibwos.blogspot.com/ (all accessed 21 August 2015).

116 Wide and detailed coverage about the ivory-bill in near-real time occurred at *Notes from Soggy Bottom*, by the late Bill Pulliam http://bbill.blogspot.com/ (accessed 21 August 2015). Pulliam's site blended an entertaining variety of science, a propensity to go where the evidence led, and a willingness to change perspective if the conditions warranted. *Notes* provided a fascinating supplement

to the historical record covering the Arkansas incident.

117 The undue dispositive weight given to birding's ***argument from authority*** was revealed unintentionally by Jackson: "Although Sibley is not an academically trained scientist … his stature within the birding community contributed greatly to public understanding of the scientific issues associated with the ivory-billed woodpecker story." See: Jackson JA (2010) *Ghost Bird* – The Ivory-billed Woodpecker: Hopes, Dreams, and Reality. *PLoS Biol* 8(8): e1000459.

118 Comment by David Sibley, October 23, 2007, at: http://sibleyguides.blogspot.com/2007/10/ivory-billed-woodpecker-status-review.html accessed 27 May 2015).

119 Additional commentary by Ned Brinkley, available at: http://listserv.arizona.edu/cgi-bin/wa?A2=ind0603c&L=birdwg01&T=0&O=D&P=2444 (accessed 18 March 2006).

120 One can scarcely imagine the acrimony (or worse) that might have occurred had the Arkansas and/or Florida ivory-bill episodes taken place over social media platforms like Twitter and Facebook. Fifteen years later, in 2021, birders continue to direct ***ad hominem*** attacks and sarcastic derision at those who in good faith offer reasonable comments about extant ivory-bill. See, for example, the 34 pages and 600+ comments of mostly snarky responses to the initial entry here: https://www.birdforum.net/threads/ivory-billed-woodpecker-continued.403334/ (accessed 30 June 2021).

121 Additional commentary by David Sibley, footnote 111. This statement is a prime example of logical fallacy arising from the ***argument from ignorance***, i.e., making the claim that the ivory-bill is extinct because it cannot be proven to be not extinct, or that the ivory-bill is not extant because it cannot be proven to be extant. One blogger argued that evidence for the woodpecker must constitute "proof of the existence of an animal thought to be extinct." Moore. Note that this utter breakdown in logic requires that ***argument from ignorance*** be the evidentiary standard used to refute a prior error caused by ***argument from authority***!

122 *Ibid.*

Chapter 13

Guillotine Fence

There is nothing new under the sun but there are lots of old things we don't know.
—Ambrose Bierce

On a moonless winter night, a compact bundle of dull green and yellow dashes low over parched earth. Shrouded in utter darkness, the winged fugitive zigzags low over rocky outcrops scattered across ochre-tinged gibber plain in the remote outback of southwestern Queensland. Driven by thirst in the seasonally arid Channel Country, the young female darts just above meager shrubs and grass hummocks scattered along her flight path over a dissected tableland. As she approaches the head of a dry creek bed, her track straightens out and accelerates. Then, running full tilt into some concealed hazard, her trajectory stops abruptly. Never sensing the slightest danger, she plummets dead. A few stray feathers drift away in the desert wind. Head severed cleanly from the squat body; she falls victim to a decapitation from barbed wire fencing.[1]

In September 2006, ranger-on-duty Robert "Shorty" Cupitt focused his attention on restoring an uneven section of wilderness track in an isolated reach of Australia's Diamantina National Park. Suddenly his vehicle's grading blade exposed the desiccated remains of a small bird. Noticing an unfamiliar yellowish wash on the bird's belly, he dismounted, picked up the headless corpse for a closer inspection, but he could not pin down this species. Cupitt noticed a few matching feathers still clinging to a strand of nearby fence wire. Later the mummified specimen would find its way to the Queensland Museum where it was identified finally as an immature female night parrot

Pezoporus occidentalis,[2] one of the planet's rarest birds.[3] For just the second time in nearly a century, tangible evidence settled reasonable doubt that the night parrot was still here.

Image 13.1 Night parrot (*Pezoporus occidentalis*), an extremely rare Australian bird that shares a number of key historical and biological traits with the ivory-bill. Like the woodpecker, the seldom-seen night parrot has the distinctive ability to prompt odd reactions, thinking errors, and eccentric behaviors in our species (illustration by Megan Messa, used with permission).

Shrouded in mystery, the night parrot has long tormented us with riddles. A species endemic to Australia's vast arid interior, practically nothing is certain about its life history. In shape and appearance, it resembles a larger, dumpy budgie *Melopsittacus undulatus*. Adults are colored overall a rich green, barred and mottled black, dark brown, and yellow, with a distinctively stubby, yellowish tail. The night parrot also bears such evocative aliases as porcupine parrot (for living in prickly grass), midnight cockatoo, nocturnal ground parakeet, and solitaire.[4] Aborigines referred to it as "myrrlumbing" for a fancied similitude of the parrot's whistling notes to the uttered sound of that word.

During the day, night parrots conceal themselves in a burrow fashioned from a tussock of spinifex (*Triodia*). Only at night do they exit dense cover to seek food and water.[5]

Cultural and biological parallels between the night parrot and ivory-billed woodpecker are so curiously alike as to prompt nothing short of an unnerving awe. For starters, and identical to ivory-bill, our historical knowledge about night parrot is anchored almost entirely on the tireless pursuit of Victorian-era taxidermy and museum-based collecting (Chapter 3). Although first taken in 1845 near Coopers Creek during the Charles Sturt expedition to South Australia, the night parrot would not be described by science until 1861, and only then from a second specimen that had been collected in 1854.[6] Delays in the night parrot's formal naming arose because the type specimen was mistakenly sent to England under the name of its closest relative, the ground parrot *Pezoporus wallicus*.[7]

Real knowledge about night parrot was gleaned still later. Frederick W. Andrews, employed by the South Australian Museum as a taxidermist, knew the enigmatic parrot better than anyone. Of the few night parrots ever destined to reach museums,[8] Andrews secured at least 16 of the initial 20. When it came to tenacious pursuit of this rare bird, Andrews was evidently the night parrot's counterpart to Arthur T. Wayne and the ivory-bill (Chapter 7). Like Wayne's fervor with the woodpecker, Andrews' zeal for night parrots was driven by their high value relative to other bird specimens that might be traded among the competing museums.[9] Apparently tipped off by a night parrot captured live at a sheep station,[10] Andrews focused his most successful quests between 1867 and 1883 in the Gawler Ranges of South Australia, then newly settled by Europeans.

Waning decades of the nineteenth century marked the zenith for night parrot, as they had for ivory-bill. By 1890, despite intense efforts, Andrews' once reliable source for these birds had dried up. During 1892 at Alice Springs, so many night parrots were brought in serendipitously by cats that enough skins were available to adorn hanging picture frames.[11] Furnishing only a false hope for its continued survival, one of four night parrots flushed during a 1912 bird hunt in Western Australia was shot mistakenly as a pigeon, but this confirmation was later lost to science.[12] Then, the night parrot vanished. For the next 78 years, no more parrots were documented by any physical evidence whatsoever.

In concert with declines in specimen records, night parrots became less and less frequently sighted. Like ivory-bill, the Australian rarity was prone to vanish from areas it had once inhabited. And once the night parrot disappeared locally, it was then widely assumed to be gone everywhere. People who had been familiar with night parrot in the 1870s and 1880s could no longer find it.[13] Period ornithologists drew attention to the species as "missing."[14] By 1917, one of the most authoritative references on Australian birds declared the night parrot extinct.[15] "Following World War I, many Australian scientists sounded [its] death knell."[16] So, even the primary decade advanced for its putative extinction (1910s) linked night parrot and ivory-bill.

Contradictions (Chapter 7) about night parrot biology flourished too. Andrews warned: "numerous mistakes were made concerning its habits and economy which I have endeavoured to correct by many years of study and observation."[17] Night parrots were described as completely silent, for example, with respect to both vocalizations and wing noise.[18] Others thought the parrot gave short, sharp repeated notes delivered during flight at night.[19] Some claimed the bird marked its presence with a faint whistle,[20] others described calls as more substantive, long-reaching, and drawn-out.[21] Night parrots supposedly even croaked like frogs.[22] When at last they were recorded, the parrot's calls resembled either a bell-like "ding-ding" or a frog-like croaking.[23]

Habitat affinities averred for night parrot diverged, too, clouding views on veracity of sightings, range extent, and the species' degree of ecological specialization.[24] Whereas the ivory-billed woodpecker was anchored tightly to old-growth bottomland hardwood forest, the night parrot was usually fixed to old-growth unburned and lightly-grazed *Triodia* grassland in stony or sandy soils.[25] Nevertheless, unconfirmed reports placed night parrots in mallee shrubland with a spinifex understory,[26] samphire and chenopod shrubland dominated by *Atriplex, Bassia,* and *Maireana,*[27] or open *Eucalyptus* woodland with understory grasses.[28] Moreover, even caves[29] and vegetated margins of salt lakes and creeks[30] were also asserted to be used by this eccentric species.

Clashing accounts were given even for something as basic as ground locomotion in the parrot, matching the inconsistent reports of flight characteristics attributed to the ivory-bill (Chapter 7). Night parrots were believed to hop around on the ground, much like tiny kangaroos.[31] Yet later on, Metcalf and Davis directly contradicted this assertion, describing the motion of birds they watched at dusk at Minga Well, in the Pilbara of Western Australia, as "a definite run, more akin to that of a scurry and not ... hop.... They seemed well adapted to the ground and ran in short waddling bursts."[32]

In grasping to explain the parrot's obscure habits, a plausible and sophisticated model was proposed in which the bird moved between spinifex grasslands (in wet periods, when that habitat provided a flush of seeds) and denser chenopod shrublands (in dry seasons).[33] But reports of the night parrot in a variety of other habitat types did not conform neatly to this model,[34] signifying error from ***illusory correlation***. Because evidence pointed to a likelihood that night parrots are nomadic and occupy extremely large home ranges, it led ornithologists to "doubt that there is much conservation value in setting [particular areas] aside as a reserve for this species."[35] Like ivory-bill (Chapter 10), land set asides were frustrated as an easy conservation remedy for protecting Australia's avian rarity.

Night parrots live in one of the driest continents on earth. Most of Australia's vast interior is arid or semi-arid. In the outback, species must adapt to conditions that are notoriously extreme and unpredictable. In such taxing environments, the region's "animals must dig in, move or perish."[36] Tellingly, the parrot's fundamental lifestyle is linked to a burrowing "strategy," as it tunnels deep into prickly vegetation to seek protection from predators. As to the birds' inclination for itinerant wandering, Andrews wrote cryptically that night parrots:

> *come and go according to the nature of the season. When the early season is wet the porcupine grass flourishes and bears large quantities of seed, on which many birds feed; but if, on the contrary, the season is a dry one the grass does not seed, and no birds are to be seen.*[37]

Andrews' frugal narrative did not divulge whether he thought night parrots waned in abundance during poor seeding years, or instead left entirely for better foraging prospects elsewhere, returning only if and when environmental conditions might improve again. Like the ivory-bill, though, ultimate fate of night parrot may hinge on whether the species is chiefly sedentary or at minimum flexibly nomadic (Chapter 1).

Futile searches were the norm after keen interest was first ignited by tantalizing but invariably ephemeral encounters with ivory-bill (Chapter 11). An identical sequel was true of the night parrot. After three birds were seen at a water source on 12 April, 2005, in Western Australia, more searches failed to find the birds on subsequent nights or on follow-up trips later that year and the next.[38] Once a road-killed night parrot near Boulia in northwest Queensland reawakened public awareness of this avian icon in 1990,[39] concerted and repeated efforts were made to locate more living individuals, all to no avail. Although a flurry of reports was generated, most of these turned out to be decades old. Despite dedicated searches in several locations backed by two nationwide publicity campaigns,[40] not even a single extant population of the night parrot was found.[41]

Whereas the woodpecker could be "rarely met with,"[42] the night parrot "might not even be rare – just rarely observed."[43] Privileged observers of the legendarily-elusive parrot were usually those who spent substantial time in the remote outback: kangaroo hunters, ranchers, long-haul truckers, cattle drovers, and caffeinated drivers equipped with powerful searchlights. So, like ivory-bill, those persons most likely to encounter night parrot lived in the native home of the rarity, yet were also socially isolated from realizing the full conservation significance of their observations (***outcome reporting bias***).[44]

On the other hand, night parrot seemed uncannily adept at avoiding scientists and birders. "Apparently the harder you seek a night parrot, the less likely you are to observe it. Sightings [consist of] … no more than a glimpse that leaves the observer doubting their own senses."[45] Like the woodpecker, night parrot sidestepped authentication standards demanded by elite bird watching, i.e., repeated confirmation by "reliable" experts sent out to confirm the claims put forth by those perceived as less qualified. "There is something about night parrots that they have a bird watcher evasion system and are so only seen by non-birdwatchers."[46] Professional ornithologists, resorting to biological reasoning instead of mystical frustration, suspected that "by the time a report comes in and an expedition is planned, the nomadic parrots have decamped and are a long way off."[47]

Given that some records were perceived as valid while others not, a hefty dose of whim and fancy shaped what was considered reliable knowledge about the night parrot. By the mid 1990s, more than 70 reports[48] of night parrot from all Australian mainland states and the Northern Territory had produced "a number ... [that] appear reliable but cannot be verified."[49] A number of reports of night parrot from the Cloncurry region of Queensland since 1990 "have [not] been verified and are not currently accepted."[50] As with ivory-bill, tension between professionals and amateurs was part of a caprice from reliance on ***argument from authority***. "While many of these [night parrot] reports have come from unskilled observers, rather than experienced birdwatchers or trained ornithologists, some of these records have been accepted as genuine."[51]

Night parrot sightings were thus treated indiscriminately by authorities of the day,[52] as usually happened with the ivory-bill. Vague reports of night parrots in mallee habitat,[53] even of a nest with five eggs in the Wimmera plains region,[54] were rejected in later syntheses of the regional avifauna.[55] Discarding this otherwise valuable information was nonetheless perceived as risky: "acceptance of only those records that have been authenticated by photographs or specimens is doing little towards our understanding of the night parrot or assisting in its conservation."[56] The same authors conceded that "sightings of night parrot require a level of proof far greater than for other [presumably more common bird] species." This historical tendency to disregard all but consensually-verified, tangible evidence wielded irresistible sway over dictating the social perceptions we came to hold for both night parrot and ivory-bill (Chapter 11).

Due to the night parrot's alleged extinction, its observers had to be defended with special endorsement or fall to suspicion (***argument from authority***). In describing sightings of night parrots made between 1954 and 1959 by Evan Walton near Ross's Spring, Victoria, Menkhorst and Isles reasoned candidly why Walton's sightings ought not to be dismissed readily as mere ***confirmation bias***:

Mr. Walton has lived in the area all his life and is an experienced

> *bushman. He is familiar with local fauna, including the mallee parrots.... He is not a member of any natural history society and is unfamiliar with ornithological literature. Thus his observations and recollections have not been biased by preconceived ideas or published accounts.... His memory of the strange parrots and the area they occupied is clear and precise.*[57]

Five persons with night parrot sightings from Queensland in the early 1990s felt obliged to defend their reports, too. Acknowledging the rare parrot could "undoubtedly be the subject of misidentification," they still argued "several factors make this explanation of the sightings untenable."[58] None of the observers were aware of an earlier and nearby bird found dead in 1990; all but one observer was unaware of anyone else's sightings. A resilient ***anchoring bias*** set on extinction tended to place any fortuitous observers of either rare species on the defensive, compelling a proof of life instead of death.[59]

Australia's ghost bird "fascinated scientists, frustrated twitchers [birdwatchers] and inspired artists, poets and novelists for more than a century."[60] Searches for night parrot leaned toward the obsessive and could take on cultish overtones. Referring to locations where two night parrot corpses were retrieved in western Queensland: "pilgrimages are made to the sites of these events by amateur and professional seekers alike."[61] Night parrot was the Tasmanian tiger of the skies.[62] Like ivory-bill, night parrot was also a Holy Grail, dooming this species to silly allusion as well: "The only uncontested evidence of the species' continued existence, perversely, comes from (with apologies to Monty Python) a couple of ex-parrots."[63]

If "the bar at Winton is always a great place for a night parrot story," then tall yarns spun about the bird could make it appear even more inscrutable. "The night parrot, in its enigmatic obscurity, remains a creature capable of invoking strong passions, controversy, and visions of the sublime."[64] Unlike the

Americans, leveler heads down under conceded that at least the night parrot was "no longer a yowie [Australia's Bigfoot we're chasing."[65] Duplicating Gallagher and Harrison's show of emotion over seeing a large woodpecker in Arkansas, however, night parrot observer Chris Tzaros conceded: "It takes a lot to get me to shake, and my eyes were actually watering as well."[66]

When at last a night parrot was photographed alive in 2013 for the first time ever, the trustworthiness of its discoverer, "wild detective" John Young, was placed under pitiless scrutiny by the armchair brigade. Young produced sound recording and seconds of video that he played initially only for audiences at closed-door sessions. Provenance of this evidence was such that:

> *it may not be easy to prove Young's claims. He won't reveal where the photos were taken nor hand over the footage.... On top of that, Young himself is a rather controversial figure in ornithology. His supposed discovery of what he called the blue-fronted fig parrot in 2006 has been disputed because it was later discovered that he digitally altered his images.*[67]

One can only imagine howls of derision erupting had such tactics been used with the Arkansas woodpecker and Luneau video (Chapter 11). Young's controversial role in night parrot history evokes somewhat that of John Dennis and ivory-bill in the Big Thicket of Texas. Just as Dennis' finds were challenged bluntly by his skeptics, Young was accused of having "assembled members of the media and told them something they should not have believed."[68] Exactly like ivory-bill, the night parrot "is a bird that is held up to exceptionally, and uniquely, high evidentiary standards."[69]

Young's detractors zeroed in on evidence seen as most vulnerable: heavily copyrighted, digitally altered photograph(s) of night parrot. Detractors did not hesitate to deploy ***ad hominem, fundamental*** or ***ultimate attribution error*** to hammer their points.[70] "As you say – we are entitled to question, especially given the man's history. As I see it, many respected scientists are questioning certain behavioural aspects of the finder, such as suppressing the call and altering images. This is scientific debate, NOT casting aspersions on the validity of the find."[71] Once again, what was chiefly an issue of bird identification got framed as high-brow science, sowing confusion and potential for conflict among different social interests (conservation, science, birding). As the most vocal critics of the Arkansas ivory-bill had done, "John Young's protagonists now try to justify their personal nasty stuff as quasi scientific discussion ... the irony of it all is that most of their comments are based on completely unsupported speculation."[72]

Rediscovery of the night parrot thereby generated much wider media attention than was customarily the case with other lost species, in line with our strong cultural reactions to the ivory-bill. After a road-killed night parrot was picked up by chance during a museum expedition to western Queensland, the phones rang off the hook for the unintentionally lucky ornithologists who made the find.[73] The 2013 announcement about finally obtaining images and sound recording of this elusive species also received wide acclaim. The story was covered by the Australian Broadcasting Company, national and international media, science-oriented magazines, and of course bloggers.

As with the woodpecker, media coverage of the night parrot's rediscovery could be pungent with fabrication, clichéd comparison, and excess literary license. Sources announced wrongly that no definitive evidence of the species had been gathered since the 1880s. The 2013 rediscovery by bushman John Young was often portrayed as the first in over a century.[74] Another false claim was so precise with its misleading as to inform readers that the night parrot had not been confirmed as a living presence for exactly 101 years.[75]

Night parrot was even "regarded as the ghost bird of the outback."[76] Detecting it at long last was like finding a needle in a haystack. "It's got this myth about it," said Max Tischler of Bush Heritage Australia, "only a handful of specimens were collected in the late 1800s and then it largely disappeared."[77] *Australian Birdlife Magazine* editor Sean Dooley could not help blurting out: "If John Young's claims prove to be correct this is the world bird-watching equivalent of finding Elvis flipping burgers in an outback road house."[78] Remarked film-maker Rob Nugent: "I like the mystery of the bird, how it lives only an inferred existence."[79] And how could the media possibly resist yet another tired metaphor for a "bird that's back from the dead"?[80]

Rediscovery of night parrot was also veiled in furtiveness. Some discretion was motivated by the same alarms expressed after finding ivory-bill – the birds' welfare.[81] Given concerns for the parrot, "fragile environments at the locality, if revealed, could be damaged by well-meaning but perhaps over-enthusiastic birdwatchers."[82] Steve Murphy explained: "For as long as we can, we'll keep it as secret as we can. It's just such a critical thing that we do everything that we can to save this species..."[83]Audio clips of the call were at first not released publicly, in order to prevent birders from using them to lure out the secretive night parrot.[84] There was worry, too, that the parrot and eggs might get poached for trading in the lucrative global wildlife market. Consequently, "the news has ruffled some feathers in the bird-watching community because the photographer [John Young] remains coy on some of the details."[85]

Indeed, night parrot sagas pitted the interests of conservation (and to a lesser extent science) against the desires of keen birdwatchers. Birding experts were especially frustrated about delays by the Queensland Environmental Protection Agency (QEPA) in announcing the 2006 find of a dead night parrot. Some described the agency's conduct as "paranoid," others faulted its "incredible secrecy." In truth the bird itself was sowing confusion – follow-up surveys had been carried out, but with no success. In response to its critics, a QEPA spokesperson called the whole controversy a "media beat-up."[86]

Rampant speculations about the ivory-bill and night parrot also led to pronouncing a premature demise for each species. When the first night parrot in almost eight decades was recovered dead in 1990, some still hazarded wildly that "the ... specimen could have been the last survivor."[87] Much the same was said about a single putative ivory-bill found in Arkansas in 2004 and 2005. That each species may have/has persisted for so long just below our perceptual radar, however, allows also for a kind of unbridled optimism to be circulated about the prospects of each species.[88]

Lack of robust, repeatable detection methods deterred our ever-gaining reliable knowledge about night parrot, the same constraint to what we never knew about ivory-bill. "The absence of a survey technique for the [night parrot] means that even when a sighting is made it is difficult to learn anything more about the species, such as preferred habitat, population density, diet and patterns of movement."[89] Despite the highly sophisticated instrumentation used to detect these two rarities,

Box 13.1. ***Pro-innovation bias*** refers to a contemporary tendency to inflate our real ability to acquire any knowledge or data at will. It is a thinking error promulgated by the mistaken belief that novel technology can overcome any obstacle to our securing accurate information about the natural world, thus neglecting to properly account either for the limitations we cannot yet fathom (meta-ignorance) or the true likelihood of some phenomenon given an unknown probability.

including autonomous recording units (ARUs), remote time-lapse cameras, and computer sound-recognition algorithms,[90] our state-of-the-art technologies have not been always successful (see ***pro-innovation bias***).[91] Studying the night parrot "has been a frustrating exercise."[92]

Night parrot disappeared during an era when social awareness of conservation threats and how to abate them was primitive – again, identical to ivory-bill. Causes for the parrot's decline can only be guessed: predation by feral cats and foxes, degradation of arid habitat by intense fires, excess grazing by stock or rabbits, including competition with these herbivores, and declines in waterhole quality. In truth, however, "there is no direct evidence to link any threatening process to the apparent decline of the species."[3]

Rarity, elusiveness, and low density of the parrot and woodpecker warn us how inept can be our attempts at forecasting a species' actual extinction (Chapter 14). The interval between mustering physical evidence for night parrot was 78 years (1912–1990), just a bit longer than the time span (69 years) between acquiring motion film evidence for ivory-bill (1935–2004). Unverified reports of ivory-bills[93] and night parrots[95] were made nevertheless every single decade of the late twentieth century. For each species, high mobility and scarce encounters hindered even a reckoning of the number of sites occupied. Expansive distributions across inhospitable terrain with sparse human settlement, cryptic or wary habits, and potential for nomadic behavior render the statistical likelihood of our finding either parrot or woodpecker exceedingly small.[96] John Young spent 17,000 hours driving 325,000 km in the outback for 15 years before ever obtaining any images and recordings of the parrot.[97]

Our cultural bonds with the ivory-bill and night parrot also remain conspicuously similar. Night parrot "triggered heated debate among conservationists, birders and scientists alike."[97]

The night parrot's historical decline is based on "contemporary population parameters [that] are essentially guesswork."[99] The parrot's apparent reduction "may have had as much to do with Andrews death in 1884 as with actual changes in abundance of the species,"[100] with the result that it is "impossible to be sure if the [disappearance] was less parrots or less observers."[101] The same can be said of the ivory-bill. Just as the vast outback guards well a sense of mystery inside remoteness,[102] so too does the wild expanse of southern swamp forest hold close its own secrets.

Like the woodpecker, the night parrot dwells in a cognitive realm where we are prone to deceive ourselves with a false confidence about how much we think we know.[103] Both chance and necessity, enabled by a peculiar brand of "cultural amnesia," prompt us to create icons out of our extinct or near-extinct species.[104] "What also makes the apparent survival of night parrots interesting is that it seems to verify the various eyewitness reports made prior to the discovery of the 1990 and 2006 specimens. People who see 'extinct' animals can't always be right, but perhaps they sometimes are."[105] Each of these two birds seduce us with ***truthiness***, leading us to form convenient story-book narratives to disguise our real impotence. Self-deception, guesswork, and failure to achieve a social consensus, however, make for mighty poor justifications to render final judgment about the obliteration of an entire species.

Endnotes

1 The bird was an immature female undergoing post-juvenile molt, indicating in this case that the night parrot also had reproduced in the very recent past. See: Cupitt, R., and S. Cupitt. 2008. Another recent specimen of the night parrot *Pezoporus occidentalis* from western Queensland. *Australian Field Ornithology* 25: 69–75; McDougall, A., G. Porter, M. Mostert, R. Cupitt, S. Cupitt, L. Joseph, S. Murphy, H. Janetski, A. Gallagher, and A. Burbidge.

2009. Another piece in an Australian ornithological puzzle – a second night parrot is found dead in Queensland. *The Emu* 190: 198–203.

2 Although it was originally placed in the monotypic genus *Geopsittacus* by Gould, evidence from molecular comparisons indicate that the night parrot is closely related to ground parrot *Pezoporus wallicus*. See: Leeton, P.RJ., L. Christidis, M. Westerman, and W.E. Boles. Molecular phylogenetic affinities of the night parrot (*Geopsittacus occidentalis*) and the ground parrot (*Pezoporus wallicus*). *The Auk* 111: 833–843. *Pezoporus* is in turn closely allied to *Neophema* and *Neopsephotus*, a biogeographically cohesive clade of platycercine parrots of Australia and the Pacific. See: Joseph, L., A. Toon, E.E. Schirtzinger, and T.F. Wright. 2011. Molecular systematics of two enigmatic genera *Psittacella* and *Pezoporus* illuminate the ecological radiation of Australo-Papuan parrots (Aves: Psittaciformes). *Molecular Phylogenetics and Evolution* 59: 675–684.

3 BirdLife International. 2015. Species factsheet: *Pezoporus occidentalis*. http://www.birdlife.org (accessed 22 September 2015).

4 Olsen, P. 2009. Night parrots: fugitives of the inland. Pp. 121–144 in *Boom and Bust: Bird Stories for a Dry Country* (Robin, L., R. Heinsohn, and L. Joseph, Eds.). Commonwealth Scientific and Industrial Research Organization (CSIRO), CSIRO Publishing, Collingwood, Victoria, Australia.

5 Andrews, F.W. 1883. Notes on the night parrot. *Transactions and Proceedings of the Royal Society of South Australia* 6: 29–30. Because they came mostly from one era and a single region, Andrews' observations were prone to later errors from ***primacy effect, anchoring bias***, and the ***availability cascade***. So few observations of the night parrot were ever made that any bits of evidence tended to take on an exaggerated importance, just as had Tanner's observations of the ivory-bill.

6 Forshaw, J.M., P.J. Fullagar, and J.I. Harris. 1975. Specimens of the night parrot in museums throughout the world. *The Emu* 76:

120–126.

7 *Ibid.*

8 As of 2012, only some 28 specimens were known to have ever existed. See: Black, A.B. 2012. Collection localities of the night parrot *Pezoporus (Geopsittacus) occidentalis* (Gould, 1861). *Bulletin of the British Ornithologists' Club* 132: 277–282.

9 Olsen, p. 129.

10 Paton, J.B. 1975. Birds of the Gawler Ranges, South Australia. *South Australian Ornithologist* 26: 180–192.

11 Olsen, p. 135.

12 Wilson, H. 1937. Notes on the night parrot, with references to recent occurrences. *The Emu* 37: 79–87.

13 Whitlock, F.L. 1924. Journey to central Australia in search of night parrot. *The Emu* 23: 248–281.

14 Campbell, A.J. 1915. Missing birds. *The Emu* 19: 167–168.

15 Mathews, G. 1917. *The Birds of Australia, Vol. 6,* Witherby and Co., London, pp. 495–498.

16 Weidensaul, S. 2002. *The Ghost with Trembling Wings: Science, Wishful Thinking and the Search for Lost Species.* Farrar, Straus and Giroux, New York, pp. 75–76.

17 *Ibid.*

18 Davis, R.A., and B.M. Metcalf. 2008. The night parrot (*Pezoporus occidentalis*) in northern Western Australia: a recent sighting from the Pilbara region. *The Emu* 108: 233–236.

19 Boucher, N.J., A. Burbidge, and M. Jinnai. 2008. Computer recognition of sounds that have never been heard before. *Australian Institute of Physics 18th National Congress* 211: 1–4.

20 Gould, J. 1867. *The Birds of Australia: Supplement.* Published by author, London, Part 4, opposite plate 66.

21 Bourgoin, as cited in Wilson.

22 Higgins, P.J. 1999. *Handbook of Australian, New Zealand and Antarctic Birds, Vol. 4, Parrots to Dollarbirds.* Oxford University Press, Melbourne.

23 Slezak, M. 2016. Australian night parrot legend lives on but bird

remains as elusive as ever. The Guardian. http://www.theguardian.com/environment/2016/apr/17/australian-night-parrot-legend-lives-on-but-bird-remains-as-elusive-as-ever (accessed 1 May 2016).

24 A full range extent for night parrot is speculatively estimated to reach up to 2–3,000,000 km². See: Department of the Environment. 2015. *Pezoporus occidentalis* in Species Profile and Threats Database, Department of the Environment, Canberra. http://www.environment.gov.au/sprat (accessed 17 September 2015). Allowing that ivory-bill might occur in some habitats adjacent to bottomland hardwood forests, a total range extent of about 750,000 km² (i.e., 1/4 of the night parrot's Australian range extent) encompasses the more suitable portions of Texas, Louisiana, Arkansas, Tennessee, Mississippi, Alabama, Florida, Georgia, South and North Carolina.

25 North, A.J. 1898. List of birds collected by the Calvert Exploring Expedition in Western Australia. *Transactions of the Royal Society of South Australia* 22: 125–192; McGilp, J.N. 1931. *Geopsittacus occidentalis*, night-parrot. *South Australian Ornithologist* 11: 68–70. If night parrot relies on habitat other than spinifex, contrary to the prevailing paradigm, the excess reliance on the first study or even a few studies (***anchoring bias***) would also constitute ***insensitivity to sample size***.

26 Menkhorst, P.W., and A.C. Isles. 1981. The night parrot *Geopsittacus occidentalis*: evidence of its occurrence in north-western Victoria during the 1950`s. *The Emu* 81: 239–240.

27 Powell, B. 1970. The night parrot. *South Australian Ornithology* 25: 208.

28 Garnett, S., G. Crowley, R. Duncan, N. Baker, and P. Doherty. 1993. Notes on live night parrot sightings in north-western Queensland. *The Emu* 93: 292–296.

29 Schodde, R., and I.J. Mason. 1980. *Nocturnal Birds of Australia*. Lansdowne Publishing, Melbourne.

30 Andrews; McGilp; Wilson. By 2021, investigators had indeed found populations of night parrot around salt lakes in Western

Australia (Michelmore, K. and L. Birch. 2020. Night parrot located by KJ rangers on Martu country in the Pilbara. https://www.abc.net.au/news/2020-08-21/elusive-night-parrot-discovered-in-the-pilbara/12581900). Consequently, a so-called "habitat specialist" does in fact use rather distinct landscapes at opposite ends of that continent, one habitat characterized by old-growth spinifex grasslands near escarpments and another dominated by vegetation surrounding flatter saline wetlands.

31 Comparti, N. 2013. DNA confirms elusive night parrot found. Western Australian Museum. http://museum.wa.gov.au/about/latest-news/dna-confirms-elusive-night-parrot-found (accessed 20 September 2015).

32 Davis, R.A., and B.M. Metcalf. 2008. The night parrot (*Pezoporus occidentalis*) in northern Western Australia: a recent sighting from the Pilbara region. *The Emu* 108: 233–236.

33 Forshaw, J.M. 1982. *Australian Parrots*. Lansdowne Editions, Melbourne, Victoria; Blakers, M., S.J.J.F. Davies, and P.N. Reilly. 1984. *The Atlas of Australian Birds*. Melbourne University Press, Melbourne, Victoria.

34 Department of the Environment 2015.

35 Boles, W.E., N.W. Longmore, and M.C. Thompson. 1994. A recent specimen of the night parrot *Geopsittacus occidentalis*. *The Emu* 94: 37–40.

36 Olsen, p. 121.

37 Andrews, p. 30.

38 Davis and Metcalf, p. 234. It would take *16 years* before additional evidence in the form of photograph and sound recordings would corroborate Davis and Metcalf's sighting at this region in Western Australia (Young, E. 2021. FMG confirms population of elusive night parrot at WA iron ore mine. The Sydney Morning Herald. https://www.smh.com.au/national/fmg-confirms-population-of-elusive-night-parrot-at-wa-iron-ore-mine-20210625-p5849l.html).

39 Boles, W., W. Longmore, and M. Thompson. 1991. The fly-by-night parrot. *Australian Natural History* 23: 689–695.

40 Garnett, S.T., J.K. Szabo, G. Dutson. 2011. *The Action Plan for Australian Birds 2010*. CSIRO Publishing, Melbourne.

41 See, for example: Blyth, J., and W. Boles. 1999. An expedition to the Murchison, Gascoyne and East Pilbora areas to search for night parrots, November 1997. *Eclectus* 6: 12–16; Blyth, J., A.H. Burbidge, and W. Boles. 1997. Report on an expedition to the Western Desert and East Pilbara areas in search of the night parrot *Pezorus occidentalis*. *Eclectus* 2: 25–30.

42 *Field & Stream* 14: 407, a comment by "Yell" when referring to the tendency of the ivory-bill to inhabit "unfrequented swamps" in Arkansas.

43 Weidensaul, p. 76.

44 ***Outcome reporting bias*** and ***sunk cost fallacy*** are offered as the sensible reaction by some Australian birders to the prospect of expending great effort in searches for the night parrot. As one keen birder confessed: "I've never gone looking for the night parrot and doubt that I'd ever have enough time, money or commitment ... to look for a bird that no-one will believe I have seen...." Gosford, B. 2010. Bird of the week – the night parrot resurfaces...again... maybe. *Crikey*. http://blogs.crikey.com.au/northern/2010/06/09/bird-of-the-week-the-night-parrot-resurfaces-again-maybe/ (accessed 22 September 2015).

45 Nugent, R. 2015. *Night Parrot Stories*. https://www.facebook.com/NightParrotStories/info?ref=page_internal (accessed 15 September 2015).

46 Comment by Cas Liber, at *Birding-Aus*, 17 February 2007. http://bioacoustics.cse.unsw.edu.au/archives/html/birding-aus/2007-02/msg00293.html (accessed 21 September 2015).

47 Boles, as quoted in Weidensaul, p. 79.

48 A rate of 70 reports of night parrots across a 78-year time span is not altogether different from the more than 100 similar reports of ivory-bill in the last 80 years. See, for example: Hunter, W.C. 2010. *Interpreting historical status of the ivory-billed woodpecker with recent evidence for the species' persistence in the southeastern United States.*

Appendix E, Recovery Plan for the Ivory-billed Woodpecker (*Campephilus principalis*), U.S. Fish & Wildlife Service, Atlanta, GA.

49 Boles et al., p. 37.

50 Department of the Environment 2015.

51 Fortescue Metal Group. 2005. Pilbara Iron Ore Project, night parrot (*Pezoporus occidentalis*) management plan., p 8. http://reports.fmgl.com.au/ENVIRO/Cloudbreak%20PER/Appendix%20L%20Night%20Parrot%20Management%20Plan.pdf (accessed 20 September 2015).

52 Menkhorst and Isles, p. 239.

53 Campbell, A.J. 1897. Ornithological notes. *Victorian Naturalist* 14: 91–92.

54 Menkhorst and Isles, p. 239.

55 Wheeler, W.R. 1967. *A Handlist of the Birds of Victoria*. Victoria Ornithological Group, Melbourne.

56 Garnett et al. 1993, p. 294.

57 Menkhorst and Isles, p. 239.

58 Garnett et al. 1993, p. 294.

59 Some Australians caught on to misguided attempts to exploit an ***argument from ignorance*** in order to discredit the re-discovery of night parrot. "Correct me if I'm wrong, but [John Young] doesn't have to prove the [night parrot] still exists, as it was never declared extinct." Blogger John Leonard at *Birding-Aus*. http://bioacoustics.cse.unsw.edu.au/birding-aus/2013-10/msg00243.html (accessed 15 September 2015).

60 Huxley, J. 2007. Twitchers cry foul in case of the deceased parrot. *Brisbane Times*. http://www.brisbanetimes.com.au/news/national/twitchers-cry-foul-in-case-of-the-deceased-parrot/2007/06/22/1182019367467.html (accessed 15 September 2015).

61 McCloy, M. 2013. Found but teetering and still enigmatic. http://citizenj.edgeqld.org.au/found-but-teetering-and-still-enigmatic/ (accessed 20 September 2015).

62 Platt, J.R. 2013. After 100 years, has the elusive night parrot

finally been discovered? *Scientific American*. http://blogs.scientificamerican.com/extinction-countdown/night-parrot-discovered/ (accessed 15 September 2015); Huxley.

63 Stafford, A. 2012. Is there life yet in these ex-parrots? The Age http://www.theage.com.au/victoria/is-there-life-yet-in-these-exparrots-20120604-1zs82.html (accessed 15 September 2015).

64 Synopsis of the film *Night Parrot Stories* (2015), by Robert Nugent. This Australian documentary film took a near identical perspective as did *Ghost Bird* for the ivory-bill in the U.S., focusing especially on the contradictory and often illogical connections that we forge with a bird flirting with (or over) the edge of extinction. http://www.screenaustralia.gov.au/find-a-film/detail.aspx?tid=31893 (accessed 15 September 2015).

65 Watson, C. 2015. Night parrot – a possible sight record from the Northern Territory. *The Grip*. http://www.chriswatson.com.au/blog/night-parrot-a-possible-sight-record-from-the-northern-territory (accessed 21 September 2015).

66 Stafford.

67 Platt. Young remained controversial in matters related to night parrot (Borrell, B. 2018. A Naturalist With a Checkered Past Rediscovered a Long-lost Parrot . . . Then Things Got Interesting. https://www.audubon.org/magazine/fall-2018/a-naturalist-checkered-past-rediscovered-long-lost). Irregularities in Young's evidence for additional populations eventually led an independent review panel to retract much of his parrot work with the Australian Wildlife Conservancy (Borrell, B. 2019. John Young Rediscovered the Australian Night Parrot, but Did He Lie About His Later Findings? https://www.audubon.org/news/john-young-rediscovered-australian-night-parrot-did-he-lie-about-his-later).

68 Lund, N. 2013. Crying wolf, and auk, and Tasmanian tiger: fake sightings of probably extinct species. *Slate Magazine*. http://www.slate.com/articles/health_and_science/science/2013/09/fake_wildlife_sightings_probably_extinct_night_parrot_thylacine_auk_and.html (accessed 2 September 2015). As the largely clueless

European bloggers did with the American ivory-bill, Lund, who writes for this magazine out of Washington, D.C., relied on ***ad hominem*** to discredit the discovery made by Young, seemingly ignorant of the fact that two dead night parrot specimens confirming the species' extant status had already surfaced just a few years before.

69 Watson.

70 Some birders in the blogging community realized, albeit unintentionally, that ***fundamental attribution error*** was being committed by critics. *Birding-Aus* blogger David Adams drew particular attention to the situational nature of the find: "For whatever reason, or combination of reasons good and/or bad, the night parrot discovery has been handled in a novel manner." http://bioacoustics.cse.unsw.edu.au/birding-aus/2013-10/msg00280.html (accessed 15 September 2015).

71 Comment by blogger Geoff Price at *Birding-Aus*, 15 October 2013. http://bioacoustics.cse.unsw.edu.au/birding-aus/2013-10/msg00251.html (accessed 15 September 2015).

72 Comment by blogger Ian May at *Birding-Aus*, 15 October 2013. http://bioacoustics.cse.unsw.edu.au/birding-aus/2013-10/msg00258.html (accessed 15 September 2015). At times the kinds and intensity of animosity that the night parrot instigated among birding bloggers was, again, nearly identical to that witnessed with the Arkansas ivory-bill. However, the scope of this debate was not nearly as large, perhaps because Australia has fewer birders than reside in the U.S. Also the debate was not as protracted, as either the moderators or the commentators tried with some success to tone down the acrimony. For example: "It is very very evident that nearly everyone, as I am, is thoroughly tired of the continued verbal assault." Comment by blogger Lloyd Nielsen at *Birding-Aus*, 15 October 2013. http://bioacoustics.cse.unsw.edu.au/birding-aus/2013-10/msg00247.html (accessed 15 September 2015).

73 Weidensaul, p. 78.

74 Ong, T. 2013. Night parrot, nocturnal bird in Australia, seen alive

for first time in over a century. http://frontiersofzoology.blogspot.com/2013/07/night-parrot-nocturnal-bird-in.html (accessed 15 September 2015).

75 McCloy.

76 http://www.wildlifeextra.com/go/news/night-parrot013.html#cr (accessed 15 September 2015).

77 Pickrell, J. 2013. Night parrot: tantalising clues revealed. *Australian Geographic*. http://www.australiangeographic.com.au/news/2013/07/night-parrot-tantalising-clues-revealed/ (accessed 15 September 2015).

78 Worthington, E. 2013. Queensland bird enthusiast presents museum with photos of elusive night parrot. Australian Broadcasting Company. http://www.abc.net.au/news/2013-07-03/man-claims-to-have-filmed-the-rare-australian-night-parrot/4796342 (accessed 20 September 2015).

79 Brain, C. 2013. Solving the night parrot mystery. http://www.abc.net.au/news/2013-05-07/ntch-mystery-night-parrot/4674266 (accessed 20 September 2015).

80 Pickrell, J. 2015. Extinct no more: protecting the night parrot. *Australian Geographic*. www.australiangeographic.com.au/blogs/austropalaeo/2015/07/protecting-the-night-parrot/ (accessed 21 September 2015).

81 Some birders tried to steer the debate about rediscovery of night parrot away from science or mere documentation, however, instead calling for a focus on its conservation. "this debate should be centred around ... the welfare of the night parrot. Too many people forget that this endangered species is not the property of any individual, however noteworthy their efforts in the field may be." http://bioacoustics.cse.unsw.edu.au/birding-aus/2013-10/msg00324.html (accessed 15 September 2015).

82 Joseph, L. 2013. Found: world's most mysterious bird, but why all the secrecy? *The Conversation*. http://theconversation.com/found-worlds-most-mysterious-bird-but-why-all-the-secrecy-18000 (accessed 20 September 2015).

83 Pickrell 2013.

84 Elusive parrot caught in Outback. 2015. *Daily Telegraph UK*. http://www.nzherald.co.nz/world/news/article.cfm?c_id=2&objectid=11495619 (accessed 20 July 2021).

85 Worthington. Although Young's "discovery proved controversial, with some critics claiming photographs had been retouched," other night parrots were later confirmed in the area of his original finding. This habitat was subsequently protected, including implementation of limits to human access and a control program to prevent the predation of night parrots by feral cats. *Daily Telegraph UK*. http://www.nzherald.co.nz/world/news/article.cfm?c_id=2&objectid=11495619 (accessed 21 September 2015).

86 Huxley.

87 Weidensaul, p. 79.

88 McKechnie, A. 2014. And along came Polly: Australia's feathered phantom returns. *African Birdlife* 2: 18.

89 Fortescue Metal Group.

90 Boucher et al.; Song, D., et al. 2008. System and algorithms for an autonomous observatory assisting the search for the ivory-billed woodpecker. *IEEE International Conference on Automation Science and Engineering* 2008: 200–205. Autonomous recording units (ARUs) miss entirely a large proportion (up to 40%) of the bird sounds that are detected and recorded by human observers. Along with the extra time and cost associated with processing this information, ARUs "would not provide a cost-effective means of gathering data." See: Hutton, R.L., and R.J. Stutzman. 2009. Humans versus autonomous recording units: a comparison of point-count results. *Journal of Field Ornithology* 80: 387–398.

91 Rogers, E.M. 1976. New product adoption and diffusion. *Journal of Consumer Research* 2: 290–301.

92 Joseph. Despite deployment of 30 camera traps at the Queensland site, a location known certainly to have a population of night parrots, one of which was mist-netted and fitted with a radio transmitter, not a single photograph of the species had been secured

with these devices during the first 15,000 hours of operation. See: http://sunshinecoastbirds.blogspot.com.au/2015/08/night-parrot-news_29.html (accessed 21 September 2015). Thus the following could just as easily be written of ivory-bill: "John's story also demonstrates that even once the presence of the bird [night parrot] is confirmed at a particular site, it is still highly unlikely that you will be able to observe it in the open. The possibility of obtaining photographic evidence of a sighting, especially a chance sighting while driving, is so remote that it needn't even be considered." Watson.

Nevertheless, once vocalizations of night parrots were conclusively identified, sound recording units placed strategically in promising habitat have expanded considerably the known distribution and/or population size of this extremely rare bird. Collins, B. 2021. New recordings of critically endangered night parrots music to ears of Kimberley rangers, scientists. https://www.abc.net.au/news/2021-05-26/biggest-night-parrot-population-discovered-great-sandy-desert/100159378

93 Department of the Environment.

94 Hunter.

95 Garnett et al. 1993, p. 292.

96 Boles et al., p. 37.

97 Worthington; McKechnie.

98 McKechnie.

99 Garnett et al. 2011.

100 Blyth, J.D. 1996. Night parrot (*Pezoporus occidentalis*) Interim Recovery Plan for Western Australia, 1996 to 1998. Department of Conservation and Land Management, Wanneroo, Western Australia.

101 Fraser, I. 2013. That benighted parrot. *Ian Fraser, Talking Naturally.* http://ianfrasertalkingnaturally.blogspot.com/2013/07/that-benighted-parrot.html (accessed 2 September 2015).

102 Until just recently, and despite broad familiarity with Australia's natural landscapes, night parrots had remained unknown even

to some of the continent's indigenous peoples. Aboriginal elders heard nocturnal noises that they later learned emanated from these parrots, but they didn't know what they were, and at first would not investigate them more closely because the sounds were "similar to an evil spirit noise." Collins.

103 McKechnie.

104 Turvey, S.T., and A.S. Cheke. 2008. Dead as a dodo: the fortuitous rise to fame of an extinction icon. *Historical Biology* 20: 149–163.

105 Naish, D. 2009. The 2006 parrot: dead, decapitated, evidence for collision with a fence … but otherwise the news is good. *TetrapodZoology Science Blog*. http://scienceblogs.com/tetrapodzoology/2009/10/06/2006-dead-night-parrot/ (accessed 31 August 2015).

Chapter 14

Twain's Rebuke

Prediction is very difficult, especially if it's about the future.
—Niels Bohr and others (apparently an old Danish proverb)

Manifest similarities in biology of night parrot and ivory-bill establish that the woodpecker's fate need not be viewed as especially remarkable contrasted to other very rare birds.[1] For two species having such parallel traits,[2] neither should we be startled to unearth such close matches in our cognitive and cultural reactions. Despite many likenesses, though, anecdotal resemblance[3] is not enough to gauge a prospect of the ivory-bill's survival. Instead, we must comprehend how factors like low population size, enigmatic habits, and nomadic distribution might dictate what we know about the woodpecker (Chapter 1), but also how certain we can be in believing that we properly grasp our true level of knowledge.

For that task, numerical corroboration may give us some insight. This rationale embraces plausible models that explore scenarios about what *could* be true about the woodpecker. We may resort to statistical consultation, including estimates for the odds of detection and survival (or extinction) in the ivory-bill. Probability[4] lends us a standard language in which to evaluate and discuss the putative legitimacy of the communal beliefs we hold about this species' fortune.[5] Yet given the woodpecker's many contradictions and mysteries, should we just trust at face value the results obtained from such models?

Only to a point. How much trust we grant such numerical appraisal depends on the data used, the model assumptions, and the intent for the model output. Models in general, statistical ones in particular, exemplify an old adage: "garbage in, garbage out." Mark Twain expressed his characteristically witty misgivings over our penchant to bend mathematics to suit our preordained conclusions: "First get your facts; and then you can distort them at your leisure."[6] And when it came to this skepticism over use of applied statistics, Twain was in outstanding company:

"Facts do not cease to exist because they are ignored."—Aldous Leonard Huxley

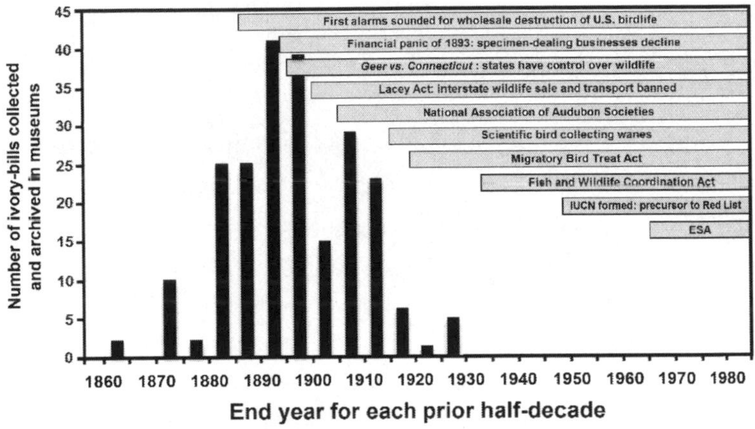

Image 14.1 Ivory-billed woodpecker specimens as collected during half-decade intervals, 1865–1930, contrasted with the dates of implementation for selected legal, financial, and social changes taking place in the American experience. Specimen numbers taken and adapted from: Jackson, J.A. 2002. Ivory-billed Woodpecker (*Campephilus principalis*), The Birds of North America Online (A. Poole, Ed.). Ithaca: Cornell Lab of Ornithology; Retrieved from the Birds of North America Online: http://bna.birds.cornell.edu/bna/species/711 (accessed 21 April 2015).

"The purpose of models is not to fit the data but to sharpen the question."—Samuel Karlin

"For every complex problem, there is a solution that is

simple, neat, and wrong." —Henry Louis Mencken

"Telling the future by looking at the past assumes that conditions remain constant. This is like driving a car by looking in the rearview mirror." —Herb Brody

"The combination of some data and an aching desire for an answer does not ensure that a reasonable answer can be extracted from a given body of data." —John W. Tukey

"Statisticians, like artists, have the bad habit of falling in love with their models." —George E. P. Box

"The statistician cannot evade the responsibility for understanding the process he applies or recommends." —Sir Ronald A. Fisher

"When evaluating a model, at least two broad standards are relevant. One is whether the model is consistent with the data. The other is whether the model is consistent with the 'real world.'" —Kenneth A. Bollen

"All models are wrong, but some models are useful." —George E. P. Box

After wide interest in the woodpecker was ignited by the Arkansas saga (Chapter 12), several studies sought to evaluate quantitatively the prospects for a still-living ivory-bill. Most claimed that based on model projections that gave very low probability for its persistence, the bird could be verified now as extinct. With consummate audacity, Gotelli and colleagues opined that their study[7] and two others[8] showed that "there is virtually no chance the ivory-billed woodpecker is extant … in the southeastern United States," and, moreover, "all point to the inescapable conclusion that the ivory-billed woodpecker is now extinct."[9] Several studies even projected a time of extinction for the woodpecker, with wild guesses for its demise ranging from as early as 1941 to "no later than 2010."[10]

Evaluating these various claims for ivory-bill extinction does not require our plunging into computational details. Rather, the decisive tests of legitimacy are had solely by dissecting the origins and quality of data used for model input. In each case it was the historical record formed by museum specimens and (sometimes) arbitrarily culled sight reports. A chronological series of ivory-bill collections, seemingly our only picture for the bird's historical abundance (***Image 14.1***), led to conclusions that the bird was extinct. Each study constituted a kind of time series in which sequential pattern(s) were identified in the data; an earlier (and arbitrarily-chosen) part of the series then generated expectations for the more recent parts of the sequence.

Box 14.1. After putative ivory-billed woodpeckers were reported from Arkansas and Florida in the twenty-first century, researchers believed they could test for statistical odds that the bird was still alive using specimen collection records collected from 100 years earlier. However, those records were a highly biased depiction of the woodpecker's real abundance, because the records failed to conform to strict requirements of statistical models that seek to understand patterns in time series like those made up of historical specimen collections.

Time series analyses should adhere to a cardinal rule. Data must conform to stationarity.[11] That is, data must be generated by the same process(es) that create the entire sequence, or the confounding influences from additional process(es) must be accounted for before analyses reach a trustworthy conclusion. A temperature gauge on my car will deceive me about the level of radiator coolant if I happen to park over a grass fire. For studies on the ivory-bill, this kind of basic common sense was ignored. Violations of stationarity were conspicuous to anyone with fluency in the nation's environmental history. Given the many economic, social, cultural, and statutory changes taking place during the Progressive Era (Chapters 3–6), we can place no trust whatsoever in a belief that a "diminishing curve of museum specimens ... be considered a proxy of total population size"[12] in the ivory-bill.

Petitioning a demise of ivory-bill after 1890 based on historical trends in the specimen record makes no more sense than using these same data to establish an *increase* in the woodpecker's population just earlier in the nineteenth century. Social impediments to collecting the woodpecker (as for all birds) began in the 1880s, with these restrictions expanding in number and scope practically unabated through the present (***Image 14.1***). Interest in protecting non-game birds took off in 1886 after *Forest and Stream*'s editor, George Bird Grinnell, tapped into popular sentiment building across American society and opposed to taking of birds for commercial profit. That sentiment was buttressed by recreational hunters led by Roosevelt who staunchly opposed the "game hogs" of market hunting. Social disapproval spanned market hunting for game birds as well as other enterprises (like the millinery trade) that targeted almost any avian species.[13]

Also, in 1886, the AOU's bird protection committee designed and then made available a Model Law for the protection of non-game birds, one tailored for adoption by state legislatures. This template prohibited killing of any but game birds as well as taking the nests or eggs of any bird unless the collection was certified for some legitimate scientific objective. Thirteen states used wording furnished straight from the Model Law. Nearly all states had passed some version of a statute that protected non-game birds by the end of the 1800s.[14] By 1900, the more stringent Lacey Act outlawed interstate shipment of wildlife, including birds, in defiance of these state laws.

Economic setbacks further curtailed the incentives propping up the taxidermy business (***Image 14.1***). Commercial taxidermy supplied Victorian decor in homes as well as content for natural history museums. Ivory-bill collections plummeted and never rebounded after the Financial Panic of 1893. This crisis bankrupted specimen dealers and ornithologists alike.[15] "Many dealers were forced to sell out or abandon their businesses. Some, like Walter F. Webb and his one-time partner Frank Lattin, one of the largest egg and skin dealers in the country, moved into other, less contentious areas of natural history."[16] Resentful taxidermists griped about losing their livelihood: "It is simply dead. If it warn't [sic] for rugs and deer heads we couldn't live. Those ... Audubon Societies and bird books and new-fangled laws are just crowding us out."[17]

Fueled by the politicized Audubon societies (Chapter 4), protection crusades launched by citizens further eroded a facility in acquiring bird specimens, even for valid scientific purposes. Natural history dealers and commercial taxidermists were only the first to be thrown under the bus. Opposition to bird killing reached such a fervor that by 1899, Witmer Stone, future president of the AOU, was mentoring his young future ornithologists to consider: "There is nothing to be gained by the collecting of series, except the extermination of the birds, which is surely not your object."[18]

By the turn of the century, then, state authorities "were increasingly suspicious of ornithologists' claims for the continued need for collecting."[19] In a 1903 editorial from *The Condor*, Walter K. Fisher protested bitterly about the ever more restrictive regulations now placed on their discipline: "Almost without exception it is a positive hardship to secure a permit to collect in states where the AOU bill has been accepted, particularly in the case of non-residents ... it is a little farfetched when a recognized student of birds must be subjected to delay, annoyance, and highway robbery if he wishes to collect for his own purposes, or for those of the Government."[20]

Fewer and fewer specimens of the woodpecker were collected as these new laws went into effect. Just 20% of known ivory-bill specimens were acquired after the Lacey Act (1900), a statute that curbed greatly any interstate movement of specimens.[21] Only about 3% of all ivory-bill specimens were collected once the nation had adopted the Migratory Bird Treaty Act (1918).[22] This fall in collections (***Image 14.1***) of the big woodpecker corresponds at least as well to historical social mores (*specimen demand*) as to a putative decline in population size (*specimen supply*). Regardless of merits to each hypothesis, an inability to differentiate between supply and demand here makes it negligent to argue that a decline in specimen records is a reliable stand-in for ivory-bill abundance, never mind decisive evidence of the bird's outright extinction.

Specimen acquisition was not the only variable sensitive to historical change.[23] Public aptitude for recognizing the woodpecker also evolved. The fight for the Singer Tract followed on the heels of Roger Tory Peterson's launch of a new popular type of bird field guide in 1934, so the resources available to field observers who might see the big woodpecker did not remain constant either. In contrast, some of today's field guides are adamant that those who report this bird in modern times are engaging in "wishful thinking."[24] Finally, even our scientific usage of words related to the term "extinction" changed markedly after about the mid 1950s.[25]

Skepticism over ivory-bill reports fluctuated across decades, too, especially after Eckleberry's portrait had concretized the belief for a "last ivory-bill" in 1944 (Chapter 9). Along with ***argument from authority*** exacted by those declaring the bird extinct, we cannot presume that all reports from eras after the 1940s were ever forwarded for consideration in the first place. As the extinction paradigm became steadily reinforced, ***outcome reporting bias*** became more and more likely, further corrupting a fundamental requirement for the stationarity needed for and from any sighting record.

Undoubtedly the various deterrents on the taking, trading,

and transport of the ivory-bill did not stop entirely the killings of this rare bird. But that is not the point: the rate of such taking was inconstant across time due to larger, ongoing shifts in American society. Exploiting the series of specimen records to justify a belief in the ivory-bill's extinction stands on an irretrievably defective premise.[26] Not one study that did so can be remedied with any amount of special pleading or grasping at surmised covariates. Rash conclusions arose from falling in love with the models, substantiating how easily our rational thinking can get hijacked by ***congruence bias*** (due to the very strong appeal of direct testing with data available from the collection record), and ***illusory correlation*** (from deceptive links presumed between the collection record and the ivory-bill's real abundance).

For a species hanging so precariously in the balance, one might wish that more vigilance had been taken to avoid such reckless abuse of numbers. After all, "the goal of all statistical analysis, ultimately, is to help people use data to rationally update their beliefs about scientific hypotheses."[27] Instead, tendering the woodpecker's collection record as corroborating evidence for extinction constitutes a ludicrous attempt to drive forward using a rear-view mirror. Years after the extinction studies had made their sweeping claims, the underlying premises to the analyses crumbled at deeper inspection. Of course, these later updates and revisions did not attract nearly the media attention (if any at all) as had the earlier work that had made the far bolder, reckless claims.[28]

If uncertain (or less certain) reports of ivory-bill are added to the reports considered to be certain, the statistical estimation does not invariably even favor the woodpecker's extinction.[29] Indeed, when models realistically portray the inherent uncertainties associated with rare species like ivory-bill, the outcome is "a much higher possibility of population persistence."[30] Early models were unduly naïve in ignoring the "inherent uncertainties in ... rare populations with weak demographics on the verge of extinction."[31] Even statistical treatments purporting to address this uncertainty persisted in culling (arbitrarily) some ivory-bill sight reports that were available,[32] limiting the sample size and skewing the results, or they gave more credence to sight reports early in the historical record (see also Chapter 7). One study using only about 30% of all ivory-bill reports available since 1944 was unintentionally frank about its resort to circularity: "prior to extinction, roughly two-thirds of the uncertain sightings were [considered] valid."[33]

Attempts to justify probability to support a hypothesis of extinction for the ivory-bill were plagued by other problems, too. Only years after the 2004 Arkansas incident did models include both search effort and uncertainty in quality of the sighting record when estimating the extinction risk.[34] Other methods required repeated area sampling to infer extinction,[35] a procedure never undertaken for the woodpecker. None of the models "account [for] the possible spatial dimension of the underlying population nor stochasticity."[36] This particular oversight then distorted "estimates of extinction probability upwards in the case of the population subsequently establishing in areas of lower detection probability,"[37] a virtual certainty for a species like the ivory-bill that occupies (or occupied) dense, inaccessible swamp forest.

Equations that estimated the extinction date were also very sensitive to any large gaps present at the end of the sighting record.[38] Exacerbating that problem, researchers openly resorted to ***argument from authority*** "by only including those [ivory-bill records] that are widely accepted."[39] In excluding most of the approximately 100 incidents[40] available for the woodpecker since 1944, estimates of extinction were blinded by researchers' particular belief system. Because "suitability of a technique is based on reliability of its results, not on majority vote,"[41] the ***availability cascade*** had once again hijacked reason.

Prejudice from these doubtful beliefs was easy to find. One study made the curious defense that it was "generally preferable" to "[base policy] decisions on imperfect data sets."[42] This justification appeared after earlier discussion about the ivory-bill specimen record had already confessed that: 1) "additional observations may have been made and not included in our sighting records," 2) "inadvertent exclusion of valid reports from after the last year in a sighting record ... would move estimated extinction dates forward in time," 3) "historical data generally were not collected with a well defined sampling scheme in mind," and 4) "many [techniques] assume that there has not been directional change over time."[43] Despite what may be an "aching desire" to seek out probabilistic answers for whether the big woodpecker survived, we are obligated nonetheless to "base ... statistical models on biological knowledge rather than mathematical convenience."[44]

Nearly all models that inferred extinction in ivory bill assumed a constant sighting rate.[45] We have no idea what the detectability[46] in the ivory-bill actually was (or is), much less whether it stayed constant, so instead we just gambled that this key factor was trivial for evaluating something as crucial as the species' survival. Yet accuracy in the projection of extinction varies enormously by search effort, search regularity, and change in search effort over time.[47] Moreover, failure to incorporate adequately the importance of detectability took place despite clear admonitions that "it [is] preferable to estimate detectability for ... [a] species before its persistence is threatened."[48]

Warnings about placing a heedless conviction in models that inferred extinction could not have been plainer in several cases, e.g., the "estimator cannot be used with most museum archives because *some years saw no collecting effort*"[49] (emphasis supplied). Despite a duty to correct for birds present but not recorded,[50] no study averring that the ivory-bill was extinct bothered to address this deficit. "It raises the question, which seems to have been missed by scientists, as to whether this species should have been declared extinct in the first place."[51] In speculating that it was "appropriate to assume a stationary Poisson process"[52] when testing for extinction in the ivory-bill, statistical estimation and science itself succumbed to its own brand of ***anchoring bias*** (see also Chapter 8).

An entirely separate approach tried to apply "stopping rules" from species-area relationships (SARs) in avian census data to force a claim of extinction on the ivory-bill.[53] As with methods that relied on the specimen record, no adjustments were made for ivory-bill detectability. Indeed, this same study did not account for the well-known detection heterogeneity observed across bird species that compose SARs.[54] Stunningly, the assertion for the woodpecker's extinction from these SARs came from avian censuses conducted during just a single season, after as little as 9 and no more than 26 search days, and from just four sites, one each in Louisiana, Mississippi, South Carolina, and Florida. Finally, the study banked on a "flawed concept of measuring extinction rates"[55] that arises from artifacts created by not distinguishing "between the underlying sampling problems when constructing a species-area relationship (SAR) and ... extrapolating species extinction from habitat loss."[56]

Not surprisingly, and mostly from ignoring search effort out of desire to seek quick closure for our cognitive disquiet, we happen to be dreadful at guessing about species' actual extinctions. Dubbed ***Romeo error***,[57] and triggered by ***base rate fallacy***, these hasty declarations of animal extinction are remarkably common.[58] Over the last 122 years no less than 351 vertebrates were "rediscovered" worldwide, at a rate that only continues to increase exponentially with time.[59] Lazarus birds went missing for around 64 years on average, although in exceptional cases as much as centuries might elapse between the consensus for a species' extant status. Of these re-found birds, 86% remain highly threatened and still have restricted distributions with low population sizes.

Box 14.2. ***Base rate fallacy*** is one of several thinking and decision blunders that arise out of ***neglect of probability***. It occurs if we are unaware of (or choose to ignore) a background rate that better captures the statistical basis for some natural phenomenon. When humans declare a status of extinction with unjustified certainty for individual species like night parrot or ivory-billed woodpecker, they overlook extensive metadata that convincingly reveal so many instances of our being premature about a species' extinctions.

Very long interludes can separate reports of extant birds. In 1992 the Cebu flowerpecker *Dicaeum quadricolor* was refound in the Phillipines after a reporting absence of 86 years.[60] In keeping with a tendency for re-discoveries to recur, the "extinct" Myanmar Jerdon's babbler *Chrysomma altirostre altirostre* showed up after 73 years.[61] India's forest owlet *Athene blewitti* was lost until 1997, a gap of 103 years.[62] Indonesia's Banggai crow *Corvus unicolor* went missing for more than a century.[63] The New Zealand storm-petrel *Fregetta maoriana* was known from just three individuals collected during the 1850s before the species reappeared in 2003.[64] After 150 years of hiding, the storm-petrel's population is still thought to number fewer than 50 individual birds.[65]

Bermuda petrel *Pterodroma cahow* may take the prize in longevity for its remarkable hiding ability. Thought extinct in 1620, then re-found in the early twentieth century, this medium-sized seabird (not much smaller than ivory-bill) concealed itself out on a small, densely populated archipelago off the southeastern U.S. for almost 300 years![66] Our predictions for individual cases of bird extinction are simply not to be trusted. Insistence on this kind of infallibility signifies hubris of the highest sort.

Ivory-billed woodpecker does not make the perfect candidate for extinction because it lacks certain traits known to greatly heighten the odds of imperilment in higher vertebrates. For example, the continental ivory-bill clearly does not bear the substantially higher risks found in island-dwelling species.[67] Although at least 10–15% of the world's birds are endangered, risk of extinction is not distributed evenly across these avian taxa. In fact, the woodpecker family (Picidae) is the *only* group worldwide containing *fewer* threatened species than expected by chance.[68] Of the traits other than taxonomic lineage (large body size, very low fecundity) that predispose birds to extinction and examined by the same study, the big woodpecker possesses only the first.[69]

Rediscovery of vertebrates that went missing during the twentieth century is more likely to occur in those species, like ivory-bill, that are threatened by habitat loss and that occupy relatively large ranges.[70] Range size of the ivory-bill was never believed to be as small as 11,000 km^2, the threshold below which forest-dependent birds have been identified as typically becoming threatened.[71] Large-bodied birds like ivory-bill tend not to have small geographic ranges.[72] And among threatened birds, larger-bodied species tend to have larger instead of smaller population sizes, although this link is comparatively weak.[73]

As a rule, extinction risk in birds from habitat loss is associated with ecological specialization and small body size, but not with generation time.[74] Conversely, extinction from human persecution (or introduced predators) is correlated with large body size and slow reproductive rates, but not with ecological specialization. This finding poses a conundrum if applied simplistically to the ivory-bill because the two primary reasons put forth for the woodpecker's demise span both drivers of extinction, yet the bird's perceived life history traits do not fit tidily into either scenario alone.[75]

Statistically-derived models can never give us reliable evidence as to whether the ivory-bill is truly extinct. "If counts are biased or imprecise, placing them in a time series does not magically transform them into something meaningful."[76] During our struggles to understand what happened to this enigmatic bird, we thought we could play fast and loose with mathematical rules demanding that our "search effort [be] constant and perfect."[77] Even "the best way to analyse sighting data to infer extinction timing is unresolved."[78] And one reviewer slammed our addiction to "current frequentist methods for fixing extinction dates [as] not only philosophically objectionable but consistently inaccurate."[79]

On the other hand, the most faulty reasons conceivable for holding a belief in extinction does not mean that the woodpecker still lives among us. "It is almost impossible to determine with any certainty whether a species is extinct."[80] Yet our history of consigning bird species into extinction much too soon, only to have them turn up later, should give us pause. As woodpeckers are among the avian groups least susceptible to extinction, at the very least we ought to exhibit some caution in expressing undue rigidity about this bird's likelihood of perseverance.

Ivory-billed woodpecker has (or had) several distinctive biological traits that could have facilitated its survival[81] right beneath our noses (Chapter 1). Given added insights available today from contemporary biology, and giving probability its due respect so as to avoid capitulating to utter guesswork, the woodpecker's own anatomy, behavior, and ecology have furnished us tips. We may never fully discern what ultimately happened to the ivory-bill, but allowing for alternatives in the bird's fundamental nature at least renders us less likely to be led wildly astray by ***congruence bias***. And bestowing a proper measure of respect to uncertainty offers a substantial benefit – it goads us into seeking out fresh interpretation.

Endnotes

1 Large historical gaps in bird sighting reports are not exceptional compared to other vertebrates either. One species of stick-nest rat, *Leporillus conditor*, found in Australia's arid interior, had not been observed for 48 years prior to its rediscovery. Burgman, M.A., R.C. Crimson, and S. Ferson. 1995. Inferring threat from scientific collections. *Conservation Biology* 9: 923–928. Spade-toothed beaked whale *Mesoplodon traversii* has never been seen alive, and only two fully intact animals have been reported in the years since the species original discovery in 1872. Thompson, K., C.S. Baker, A. Van Helden, S. Patel, C. Millar, and R. Constantine. 2012. The world's rarest whale. *Current Biology* 22: R905–R906. Note that even the "rarity" of this whale is driven solely by scarce reports, not on the species actual known abundance.

2 Shared traits between these two birds include (or may include): 1) very low population size, 2) relatively large original or historical range, 3) high mobility, 4) apparent lack of territoriality, 5) nomadic habits associated with changing habitat or feeding conditions, and 6) behaviors (e.g., wariness, nocturnal movements, or inhospitable habitat) that make human detection exceedingly difficult.

3 "The plural of anecdote is not data." —Roger Brinner.

4 Broad agreement exists that the probability of survival and/or extinction in imperiled species like the ivory-bill should be stated in quantitative rather than absolute terms. "Any statement of extinction is probabilistic by nature." Roberts, D.L. 2006. Extinct or possibly extinct? *Science* 312: 997–998.

5 In doing so, we are better able to avoid a variety of profound errors in our reasoning, including falling victim to the ***conjunction fallacy***, ***neglect of probability***, ***Bayesian conservatism***, and forgetting about a need to factor ***Bayesian caution*** into our

assessment of the ivory-bill's real fate.

6 Twain seems to have popularized rather than coined the famous expression "lies, damned lies, and statistics." This expression's exact sourcing remains obscure, although it does bear considerable similarity to Thomas H. Huxley's club minutes from 1885 "three classes of witnesses – liars, damned liars, and experts." See: MacLeod, R.M. 1970. The X-club as a social network of science in late-Victorian England. *Notes and Records of the Royal Society of London* 24: 305–322.

7 Gotelli, N.J., A. Chao, R.K. Colwell, W-H. Hwang, and G.R. Graves. 2012. Specimen-based modeling, stopping rules, and the extinction of the ivory-billed woodpecker. *Conservation Biology* 26: 47–56.

8 Elphick, C.S., D.L. Roberts, and J.M. Reed. 2010. Estimated dates of recent extinctions for North American and Hawaiian birds. *Biological Conservation* 143: 617–624; Roberts, D.L., C.S. Elphick, and J.M. Reed. 2010. Identifying anomalous reports of putatively extinct species and why it matters. *Biological Conservation* 24: 189–196.

9 This kind of unmerited certitude invites cognitive blunders from ***false consensus effect, overconfidence effect, base rate fallacy***, and ***neglect of probability***, while also exposing a deficit of ***Bayesian caution***.

10 More than a half century of divergence in the dates projected by these different models should arouse deep misgivings about their accuracy. Suspicion in a putative extinction date as early as 1941 also arises given its obvious conflict with Eckleberry's "last" sighting date in 1944. See also Jarić, I., & D. Roberts. 2014. Accounting for observation reliability when inferring extinction based on sighting records. *Biodiversity and Conservation* 23: 2,801–2,815. Whether or not recent "controversial" sightings of ivory-bill are used in the models at all leads to estimates for an extinction date for the woodpecker that range from 1965 to 2032; see Rout, T.M., D. Heinze, and M.A. McCarthy. 2010. Optimal allocation of

conservation resources to species that may be extinct. *Conservation Biology* 24: 1,111–1,118.

11 In formal statistical terms, stationarity requires that such fundamental data properties as the mean, variance, and autocorrelation structure do not change over time. Such basic properties were demonstrably inconstant, and so the time series was irretrievably compromised, due in large part to the social and cultural changes in American society that affected bird collection and reporting rates.

12 Gotelli et al., p. 50.

13 Dorsey, K. 1998. *The Dawn of Conservation Diplomacy: U.S.-Canadian Wildlife Protection Treaties in the Progressive Era*. University of Washington Press, Seattle, p. 175.

14 Farnham, T.J. 2007. *Saving Nature's Legacy: Origins of the Idea of Biological Diversity*. Yale University Press, New Haven, CT, p. 50.

15 Arthur Cleveland Bent was one ornithologist whose business interests were wiped out by the Panic. See Taber, E. 1955. In memoriam: Arthur Cleveland Bent. *The Auk* 72: 332–339. Another amateur naturalist, Daniel Hector Talbot, lost his fortune during the Panic, too, barely donating a beloved natural history collection to the University of Iowa before his creditors arrived to enforce a legal judgment against him for almost $95,000.00. http://www.mnh.uiowa.edu/talbot-collection (accessed 7 October 2015). The Panic of 1893 directly affected key operations at the nation's natural history museums. Robert Ridgway's salary at the Smithsonian Institution for years after the Panic was 3 to 4 times lower than at comparable institutions, with financial effects so dire that Ridgway called it a "miserable existence of a 'Government Pauper.'" Lewis, D. 2012. *The Feathery Tribe: Robert Ridgway and the Modern Study of Birds*. Yale University Press, New Haven, CT, pp. 56–57.

16 Barrow, M.V., Jr. 1998. *A Passion for Birds: American Ornithology after Audubon*. Princeton University Press, Princeton, New Jersey, p. 139.

17 Barrow, p. 136.

18 Stone, W. 1899. Hints to young bird students. *Bird-Lore* 1: 125–127. Stone's advice was co-signed by such natural history dignitaries of the age as J.A. Allen, Frank M. Chapman, Robert Ridgway, C. Hart Merriam, T.S. Palmer, and William Brewster, among others.

19 Barrow, p. 135.

20 Fisher, W.K. 1903. Editorial notes. *The Condor* 5: 136. Fisher was especially irate over ornithologists having to post bond as high as $200 for the privilege of being able to collect birds for purposes of science. This fee was additional to costs for the required permit(s).

21 Approximate percentages were calculated by taking the number of specimen reports archived for the ivory-bill after 1900 (for the Lacey Act) and 1918 (for the Migratory Bird Treaty Act), then dividing by 400. The total number of ivory-bill specimens believed to exist world-wide is "slightly more than four hundred." Jackson, p. 74.

22 At least one taxidermist was in fact arrested for trading in ivory-bill specimens. Jackson, J. A. 2006. *In Search of the Ivory-billed Woodpecker*, 2nd ed. HarperCollins, New York, p. 76.

23 Investigators who naively used the ivory-bill's collection record to infer extinction were subscribing also to distortions caused by historical ***presentism***, i.e., they juxtaposed past events (the collection record) with both modern values (e.g., desire to apply information for today's conservation decisions) and current concepts about quantitative rigor (e.g., key assumptions behind cutting-edge statistical modeling).

24 Dunn, J.L., and J.K. Alderfer. 2011. *National Geographic Field Guide to the Birds of North America*. National Geographic Books, Washington, DC, p. 324.

25 Figure 2 in Smith-Patten, B.D., E.S. Bridge, P.H.C. Crawford, D.J. Hough, J.F. Kelly, and M.A. Patten. 2015. Is extinction forever? *Public Understanding of Science* 24: 481–495.

26 Attempts to apply statistical inference for extinction in the ivory-bill proceeded anyway, despite the fact that "the most serious

problem in … application is likely to be that collection frequencies will reflect changing trends in museum … collections" (Burgman et al., p. 927). Thus, an unjustified blind faith was placed into "methods … [that] assume that collections represent a stationary, random process (a Poisson process), so the average chance of collecting a species does not change over time" (Burgman et al., p. 925). Because studies that violated this assumption with the woodpecker were all published after such pointed warnings, often in the very same journal, serious collapses in the quality of earlier peer review were exposed.

27 Anderson, J.L. 1998. Embracing uncertainty: the interface of Bayesian statistics and cognitive psychology. *Conservation Ecology* http://www.consecol.org/vol2/iss1/art2/ (accessed 31 December 2015).

28 Early claims for the ivory-bill's extinction using statistical estimation with the collection record were featured prominently in the media, accompanied by biased captions such as "only a memory" and what by now was the near compulsory nod to Elvis (e.g., *The Economist*. 2011. Dead or alive. 15 October print edition). Later studies finding that there was weakened or no sound statistical evidence for the woodpecker's extinction went largely or entirely ignored by the media.

29 A notable feature of the sighting record typically used as the proxy for ivory-bill abundance is that all "uncertain" sightings occurred after the most recent "certain" sighting, which has been identified to be purely an artifact of these data. See also Chapter 7, and Solow, A.R., and A.R. Beet. 2014. On uncertain sightings and inference about extinction. *Conservation Biology* 28: 1,119–1,123. As a result, accounting for uncertain sightings (i.e., giving them some weight) leads to very different probabilities for extinction than simply excluding them altogether, or including them all as if they were certain. See also: Lee, T.E., M.A. McCarthy, B. Wintle, S.H.M. Butchart, M. Bode, D.L. Roberts, and M.A. Burgman. 2014. Inferring extinction from sighting records of variable reliability.

Journal of Applied Ecology 51: 251–258.

30 Caley P., and S.C. Barry. 2014. Quantifying extinction probabilities from sighting records: inference and uncertainties. *PLoS ONE* 9(4): e95857. doi:10.1371/journal.pone.0095857. This study, too, urged strong caution about giving in to unfettered inferences from extinction models. "The implicit assumption that all other processes have zero plausibility is quite stringent and represents strong knowledge of the system in question...why the simple model is a natural starting point is not considered in the literature" (p. 10).

31 *Ibid.*

32 Roberts, D.L., and I. Jarić. 2016. Inferring extinction in North American and Hawaiian birds in the presence of sighting uncertainty. *PeerJ* 4:e2426; DOI 10.7717/peerj.2426. Authors drew particular attention to the conservation risks of ***Romeo error*** (false conclusions of extinction) when the sighting uncertainty is not addressed properly. Few studies attempt to quantify this level of uncertainty, test for the level of accuracy experimentally, or incorporate these parameters into the analyses. None of these issues were ever addressed in the ivory-bill.

33 Solow, A., W. Smith, M. Burgman, T. Rout, B. Wintle, and D. Roberts, D. 2012. Uncertain sightings and the extinction of the ivory-billed woodpecker. *Conservation Biology* 26: 180–184.

34 Lee, T.E. 2014. A simple numerical tool to infer whether a species is extinct. *Methods in Ecology and Evolution* 5: 791–796. This method apparently has yet to be applied to the ivory-bill, however.

35 Solow, A.R. 1993. Inferring extinction from sighting data. *Ecology* 74: 962–964.

36 Caley and Barry, p. 10.

37 *Ibid.*

38 Collen, B., A. Purvis, and G.M. Mace. 2010. When is a species really extinct? Testing extinction inference from a sighting record to inform conservation assessment. *Diversity and Distributions* 16: 755–764.

39 Elphick et al., p. 622.

40 Hunter, W.C. 2010. *Interpreting historical status of the ivory-billed woodpecker with recent evidence for the species' persistence in the southeastern United States*. Appendix E, Recovery Plan for the Ivory-billed Woodpecker (*Campephilus principalis*), U.S. Fish & Wildlife Service, Atlanta, GA.

41 Thompson, W. L. 2002. Towards reliable bird surveys: accounting for individuals present but not detected. *The Auk* 119: 18–25 (p. 23).

42 Elphick et al., p. 622.

43 *Ibid.*

44 Smith, W.K., and A.R. Solow. 2011. Missing and presumed lost: extinction in the ocean and its inference. *ICES Journal of Marine Science* doi:10.1093/icesjms/fsr176.

45 A method might also model a systematic decline in detectability as extinction was approached. See Solow, A.R. 1993. Inferring extinction in a declining population. *Journal of Mathematical Biology* 32: 79–82. Nevertheless, "the assumption of a constant or declining sighting rate may be hard to justify" (Caley and Barry, p. 1). Moreover, in the case of ivory-bill we don't know when extinction was approached nor if the putative population *is* declining. Detectability of the woodpecker could have increased (e.g., due to wider use of field guides), or detectability could have fluctuated in some inconstant, non-linear fashion. Finally, if and when the probability of detecting "a species declines over time, many of the methods perform poorly" (Rivadeneira, M.M., G. Hunt, and K. Roy. 2009. The use of sighting records to infer species extinctions: an evaluation of different methods. *Ecology* 90: 1,291–1,300).

46 Detection probabilities correct for the bias that arises from birds that are in fact present but not seen or heard by the researcher. This ratio of birds not detected, but nevertheless present, can be substantial. Even in open habitats with greater visibility, as many as 60% of individual birds are missed entirely at distances beyond

50 m from the observer's position. See Diefenbach, D.R., D.W. Brauning, and J.A. Mattice. 2003. Variability in grassland bird counts related to observer differences and species detection rates. *The Auk* 120: 1,168–1,179. Even the relatively large, territorial, and loud pileated woodpecker *Dryocopus pileatus* may have detection probabilities of only 18–36%. See: Farnsworth, G.L., K.H. Pollock, J.D. Nichols, T.R. Simons, J.E. Hines, and J.R. Sauer. 2002. A removal model for estimating detection probabilities from point-count surveys. *The Auk* 119: 414–425; Watson, K. 2013. A comparison of single-day versus multiple-day sampling designs using occupancy modeling. Bachelor of Science thesis, College of William and Mary.

47 Clements, C.F., N.T. Worsfold, P.H. Warren, B. Collen, N. Clark, T.M. Blackburn, and O.L. Petchey. 2013. Experimentally testing the accuracy of an extinction estimator: Solow's optimal linear estimation model. *Journal of Animal Ecology* 82: 345–354.

48 Reed, J.M. 1996. Using statistical probability to increase confidence of inferring species extinction. *Conservation Biology* 10: 1,283–1,285. Reed calculated that for a species having a comparatively low probability of detection (around 20%), failure to find individuals required 14 to 20 repeated visits in the area(s) of interest before one could make a statistically valid decision (i.e., justified by probability having sufficient power) in support of the hypothesis for extinction. Nowhere near that level of survey effort has been applied across the entire geographic area for which specimen or sighting records are available for the ivory-billed woodpecker.

49 Collen et al., p. 761.

50 Thompson provides a succinct overview of the need to address detectability in avian surveys.

51 Roberts, p. 997.

52 Elphick et al., p. 622. See also: Vogel, R.M., J.R.M. Hosking, C.S. Elphick, D.L. Roberts, and J.M. Reed. 2009. Goodness of fit of probability distributions for sightings as species approach extinction. *Bulletin of Mathematical Biology* 71: 701–719.

53 Gotelli et al.

54 Boulinier, T., J.D. Nichols, J.R. Sauer, J.E. Hines, and K.H. Pollock. 1998. Estimating species richness: the importance of heterogeneity in species detectability. *Ecology* 79: 1,018–1,028.

55 Beck, J. 2011. Species-area curves and the estimation of extinction rates. *Frontiers of Biogeography* 3: 81–83.

56 He, F. and S. P. Hubbell. 2011. Species-area relationships always overestimate extinction rates from habitat loss. *Nature* 473: 368–371.

57 ***Romeo error*** arises when we abandon the conservation of a species based on the false premise that it is already extinct when in fact it is still alive.

58 Collar, N.J. 1988. Extinction by assumption: or, the Romeo error on Cebu. *The Oryx* 32: 239–244.

59 Scheffers, B.R., D.L. Yong, J.B.C. Harris, X Giam, and N.S. Sodhi. 2011. The world's rediscovered species: back from the brink? *PLoS ONE* 6(7): e22531. doi:10.1371/journal.pone.0022531.

60 Dutson, G.C.L., P.M. Magsalay, and R.J. Timmins. 1993. The rediscovery of the Cebu flowerpecker *Dicaeum quadricolor*, with notes on other forest birds on Cebu Philippines. *Bird Conservation International* 3: 235–243.

61 Dell'Amore, C. 2015. "Extinct" bird rediscovered in Myanmar, surprising scientists. *National Geographic News*. http://news.nationalgeographic.com/news/2015/03/150305-birds-extinct-rediscovered-myanmar-burma-animals-science/ (accessed 4 January 2016).

62 Butchart, S.H.M., N.J. Collar, M.J. Crosby, and J.A. Tobias. 2005. Lost and poorly known birds: targets for birders in Asia. *Birding Asia* 3: 41–49.

63 Indrawan, M., Y. Masala, D. Dwiputra, F.N. Mallo, A. Maleso, A. Salim, F. Masala, I. Tinulele, L. Pesik, and D.S. Katiandagho. 2010. Rediscovery of the critically endangered Banggai crow *Corvus unicolor* on Peleng Island, Indonesia, part 1: ecology. *Bulletin of the British Ornithologists' Club* 130: 154–165. Interestingly, rediscovery

of the Banggai crow was compromised and delayed in some part because its identity was confused with the very similar and far more common slender-billed crow *Corvus enca*.

64 Kirby, A. 2003. NZ bird returns 150 years on. *BBC News* http://news.bbc.co.uk/2/hi/science/nature/3344917.stm (accessed 8 January 2016).

65 BirdLife International. 2016. Species factsheet: *Fregetta maoriana*. http://www.birdlife.org (accessed 8 January 2016).

66 Madeiros, J., B. Flood, and K. Zufelt. 2014. Conservation and at-sea range of Bermuda petrel (*Pterodroma cahow*). *North American Birds* 67: 546–557. Extensively harvested for food beginning with the arrival of Spanish and English mariners, additional threats from introduced rats, hogs, cats, and dogs all served to diminish the petrel. After single petrels were encountered in 1906, 1935, and 1945, Louis S. Mowbray, Robert Cushman and Grace Murphy, and David B. Wingate found 7 pairs on a 1950 expedition to a few tiny offshore rocks near Castle Harbour.

67 Extinction risks for the island-dwelling Cuban ivory-bill are another matter entirely. See also: Johnson, T.H., and A.J. Slattersfield. 1990. A global review of island endemic birds. *The Ibis* 132: 167–180.

68 Bennett, P.M., and I.P.F. Owens. 1997. Variation in extinction risk among birds: chance or evolutionary predisposition? *Proceedings of the Royal Society of London* B 264: 401–408. Eight of 215 (slightly fewer than 4%) species of woodpecker are threatened worldwide. Later estimates using more relaxed IUCN criteria along with new taxonomic revisions listed 42 of 254 woodpeckers as vulnerable. Most of these were threatened by deforestation, including that from more selective logging practices. See: Lammertink, M. 2014. Trends in threat status and priorities in conservation of the woodpeckers of the world. *Acta Ornithologica* 49: 207–219.

69 After accounting for other factors, the relation between body size and extinction risk within bird taxonomic lineages still holds. See: Gaston, K.J., and T.M. Blackburn. 1995. Birds, body size and the

threat of extinction. *Philosophical Transactions of the Royal Society of London B* 347: 205–212.

70 Direct persecution, however, predisposes animals to an even greater risk of extinction than does habitat loss. See: Fisher, D.O., and S.P. Blomberg. 2010. Correlates of rediscovery and the detectability of extinction in mammals. *Proceedings of the Royal Society of London* B 278: 1,090–1,097.

71 Harris, G., and S.L. Pimm. 2008. Range size and extinction risk in forest birds. *Conservation Biology* 22: 163–171.

72 Brown, J.H., and B.A. Mauer. 1987. Evolution of species assemblages: effects of energetic constraints and species dynamics on the diversification of the American avifauna. *American Naturalist* 130: 1–17; Gaston, K.J. 1990. Patterns in the geographical ranges of species. *Biological Review* 65: 105–129.

73 Gaston and Blackburn, p. 210.

74 Owens, I.P.F., and P.M. Bennett. 2000. Ecological basis of extinction risk in birds: habitat loss versus human persecution and introduced predators. *Proceedings of the National Academy of Sciences* 97: 12,144–12,148.

75 Threatened taxa are rarely subjected to extinction threats posed by direct killing and habitat loss simultaneously. *Ibid.*

76 Thompson, p. 23.

77 Clements, C., T.E. Lee, and M.A. McCarthy. 2014. An experimental test of a Bayesian method for inferring extinction with varying search efforts. *PeerJ* e466v1 https://peerj.com/preprints/466.pdf (accessed 7 January 2016).

78 Caley and Barry, p. 10.

79 Alroy, J. 2014. A simple Bayesian method of inferring extinction. *Paleobiology* 40: 584–607. From a Bayesian rather than frequentist statistical perspective "there is no clear way to put a number on the chance [an imperiled species] still exists." *Ibid.*, p. 584.

80 Roberts, p. 997.

81 Surviving does not equate to thriving, however.

Chapter 15

The Last Laugh

Somewhere lives a bad Cajun cook, just as somewhere must live one last ivory-billed woodpecker. For me, I don't expect ever to encounter either one.

—William Least Heat Moon, Blue Highways: A Journey into America (1982)

Even cartoon woodpeckers get press-ganged into hauling around the abstract burdens imposed by our covert agendas and social projections.

While the ivory-bill was going silent down in the Singer Tract, cackles of maniacal laughter were unleashed from a brand-new wood-drilling species. Zany prankster and irreverent imposter, Woody Woodpecker was birthed inside a realm of the ludicrous. Woody flaunted an energetic brand of 1940's comic horror, although some of his antics verged on the criminally insane, at times veering over to the dangerously psychopathic.[1]

Woody was a composite of woodpecker mannerisms. His nature embodied vocal parallels to the pileated woodpecker, a quarrelsome nature of acorn woodpecker (from which producer Walter Lantz and his wife Gracie got their original inspiration), and a forward-leaning red crest rented from, yes indeed, the male ivory-bill. Much like Woody, my parallax ivory-bill (Chapter 1) is also a fusion. Unlike Woody, though, the biological traits from that parallax ivory-bill exist literally within woodpeckers or in other birds. A good proportion of these traits were detailed by Tanner himself.

Overall, my appraisal places a weightier emphasis on the ivory-bill's versatility, grants less prominence to its alleged, narrow specializations, yet leans toward agnostic for what any elastic behavior and ecology might have portended for the bird's ultimate fate.[2] Just because ivory-bill may have owned flexible traits does not mean it survived. Rather, *if* it survived long past the 1930s it would have been likely because certain overlooked or downplayed traits in the bird functioned as ***probability levers***[3] that lifted its prospects over the gloomy fortune we have habitually assigned to this species.

> **Box 15.1.** ***Probability levers*** refer to what are normally very small influences, but ones that can have out-sized impacts or unforeseen outcomes. Examples of conceivable probability levers for the ivory-billed woodpecker include a highly nomadic lifestyle (a trait which makes a rare species even harder to detect), or strong, year-round pair bonding, a trait making a species less vulnerable to the negative consequences of very small population size.

What seem to be such glaring paradoxes in how the bird lived (Chapter 7) instead exemplify the very plasticity expected from a species living in variable environments. Ivory-bill was a fixture of the North American avifauna for hundreds of thousands of years.[4] Over that span, it is difficult to imagine a large woodpecker so inflexible that it relied on just one or two forest types remaining in a condition of perpetual old-growth.[5] During geological upheavals over long durations, dramatic changes to forest extent, composition, age, and connectivity would have posed insuperable risks to a narrow habitat or feeding specialist.[6] Conversely, a nomadic bird with substantial flight capacity and opportunistic foraging habits might adjust readily to ecological disruptions in the evolving New World landscapes.

How strong is "modern" evidence for this nomadism?

Purely on morphological grounds, the wing structure and flight capacity cannot be disputed reasonably. Like so much about ivory-bill, however, the supplemental evidence is exceedingly sparse and largely anecdotal. Still, the best numbers inform us that the big woodpecker's tendency was to be present in one locality, later abandon it, only to pop up elsewhere.[7] This pattern predated the frontier's end. Hunter affirms an erratic nature to its distribution over the long term:

> *There is no evidence that the ivory-billed woodpecker was ever widely or consistently relocated in the same areas from year to year or from decade to decade prior to 1940.... What is clear is that the present pattern of reports that do not effectively document occurrence of the species has been repeated from decade to decade for more than a century and that the number of locations with potential encounters within the same decade has varied little since the 1870s.*[8]

In other words, without arbitrarily culling the evidence that is available to us,[9] the ivory-bill has not "acted" much differently in the most recent decades than it did more than a century ago.

Despite our manifold obsessions with it, the ivory-bill was after all just a large, powerful woodpecker. From an avian point of view, it was not that exceptional. Some of its biological flexibility came from sizeable resistance to Allee effects as a result of nomadism, lengthy pair-bonds,[10] sociality, and covert behavior. Such traits would have granted both the adults and year-one birds higher survival than otherwise. The species' potential for variable and potentially high fecundity would, over the relatively long lifespans expected, also complement any fluctuations to forest mosaics yielding episodic and patchy food from one year to the next. The ivory-bill could thus better avoid the ecological traps that can waylay more sedentary birds in deteriorating habitats.

At times ivory-billed demography might resemble that of *K*-selected species. A large investment in just one offspring during some years was more in line with life history strategies of certain marine as opposed to forest-dwelling birds. If the woodpecker dispersed, especially outside the breeding season, pairs could function as a broadly-distributed meta-population interacting regionally instead of as isolated groups. If we could have measured the spatial scales over which this bird lived, the geographic enormity might well dazzle us.[11] Much mystery about this bird could be dispelled by high vagility alone, e.g., trying to find moving targets in dense forest vegetation across some 10 large southern states is an intimidating prospect, to say the least.

My exploration does not settle the proximate or the ultimate cause(s) behind the ivory-bill's imperilment or its putative extinction. I find utterly unpersuasive the premise that the bird became endangered due to over-reliance on large wood-boring beetle larvae confined to a few species of old trees in forested wilderness.[12] Conversely, considering the weight of evidence from all accounts, I give more credence to human persecution in steering the ivory-bill's fate, conceding that this factor may have played out more in the bird's behavioral expression rather than leading directly (or inexorably) to its extinction.

Reflections on a parallax ivory-bill did unmask a troubling auxiliary threat – human interference with so many natural disturbance cycles vital to the bird.[13] Beginning in the nineteenth century, and extending practically to the present day, we interrupted and truncated natural disturbance regimes once prevalent in the forests exploited by ivory-bill. We all but eliminated a role for the American beaver *Castor canadensis* as an engineer of patchy tree mortality in eastern forests.[14] We drained, re-channeled and leveed southern rivers.[15] Besides land conversion and deforestation, modern forestry's short rotations left behind younger and more supple trees that were less vulnerable to wind damage by tornados or hurricanes. Most of all, fire suppression greatly reduced the large burn-outs that early writers said had been so favored by the woodpecker. All of these actions diminished the ecological complexity found in the large landscapes that once supported ivory-bill.

If indeed ivory-bill lived much past the 1930s, why has it been so extraordinarily difficult for us to find and to authenticate? Earlier I illustrated how difficult it is to detect over vast areas individual birds in a population of just tens to low hundreds (Chapter 1). In doing so, I assumed a generous detectability for ivory-bill. In field settings, the ivory-bill's true discernibility could be far lower than what we can portray straightforwardly.[16] In substantial part this is because ivory-bill apparently did not rely on loud broadcast vocalizations to find mates or to defend territories from conspecifics. Its calls were notoriously weak throughout the year.[17] And, unlike many birds, it did not or would not consistently respond to call play-back or other auditory lures.[18]

Mine is hardly the first comparison to accentuate how very low probability so challenges our finding (or photographing) an ivory-bill. In Arkansas, the odds of detecting just one ivory-bill with the huge search effort expended there were estimated to be only 12%.[19] Conceding to birders rather generous levels of field effort given the highly unappealing nature of searching in bottomland swamp forests, Pulliam illustrated that with even 100 ivory-bills, we might expect to get one poor sighting every 2 years, and a much better sighting or bad photo about every 20 years.[20] Pulliam rightly avowed that it is practically impossible in a statistical sense to distinguish between 10 living ivory-bills and absolutely no birds at all. "Thinly spread forest creatures, with very large areas within which they might be scattered, are destined to remain phantoms."[21]

In line with expectations of logic, greater shared effort hunting for the ivory-bill does generally pay off with more (albeit still contested) encounters when compared to the historical norm (Chapter 1). One such interval was the specimen collection era of the 1880s–1890s, another the time span bracketed by Tanner's study and the fight to save the Singer Tract (Chapter 10), and a more recent epoch in the 2000s when search effort again ramped up after the Arkansas and Florida incidents (Chapter 12). In fact, by 2010, those stepped-up efforts had found twenty-first-century evidence consistent with the identity of the ivory-bill (sounds,[22] foraging sign, cavities, sightings, and so forth) in Tennessee, Arkansas, Louisiana, Mississippi, Florida, and both South and North Carolina.[23]

Given such long odds against success, though, is it *rational* for individual searchers outside large organized teams to encounter ivory-bill? Yes, it is logical, though highly unlikely because the bird is so "hard to find and harder to follow."[24] Prolonged searching is rational since those who repeat searches over days, seasons, or locations essentially rely on a modest odds lever furnished by the additive rule of probability. Odds of finding a rare object on day 1 *or* day 2 is equal to the probability on day 1 *plus* the probability on day 2. Critics oblivious to the statistical nature of finding rare, highly mobile species often ridiculed searchers (Chapter 11), belittling such efforts as a quixotic fixation. But while indeed the overall odds remain exceedingly low (providing, of course, that the ivory-bill still does survive), more search effort delivers a greater chance at payoff. What is decidedly *irrational* is to expect the ivory-bill to perform for us like a movie-on-demand, appearing at and for the convenience of our quirky perceptual frames.[25]

Perseverance even by solo searchers who look for ivory-bill at times furnishes evidence consistent with the identity of this species, signs that corroborate other aspects of the bird's perceived biology.[26] This validation includes a struggle to obtain repeat detections,[27] lack of long-term occupancy at one site (for more than a few years), even if found,[28] and a level of wariness[29] and/or use of dense habitat that makes unobstructed, high-quality photography so arduous[30]. Low rates of capturing evidence and so many failures to detect the bird at all, however, also refute a cognitive priming theory so beloved by skeptics who think observers just go out and see what they want to see. Why spend years and years toiling in hardship before divulging one's hasty lack of prudence?

The search for the ivory-bill can appear like a cult-like obsession, then, due to the extraordinarily low probability of it being fruitful. One person's quest to persevere long enough to overcome huge mathematical odds is just another's pointless waste of time. Such disparate perceptions among us can be attributed to diverging cognitive reactions to the strict dictates of probability. A searcher counts on the addition rule of probability to serve as odds lever; a skeptic (or denier) relies upon the likelihood (or reality) of failure to never bother with such nonsense in the first place. Both viewpoints contain arithmetic truth, and either view may be absolutely correct in its ultimate verdict.

Invented out of liberal doses of ornithological and artistic license, Woody eventually came to denote more than a screwball bird. Like the ivory-bill, there was far more to Woody than met the eye. Hidden deep in Woody cartoons, his animators embedded various gags, in-jokes, and visual or auditory pranks. Animator Shamus Culhane even used Woody for experiments in avant-garde cinema. Inside the cartoon's explosive montages he ingeniously disguised short films of abstract art inspired by Soviet theorists Sergei Eisenstein and Vsevolod Pudovkin.[31] In later spectacles of allegory, Woody served as a tattoo icon for skinheads and prison gangs.[32] Woody even provided a standard for measuring levels of aggression during psychological testing of children.[33]

Woody never messed about with our heads so callously, though, as has the real ivory-bill. My exploration uncovered at least 50 cognitive errors, psychological biases, and logical failures incited by our cultural and social fixations with a big woodpecker.[34] Most of these critical thinking and decision gaffes can be ascribed to a historical uncertainty tethered to the bird, followed by our haste to find rash closures for all the ambiguity. Vagueness is profoundly trying for the human species to hold in check.[35] We'd rather be firm about a wrong decision than hesitate too long in getting it right. Yet so very many decision and thinking errors appended to the status and identity of just this one bird plead a deep grilling about what else we routinely get wrong about the world in which we live.

Thinking errors are unsettling to smart practices in conservation. Our cognitive ***framing*** of extinction matters greatly for how we act.[36] Conservation early on ventured that "species must be presumed extinct or endangered unless shown to be extant and secure."[37] This turn of logic stemmed from a fear the public would otherwise misjudge the real urgency of species in peril. To better frighten ourselves into what was believed to be right action, we fell to using a faulty mindset to brand extinction.[38] Realizing at last that this exaggeration was not helping with a conservation intent, the IUCN Red List changed the burden of proof to evidence that showed that a species was *not* present.[39] The IUCN would eventually heed calls for more transparent labels that accommodated changes to our knowledge and wider ranges of certainty.[40] Conservation can better achieve its aims if it follows guidelines that are informed by this cognitive vigilance.[41]

We give ourselves far too much credit, and the ivory-bill not nearly enough, for our respective levels of cognitive talent. We think ourselves far more sagacious about this bird than we are. To soothe our ignorance, we often portrayed the woodpecker as little more than a bark-peeling robot, addicted to gourmet foods it could find nowhere else but recently perishing trees in the virgin forests of remote wilderness. We habitually saw the ivory-bill as so incapable of amending its own behavior in reaction to our malfeasance that it just lined up timidly in front of the loggers' saws and the collectors' guns.

After one hundred years, then, there ought to be no real surprise that the ivory-bill still mocks us. Like Wally Walrus, the slow-witted foil for Woody, we allow ourselves to get set up again and again by this bird. Our desperation for closure even causes us to over-conjecture about what oblique signs of its material presence might tell us of its exact whereabouts. "So many" large cavities suitable for ivory-bill were found in Florida "that it would make finding an active roost hole daunting."[42] And, if the ivory-bill's foraging was as diverse as early naturalists depicted, and what was once thought to be the bird's signature foraging style[43] (bark peeling) can be confused with feeding squirrels,[44] what makes us think we can track easily the big woodpecker's whereabouts using such roundabout clues? Here a ***congruence bias*** boxes us in at an overly narrow hypothesis while ***illusory correlation*** ambushes us from our addiction to unreliable proxy.[45]

Nowhere have we been so irrational as when we declared the woodpecker extinct too soon.[46] After 100 years of committing such error, why should we think we are right about the ivory-bill's status now, this time? More bluntly, neither the real nor the hypothetical ivory-bill has behaved like a properly extinct bird. It puzzles me no end why we didn't gather for decades and decades after their passing similar wild hearsay about lingering survivors in the passenger pigeon and Carolina parakeet. Though perhaps not as charismatic, or altogether condemned as surely extinct, we also don't have many tall tales about re-found Eskimo curlew or Bachman's warbler *Vermivora bachmanii*. Yet time and again, the ivory-bill materializes out of obscurity, often to those who were *not* looking for it, only to agitate our social conventions and sow chaos inside our cognitive networks.

Belief in ivory-bill extinction was taking a beating from bonus testimony during the *same decade* in which ensued the celebrated Big Woods and Choctawhatchie River escapades (Chapter 12). Along the Coosawhatchie River, in South Carolina, Hamilton reported *three* ivory-bills in June 2005.[47] During an aerial waterbird survey in April 2008 for the North Carolina Wildlife Resources Commission, Houston spotted a large woodpecker from above sporting bold black and white, the latter color on the wing's trailing edge. He reported this later to the Audubon office in North Carolina.[48] Testimony for two ivory-bills at close range came from federal refuge staff in 2007 near Felsenthal National Wildlife Refuge in Arkansas.[49] In the Wambaw Creek Wilderness of Francis Marion National Forest, South Carolina, Cork watched a "well described" pair of ivory-bills in direct contrast with two nearby pileated woodpeckers in September 2007.[50] In January 2007, Hodge saw a woodpecker in straight flight with much white on the top of the wings during an aerial waterbird survey conducted for Everglades National Park.[51]

Lesser-known challenges to our wild guessing at the bird's demise occurred in earlier decades, too. Near Toledo Bend Reservoir, Sabine River, Texas, Hyde, an observer familiar with the pileated, watched in the summer of 1985 a presumed family group of 4 to 5 ivory-billed woodpeckers for 10 minutes, noting in the birds a gray-ivory bill and trailing white edges to the wings both above and below.[52] A male ivory-bill was observed in 1986, a female in 1987, by biology instructor Higginbotham near the Pearl River, St. Tammany Parish, Louisiana.[53] "Possible multiple" sightings by the Rupps were made of the big woodpecker along the lower Suwanee River, Florida, in May 1995.[54] Sometime in the "late 1970s," Stevenson noted a pair, including a female poking her head out of a cavity, in the Apalachicola National Forest, Florida.[55] All of these form a series of some 100 incidents amassed for the ivory-bill since the late 1940s.[56]

Not just purely theoretical factors (Chapter 1) point to

survivability in very small populations of birds. For a medium-sized bird like the ivory-bill, the number of years it takes to put off extinction (i.e., the population lifetime) in a small population jumps from 50 to more than 100 years if the number increases from just 5 to 10 nesting pairs.[57] Even for small passerine birds, a group expected to have lower survival and much shorter lifespans than does the ivory-bill, the probability of extinction over 100 years rapidly declines to 0–10% in populations with 20 to 60 individuals.[58]

Very small populations of birds can thus hang on for quite some time. In North America, whooping crane *Grus americana* fell to as few as 15 adult birds, and California condor *Gymnogyps californianus* bottomed out at only 22 birds.[59] The single-island endemic Seychelles warbler *Acrocephalus sechellensis* once numbered just 26 individuals.[60] The population of orange-bellied parrot *Neophema chrysogaster* in Australia is now thought to be less than 50 birds.[61] As few as 20 to 30 Madagascar pochards *Aythya innotata* were thought to remain, all apparently found at a single location.[62]

What, then, lies behind our pigheaded death wish for this bird? After decades of unchecked ***anchoring bias***, a living ivory-bill insults us immeasurably more than the dead version. A deceased bird is easy to fathom. It's gone, we fall back on truthy dogma for consolation, and our leftover ignorance is exploited for ulterior aims: legend, trigger, nostalgia, quest, weapon, myth, obsession, enforcement, motivation, litmus test, allegory, even cartoon. But a live edition of this woodpecker spotlights too many disquieting limits to human reason. A living bird also threatens hallowed assumptions we use to preserve a shaky order inside those social conventions most heavily vested in the ivory-bill's plot line. Proprietorship over the ivory-bill's "rediscovery" still remains hotly contested.[63]

Imagine how conservation practice and wildlife management (Chapter 5) might react at learning that the ivory-bill had for 70 years needed absolutely no "help" for its survival. A living bird brings up the prospect of our abject failure.[64] Each profession might be disposed to runaway ***hindsight bias*** to explain away this impotence. Both disciplines would insist that land-set asides, afforestation programs, the Migratory Bird Treaty Act, duck stamps, and hunter-generated revenue had really protected the ivory-bill all along. This is not a cynical verdict, just a realistic one. Applied forestry pounced instantly on the Arkansas saga, insisting on offering its learned treatments for stand thinning and tree girdling on behalf of the ivory-bill's welfare.[65] Americans are a hands-on people. We can't help ourselves from a rash impulse to do something, even if it's just anything.

Scientific ornithology could fare little better with a surviving ivory-bill. "Belief that this bird is extinct has been held … as rigidly and dogmatically as the tenets of the most fundamentalist religious sects."[66] Ornithology would need to confront its tight ***anchoring*** on and rigid adherence to (***bandwagon fallacy***) the meager sampling (***insensitivity to sample size***) that so narrowly delimited the bird's presumed ecology and behavior (Chapter 8). That reckoning would need to quiz how our technical invention, sampling protocol, bio-statistical savvy, and modeling know-how all missed a large woodpecker. Ornithology might blame its overconfidence on ***pro-innovation bias***, and its rational hesitancy to invest in more search effort on ***sunk cost fallacy***. After all, "no amount of time or money will bring back the Zombie woodpecker."[67]

A feast of very bitter crow awaits sport birding at such a revelation, especially those who were most obdurate in insisting on a dead ivory-bill.[68] Competitive bird watching is not likely to ever take a genuine delight in the living bird. Though embodying only part of our collective interest in the ivory-bill, no other social identity so epitomizes human grasping at infallibility. Birding cannot extricate itself from rigid ***framing*** doctrine that insists rare birds be repeatedly found and verified,[69] if not from the masses, then by its designated experts.[70] The sport is predisposed to snub ivory-bill sightings by "outsiders," thus dismissing those possibly-valid observations made by lay persons.

Adamant insistence by the sport on indisputable "proof" is moreover curiously anti-science, exposing a rigid conceit that blinds its adherents to the inexorable canons of probability.[71] Even if incontestable evidence were acquired, blogging chatter inside the hobby would likely just pivot to how the species gets counted again according to the ABA rules. I hold little optimism that birding's cognitive flaws (Chapter 12) would get exhumed, its homophily and group contagion[72] inspected, or a measure of redemption sought for its shocking apathy[73] towards conservation.[74]

What makes hijacking of the ivory-bill narrative by birding so loathsome is not the hobby's immutable presumption that others conform to its doctrine. It is rather the past-time's calculated belittling of those who don't fall in line with extinction paradigm. In a stunning act of political manipulation for something as innocuous as bird identity, authors of the *National Geographic Field Guide* denigrate all ivory-bill incidents from recent decades as "not possible in an age when most rarities discovered are photographed and those images are posted on the Internet the same day."[75] Past and future observers alike are then insulted by an avowal that "sightings that lack provable evidence more likely represent wishful thinking," topped off by "the finality of a species' extinction is difficult for many to accept." With such brazen ***ultimate attribution error*** embedded in the resource used to verify any ivory-bills that were to be still alive, birding now insures that its cognitive blunders are spread more widely to future generations.

An extant ivory-bill may settle uneasily more broadly across our social networks. The woodpecker's rediscovery heralds

Box 15.2. *Ultimate attribution error* misleads us to brand entire classes of events or people with sweeping stereotypes. To say that everyone who has ever reported a putative ivory-billed woodpecker is deluded by wishful thinking stems from an ego-centric blunder linked to our making a clichéd distinction between in-group and out-group members of two or more social identities. We simply cannot judge the interior motives or intentions of other people accurately, especially in the collective sense. This error type also may be either negative or positive in its tone.

sobering lessons about stark limits to research method.[76] A huge bird sleuthing about for decades right under our noses doesn't auger well for placing a blind faith in an omnipotent security apparatus during this brand-new era of all-seeing drones. Turnabouts for something as glaring and basic as whether an entire species is alive or dead may threaten public trust in how accurately formal science can always represent a legitimate way of knowing.[77]

Getting extinction so utterly wrong also may lead to "accusations of hype and overblown ... rhetoric"[78] fired at the practice of conservation. A living woodpecker means that this species no longer serves as a legitimate exhibit in our "told-you-so" warnings over the inevitable penalties of failure to act, especially if set-asides of conservation land are the rescue on offer. Earlier reputed occurrences of the ivory-bill were a major rationale for setting aside Congaree National Park, South Carolina, and Big Thicket National Preserve, Texas.[79] Given the (apparent) lack of persistence in the species there, conservation would be rightly scrutinized for its own overindulgence in the sly deceptions of myth.[80]

A persevering ivory-bill also sets us up for a latent and unduly impactful ***black swan event***[81] – the bird's survival superficially impossible and at first totally unforeseen,[82] yet upon deeper reflection later understood as a one-off incident practically bound to happen (***hindsight bias***). Then, after the initial excitement wears thin, we would come to realize sheepishly that every bit of human agency vis-à-vis this species had been irretrievably foiled. We were stymied from killing it off. We were frustrated from rescuing it. That inimitable notion of our American exceptionalism[83] takes a flogging. Credence in our ultimate dominion over nature gets sullied.

Can we say with confidence that ivory-bill is either extinct or that it still persists? No, we cannot, not with anything resembling an unqualified certainty. The more rational stance is to shun the ***congruence bias*** and respect each possibility in play. Since the 1930s, the bird followed one of four paths. A remnant number may have postponed the inevitable, tantalizing a few of us intermittently, yet mostly hiding woeful details of their inexorable spiral into extinction. Possibly, this remnant is still heading towards a disappearance, but the final date is not yet set. Maybe the bird treaded demographic water, neither much increasing nor decreasing. Perhaps the ivory-bill squeezed through bottlenecks exacted by our greed, then edged slowly back from the abyss of species death. After never having had any robust clues about the woodpecker's real population size or trend, I see no way to discriminate forcefully among such utterly different pathways.

So, if another ivory-bill does ever materialize, we may not be able to rule out entirely that this next bird, too, isn't auditioning for a title role as last of its kind.[84] Even if we could dodge all of our mental conceits and then elude the bird's phoenix-like chicanery (Chapter 6), we still would be left to ponder. Was it just luck? Are the odds finally turning around for a convincing encore in the bird's appearance? Instead, our social illiteracy about the intricate workings of probability condemns us to force a plausibly living bird to go extinct over and over again, a charade that obliges us always to prove life instead of death.[85]

Whatever its plight, the ivory-bill commands the last laugh. Like a scout vanished into some lost frontier, the woodpecker's absence evinces a kind of poignant dignity. For never yielding to man's deliverance, and not dying in plain sight to appease our morbid curiosity, the ivory-bill was typecast purely by accident as self-reliant, private, even noble.[86] That archetype fixed an already magnificent species into our national psyche, enshrined it inside our native mythology.[87] Like other protagonists from bygone eras, the ivory-bill long fended off daunting adversaries – practically all of us,[88] and our eccentric social pacts, too. Refusing to surrender, snubbing new order we tried to impose upon it,[89] the ivory-bill lived out some unobtrusive destiny on its own terms, wild and free.[90]

Our own species does not assent very well to getting so totally vexed and vanquished by a mere bird. But no matter – dead or alive, the ivory-bill has bested us.

Endnotes

1 Klein, T. 2010. Woody abstracted: film experiments in the cartoons of Shamus Culhane. *Animation* 6: 39–53.

2 In my conception, the ivory-bill was sufficiently resourceful in a biological sense to have survived for some period of time past the 1930s, but with no guarantee that such versatility was adequate to insure long-term survival to any particular date certain, including (and especially) the present day.

3 Hand, D.J. 2014. *The Improbability Principle: Why Coincidences, Miracles, and Rare Events Happen Every Day*. Scientific American/ Farrar, Straus and Giroux, p. 142.

4 Fleischer, R.C., J.J. Kirchman, J.P. Dumbacher, L. Bevier, C. Dove, N.C. Rotzel, S.V. Edwards, M. Lammertink, K.J. Miglia, and W.S. Moore. 2006. Mid-Pleistocene divergence of Cuban and North

American ivory-billed woodpeckers. *Biology Letters* 11: 466–469.

5 "It seems to me entirely possible that no forest, virgin or otherwise, normally has a large enough, concentrated enough and continuous enough food supply for them to be permanent breeding residents." Pough, R.H. 1944. Report to the executive director, National Audubon Society: present condition of the Tensas River forests of Madison Parish, Louisiana and the status of the ivory-billed woodpecker in this area as of January, 1944. 9 pp. See also discussion of putative habitat specialization in Appendix B.

6 Land birds in North America are recognized as generally being quite adaptable and resilient to environmental change. "These species are the survivors of massive landscape changes during the Pleistocene, when the distribution of plants and animals shifted and re-shifted across hundreds of miles with the advance and retreat of continental glaciers." Askins, R.A. 2002. *Restoring North America's Birds: Lessons from Landscape Ecology*. Yale University Press, New Haven, CT, p. 232. See also Appendix B.

7 Chapman, F.M. 1898. *Handbook of Birds of Eastern North America: With Keys to the Species*. Forgotten Books, London, 2013 reprint, pp. 267–268.

8 Hunter, W.C. 2010. *Interpreting historical status of the ivory-billed woodpecker with recent evidence for the species' persistence in the southeastern United States*. Appendix E, Recovery Plan for the Ivory-billed Woodpecker (*Campephilus principalis*), U.S. Fish & Wildlife Service, Atlanta, GA, p. 71.

9 Hunter's analysis is one of few that used all sighting information available to reconstruct a temporal trend in reports of ivory-bill. Consequently, his treatment did not deliberately exclude or include reports arbitrarily based on ***argument from authority*** or some other fallacy that suppressed a more complete sighting record.

10 Because ivory-bill maintained prolonged pair-bonds, it was less vulnerable to population imbalances caused by skewed sex ratios. Risks of extinction due to environmental or demographic

stochasticity, genetic problems, or insufficient immigration are higher in small, isolated populations, an effect magnified when the natal dispersal is primarily by females, the usual condition in birds. Female-biased dispersal exaggerates the Allee effect and thus the extinction risk. See: Dale, S. 2001. Female-biased dispersal, low female recruitment, unpaired males, and the extinction of small and isolated bird populations. *Oikos* 92: 344–356.

11 Data loggers, radio, and satellite tags now show that birds routinely cover immense distances once considered unthinkable. Bermuda petrel, a bird smaller than ivory-bill, transits between Bermuda and the Azores, a distance of about 4,000 km. Madeiros, J., B. Flood, and K. Zufelt. 2014. Conservation and at-sea range of Bermuda petrel (*Pterodroma cahow*). *North American Birds* 67: 546–557.

12 "The ivory-bill problem puzzles me exceedingly, and I do not feel that Tanner's report begins to explain the reasons for the drastic declines in this species." Pough, p. 4. "It is hard to believe that this fact alone was sufficient to affect them so seriously." Pearson, T.G. 1917. *The Bird Study Book*. Doubleday, Page & Company, p. 129.

13 Intensities, frequencies, and scales of natural disturbance strongly influence the population viability in a variety of birds. Brawn, J.D., S.K. Robinson, and F.R. Thompson, III. 2001. The role of disturbance in the ecology and conservation of birds. *Annual Review of Ecology and Systematics* 32: 251–276.

14 Müller-Schwarze, D., and L. Sun. 2003. *The Beaver: Natural History of a Wetlands Engineer*. Cornell University Press, New York. In line with the overall timing of ivory-bill imperilment, a large proportion of the North American population of beaver was depleted in southern latitudes before 1900. Rosell, F., O. Bozser, P. Collen, and H. Parker. 2005. Ecological impact of beavers *Castor fiber* and *Castor canadensis* and their ability to modify ecosystems. *Mammal Review* 35: 248–276. Beaver girdling and damming resulted in patches of rapidly dying trees; each process is remarkably similar in scale to the deliberate girdling of trees by humans said to have

been so effective at luring ivory-bills to particular locations.

15 Jackson, J. A. 2006. *In Search of the Ivory-billed Woodpecker*, 2nd ed. HarperCollins, New York, p. 65.

16 Several distinctive aspects of ivory-bill behavior and distribution *greatly* reduce its detectability. Survey efforts are much less effective if birds are spatially distributed in patterns other than uniform, or if a relict population is confined to a small sub-region of the entire area being targeted by surveys, both of which are likely to be true with this species. See: Scott, J.M., F.L. Ramsey, M. Lammertink, K.V. Rosenberg, R. Rohrbaugh, J.A. Wiens, and J.M. Reed. 2008. When is an "extinct" species really extinct? Gauging the search efforts for Hawaiian forest birds and the ivory-billed woodpecker. *Avian Conservation and Ecology* 3: 3. http://www.ace-eco.org/vol3/iss2/art3/ (accessed 5 February 2016).

Searchers generally assume ivory-bills "are neither attracted to nor repelled by observers." *Ibid.*, p. 13. But early naturalists drew attention to its evasive tactics around humans. For example: "We had hunted for three days for this particular pair of birds without ever hearing them, even though we were frequently within three hundred yards of the nest." Allen, A.A., and P.P. Kellogg. 1937. Recent observations of the ivory-billed woodpecker. *The Auk* 54: 164–184. "The nearest bird coming to within a hundred feet of me but very carefully staying out of my sight, although I am sure that I did not stay out of the bird's sight." Crompton, D.H. 1950. My search for the ivory-billed woodpecker in Florida. *Bulletin of the Massachusetts Audubon Society* 34: 235–237. Some anecdotal accounts infer that ivory-bill may (have) approach(ed) only to within a minimum distance of observers who were actively trying to attract it closer. A "rough distance of 80 meters or so has been reported as an [ivory-billed woodpecker] stopping distance by more than one researcher." See: Michaels, M., and J. Williams. Guest Trip Report, March 22–24, 2016 by John Williams. https://projectcoyoteibwo.com/2016/04/21/guest-trip-report-march-22-24-2016-by-john-williams/ (accessed 24 April 2015). Modern-day surveys have been criticized as "not secretive enough," or "not sufficiently stealthy." Steinberg, M.K. 2008. *Stalking the Ghost Bird: The Elusive Ivory-billed Woodpecker in Louisiana.* Louisiana State University Press, Baton Rouge, pp. 59, 100.

17 "The woodpeckers are very silent at all times so far as their voices are concerned." Ridgway, R. 1898. The home of the ivory-bill. *Osprey* 3: 35–36. The ivory-bill's call "lacks carrying capacity and can rarely be heard over 100 yards [90 m] on a still morning." Hoyt, R.D. 1905. Nesting of the ivory-billed woodpecker in Florida. *The Warbler* 1: 53–55. In thick vegetation, the ivory-bill's "conversational notes might hardly carry 200 yards" (~180 m). Tanner, J.T. 1942. *The Ivory-billed Woodpecker*. Dover Publications, Inc., Mineola, NY, pp. 61–62.

18 Despite no strong historical evidence for its effectiveness, modern-day researchers still attempt call play-back and simulated double knocks to lure ivory-bills to respond. Call-backs elicited from ivory-bills were tried with a saxophone mouthpiece, but "the birds paid no attention to it." Tanner, p. 61. "I tried to devise some imitation of their call that would make an ivory-bill answer, but nothing was successful." *Ibid.*, p. 22. Tanner was only "occasionally able to make ivory-bills answer by imitating their ... single or double raps as loudly as [he] could." *Ibid.*, p. 62. In modern times, just one of 13 double-knock encounters with putative ivory-bill in South Carolina occurred in response to an imitation. Moskwik, M., T. Thom, L.M. Barnhill, C. Watson, J. Koches, J. Kilgo, B. Hulslander, C. Degarady, and G. Peters. 2013. Search efforts for ivory-billed woodpecker in South Carolina. *Southeastern Naturalist* 12: 73–84. Playback or mimicked double knocks did not consistently elicit responses from ivory-bills during the Arkansas searches, either. To minimize confusion with the data recorded by remote sound stations, the Florida search team avoided entirely such human-generated mimicry. Hill, G.E. 2007. *Ivorybill Hunters: The Search for Proof in a Flooded Wilderness*. Oxford University Press, London, p. 107. Lack of consistent response by ivory-bill to

auditory stimulus from imitated double-knocks and/or recorded playback of vocalizations instead may be perceived by the birds as aggressive signals, serving as a repellent instead of an attractant. "They decided to play the ivory-bill tape – and the birds shut up instantly." Gallagher, T. 2006 *The Grail Bird: The Rediscovery of the Ivory-billed Woodpecker*. Houghton Mifflin Harcourt, Boston, MA, p. 23. This oversight again reveals our liability to ***congruence bias***.

19 If the total population sampled in Arkansas was actually 5 or more birds, however, then the odds of detecting one or more ivory-bills rose to greater than 90%. Scott et al., p. 9.

20 Pulliam, B. 2010. Shroedinger's woodpecker. *Notes from Soggy Bottom*, pp. 3–5. http://bbill.blogspot.com/2010/04/schroedingers-woodpecker.html (accessed 5 February 2016). Pulliam granted quite charitably that up to 10,000 person-hours per year of birding effort might be expended in 20,000 km^2 of bottomland habitat spread across the southeastern U.S., with only 10% of all detections constituting "seen" (rather than heard) ivory-bills. In reality, bottomland habitats are usually shunned by birders due to the logistical impediments in obtaining suitable gear, plus bother from biting insects, inaccessibility from flooding, poor viewing conditions, low species diversity, and so on.

21 *Ibid.*, p. 5.

22 Ivory-bill generated at least two sound types. Its vocalizations were usually rendered phonetically as some form or pattern of nasal *kent* calls. In addition, the ivory-bill used its bill to create very loud and resonant single- and (more commonly) double-knocks (DKs) to communicate with conspecifics. However, high-pitched alarm calls and popping wing noises have been described and linked putatively to this species, sound types that do not match the woodpecker's earlier known (or indisputably documented) repertoire. See: Collins, M.D. 2010. Putative audio recordings of the ivory-billed woodpecker (*Campephilus principalis*). *Journal of the Acoustical Society of America* 129: 1,626–1,630; Collins, M.D. 2012. Acoustic detection of the ivory-billed woodpecker (*Campephilus*

principalis). *Proceedings of the Meetings on Acoustics* 14: 010004. Collins, M.D. 2021. The role of bioacoustics in the conservation of the ivory-billed woodpecker (*Campephilus principalis*). *Journal of Theoretical and Computational Acoustics.*

23 Hunter, pp. 77, 85–87. See also: Rosenberg, K.V., R.W. Rohrbaugh, and M. Lammertink. 2005. An overview of ivory-billed woodpecker (*Campephilus principalis*) sightings in eastern Arkansas in 2004-2005. *North American Birds* 59: 198–206; Hill, G.E., D.J. Mennill, B.W. Rolek, T.L. Hicks, and K.A. Swiston. 2006. Evidence suggesting that ivory-billed woodpeckers (*Campephilus principalis*) exist in Florida. *Avian Conservation and Ecology* 1: 2 http://www.ace-eco.org/vol1/iss3/art2

24 Tanner, J.T. 2000. A postscript on ivorybills. *Bird Watcher's Digest* July/August: 52–59.

25 Sibley's rigid opposition to the identity and persistence of ivory-bill in Arkansas arose from ***neglect of probability***, and after he only "spent eight days hiking and paddling in the Arkansas marshes searching in vain for the ivory-billed." McCarty, J.F. 2010. Leading birders voice skepticism over reports of ivory-billed woodpecker sightings; most likely pileateds, they say. *The Plain Dealer* http://www.cleveland.com/neobirding/index.ssf/2010/11/leading_birders_voice_skeptici.html (accessed 3 July 2016).

26 Most putative encounters with ivory-bill during the modern era, for example, originate from areas well within its historical range and in habitat types linked to earlier reports of this species. Some modern, putative encounters, however, have been in habitat containing suitable foraging trees in more upland forest types away from extensive bottomland hardwoods, e.g., Gillum, P. 2015. Conclusion – winter of the ivory billed woodpecker. http://foreverahillbilly.blogspot.com/2015/11/conclusion-winter-of-ivory-billed.html (accessed 30 April 2016).

27 To illustrate, Collins spent parts of at least eight years searching to obtain some dozen total encounters in the Pearl River area of Louisiana and Mississippi (M. Collins, *pers. comm.*, 11 February

2016).

28 Expectations that ivory-bill(s) would persist indefinitely at sites judged initially as suitable got dashed again and again during the 2000s. Although he thought that the "evidence suggests that what we have discovered is a resident population of ivory-bills" (Hill, 2007, p. 238), those birds in fact only lingered in the Choctawhatchee River study site for some 3 calendar years, leading to "a flurry of encounters followed by an inability to detect any birds." *Ibid.*, p. 101. "Our experience over the past year indicates that ivorybills have moved out of the areas where we encountered them from 2005 to 2008." See Hill, G.E. 2009. Ivory-billed woodpeckers in the Florida Panhandle: updates from Florida. http://www.auburn.edu/academic/science_math/ivorybill/Updates.html (accessed 11 February 2016).

Encounters by Collins in the Pearl River were also temporally clustered, e.g., five sightings in 2006 were within a single localized area over just five days (M. Collins, *pers. comm.*, 11 February 2016). Repeat encounters with the apparently single Arkansas woodpecker lasted little more than a calendar year. Searchers in Louisiana have looked since 2009 for ivory-bills, acquiring recordings, sightings, and other evidence consistent with identity of this species. However, "while we believe that ivorybills are still present in the vicinity ... the birds do not seem to be frequenting our old hot zone. For this reason, we have shifted our focus to other parts of the state." Michaels, M., and F. Wiley. 2013. About Project Coyote. http://projectcoyoteibwo.com/2013/04/24/about-project-coyote/ (accessed 11 February 2016).

29 "Either the excitement of the ivorybill hunt causes competent birders to see and hear things that do not exist and leads competent sound analysts to misidentify hundreds of recorded sounds, or the few ivorybills in the Choctawhatchee River Basin are among the most elusive birds on the planet." Hill, 2009. This reaction showcases how cognitive friction arises from our trying to process sparse evidence consistent with the ivory-bill, but doing so after

decades of ***anchoring bias*** already favor overwhelmingly an insistent belief in the bird's putative extinction.

30 Due to extreme difficulties of detection, the expected waiting time for securing a clear photograph of ivory-bill is estimated to be several orders of magnitude longer than for other species of comparable rarity. See: Collins, M.D. 2019. Statistics, probability, and a failed conservation policy. *Statistics and Public Policy* 6: 67–79.

31 Cieply, M. 2011. That noisy woodpecker had an animated secret. *New York Times*, April 10. http://www.nytimes.com/2011/04/11/arts/design/woody-woodpecker-and-shamus-culhanes-animation.html?_r=0 (accessed 1 February 2016).

32 Anti-Defamation League. 2016. Symbol guide for law enforcement: peckerwood. http://www.adl.org/mobilehatesymbols/symbol-50.html (accessed 1 February 2016).

33 Björkqvist, K., and K. Lagerspetz. 1985. Children's experience of three types of cartoon at two age levels. *International Journal of Psychology* 20: 77–93.

34 See Glossary.

35 At its core our intolerance to uncertainty stems from apprehension. Berenbaum, H., K. Bredemeier, and R.J. Thompson. 2008. Intolerance of uncertainty: exploring its dimensionality and associations with need for cognitive closure, psychopathology, and personality. *Journal of Anxiety Disorders* 22: 117–125. Those having greatest needs for closure are the most resistant to persuasion. Webster, D.M, and A.W. Kruglanski. 1994. Individual differences in need for cognitive closure. *Journal of Personality and Social Psychology* 67: 1,049–1,062. Moreover, need for closure leads to our seeking consensual validation from others, preferring more conventional social norms, wanting to preserve stability, and conforming to and maintaining the status quo. Cologero, R.M., A. Bardi, R.M. Sutton. 2009. A need basis for values: associations between the need for cognitive closure and value priorities. *Personality and Individual Differences* 46: 154–159.

36 Muddled semantics can easily cloud messaging so that the public may perceive that extinction isn't really forever, or that the crisis is being invented. "We cannot ensure that our words will be disseminated accurately, but we can rein in our egos, our bleeding hearts, and our best intentions before we communicate with the public." Smith-Patten, B.D., E.S. Bridge, P.H.C. Crawford, D.J. Hough, J.F. Kelly, and M.A. Patten. 2015. Is extinction forever? *Public Understanding of Science* 24: 481–495.

37 Diamond, J.T. 1987. Extant unless proven extinct? Or, extinct unless proven extant? *Conservation Biology* 1: 77–79. Note that subscribing to the latter stance requires an ongoing and presumably never-ending proof of life.

38 Making us prone to both Type I decision error (concluding extinction when a species actually persists, i.e., the ***Romeo error***) and Type II decision error (continuing to search for a species after it is gone). See: Reed, J.M. 1996. Using statistical probability to increase confidence of inferring species extinction. *Conservation Biology* 8: 309–319.

39 Mace, G.M., N.J. Collar, K.J. Gaston, C. Hilton-Taylor, H.R. Akcakaya, N. Leader-Williams, E.J. Milner-Gulland, and S.N. Stuart. 2008. Quantification of extinction risk: IUCN's system for classifying threatened species. *Conservation Biology* 22: 1,424–1,442.

40 Collen, B., A. Purvis, and G.M. Mace. 2010. When is a species really extinct? Testing extinction inference from a sighting record to inform conservation assessment. *Diversity and Distributions* 16: 755–764.

41 A succinct overview of individual and social cognition, including a synthesis of mankind's most prevalent thinking biases and errors during decision making, is given by Pronin, E. 2008. How we see ourselves and how we see others. *Science* 320: 1,177–1,180.

42 Hill, 2007, p. 149. *Campephilus* woodpeckers do tend to return to trees visited before, but the number of foraging trees is inversely related to flight distance, in keeping with radial distribution functions that greatly lower the probability of detection with

increasing distance from any central place. Vergara, P., and R.P. Schlatter. 2004. Magellanic woodpecker (*Campephilus magellanicus*) abundance and foraging in Tierra del Fuego, Chile. *Journal of Ornithology* 145: 343–351.

43 "I've also been somewhat obsessed with finding a category of feeding sign that is beyond the physical capacity of a pileated woodpecker to produce and is therefore diagnostic." Michaels, M. 2013. Encounter rate. http://projectcoyoteibwo.com/2013/10/13/encounter-rate/ (accessed 11 February 2016).

44 Michaels, M. 2016. Squirrels stripping bark. http://projectcoyoteibwo.com/2016/01/05/squirrels-stripping-bark (accessed 9 February 2016). It may well be that *some* ivory-bill foraging sign is entirely unique to this species. However, whatever foraging sign that overlaps with other woodpeckers (or other animals) is obviously not so, with the consequence that ivory-bill is more cryptic and less easily detected or tracked through proxy in the field than is often assumed, a condition that sets us up for decision errors from ***illusory correlation***.

45 Sibley's "mounting doubt" to the prospect of living ivory-bill in Arkansas was premised in substantial part on his failure to find "any signs of bark peeled from decaying oak trees – a key indicator of the presence of ivory-billeds feeding on their favorite food, giant grubs." McCarty. However, Sibley's logic was compromised by the falsifiability of its fundamental premise – ivory-billed woodpeckers were known to use foraging tactics other than bark-peeling. See, e.g., Allen and Kellogg.

46 "We declare species extinct too soon, sealing them off from further investigation and only realizing our mistake when evidence … emerges from some unexpected quarter." Collar, N.J. 1988. Extinction by assumption: or, the Romeo error on Cebu. *The Oryx* 32: 239–244.

47 Hunter, p. 77.

48 *Ibid.*

49 *Ibid.*, p. 87.

50 *Ibid.*, p. 77.

51 *Ibid.*, p. 83.

52 *Ibid.*, p. 88.

53 Steinberg, pp. 76–78.

54 Hunter, p. 79.

55 *Ibid.*

56 *Ibid.*, pp. 75–89.

57 Pimm, S.L., J. Diamond, T.M. Reed, G.J. Russell, and J. Verner. 1993. Times to extinction for small populations of large birds. *Proceedings of the National Academy of Sciences* 90: 10,871–10,875.

58 Legendre, S., J. Clobert, A.P. Møller, and G. Sorci. 1999. Demographic stochasticity and social mating system in the process of extinction of small populations: the case of passerines introduced to New Zealand. *The American Naturalist* 153: 449–463.

59 BirdLife International. 2016. Species factsheet: *Neophema chrysogaster*. http://www.birdlife.org (accessed 4 February 2016).

60 Komdeur, J. 1994. Conserving the Seychelles warbler *Acrocephalus sechellensis* by translocation from Cousin Island to the islands of Aride and Cousine. *Biological Conservation* 67: 143–152.

61 BirdLife International. 2016. Species factsheet: *Neophema chrysogaster*. http://www.birdlife.org (accessed 4 February 2016).

62 BirdLife International. 2016. Species factsheet: *Aythya innotata*. http://www.birdlife.org (accessed 4 February 2016).

63 "Who determines when something is rediscovered?... Is it a group of elite birders?... Or can local people or state wildlife officials be credited with rediscovery?" Steinberg, p. 7.

64 "It's going to be rather hard to implement a recovery plan for a bird that by and large we cannot locate, cannot monitor, cannot even begin to count, but yet we know it's out there." *Ibid.*, p. 73.

65 Given that "forestry" destroyed the presumed last redoubt of the ivory-bill in the Singer Tract, the irony here needs little comment. See: Shoch, D.T. 2005. Forest management for ivory-billed woodpeckers (*Campephilus principalis*): a case study in managing an uncertainty. *North American Birds* 59: 214–219. Among

various treatments, artificial inputs to standing dead wood were promoted, either through selective girdling or prescribed fire. These and other prescriptions were offered at least with an admission of the inevitable limitations to "an approach to deriving practical management applications from incomplete information, a situation with which land managers continue to be confronted" (p. 216).

66 Gallagher, p. 237.

67 http://www.economist.com/node/21532341/comments#comments (accessed 8 October 2015).

68 "Sibley and Kaufmann were in error in concluding that the extinction of this species (and two others) was certain enough that it did not [need] to be included in 'comprehensive' North American field guides." Pulliam, p. 5.

69 "Dunn commented that with many rare birds now being photographed, and with the images posted to the internet often within hours, it seems inconceivable that populations of ivory-billed woodpeckers in the eastern U.S. could completely evade the efforts of birders and photographers to document their existence." Pranty, B., J.L. Dunn, D.D. Gibson, M.J. Illiff, P.E. Lehman, M.W. Lockwood, R. Pittaway, and K.J. Zimmer. 2011. 22nd Report of the ABA checklist committee 2011. *Birding November 2011*: 26–33.

70 Indeed, birding's ***argument from authority*** is preserved indefinitely through unchecked ***system justification***. Some birders admit their acquiescence to this rigid internal authority with a surprising candor. Enforcement of the status quo is seen as "the way in which birders ... defer to each other, respond to authority and reputation, stake out territory, and all of that ... birdwatching social structures are very feudal in their nature.... The ABA area, like Medieval Europe, has ... several kings ... [who] still roam North America with regal acclaim, and there are more with equal powers who don't have their names attached to field guides.... They post often and thoroughly to the listserv and they police it with equal vehemence – fulfilling the lords'

obligation of administering justice among his subjects, no doubt. They also are often eBird moderators ... for their individual fiefs, controlling what and how the knights and peasants get to bird in their domain, what is accepted ... the surest way for one to rise and fall in the birding world is based on reputation. Reputation is immortal. It takes a long time to build and can quickly be lost. It affects every bird-related part of the birder's life, the birder's family, the birder's capital, the birder's place in heaven."

Of course, not even feudal systems were entirely secular, so birding too has its spiritual authority: "the birder's church is principally the ABA." See: Butler, L. 2015. Birder sociology revisited: birding in a feudal world. Butler's Birds http://butlersbirdsandthings.blogspot.com/2015/09/birder-sociology-revisited-birding-in.html (accessed 7 February 2016).

71 Errors from the ***neglect of probability*** are rampant inside birding. By relying entirely on a ***framing effect*** that equates reliable information with "proof," that is, only material that is unassailable or irrefutable, birding makes itself vulnerable to thinking and decision errors that arise from ***base rate fallacy***, lack of ***Bayesian caution***, as well as ***Bayesian conservatism***, ***insensitivity to sample size***, and ***zero risk bias***.

72 Mark, N.P. 2003. Culture and competition: homophily and distancing explanations for cultural niches. *American Sociological Review* 68: 319–345; Loersch, C., H. Aarts, B.K. Payne, and V.E. Jefferis. 2008. The influence of social groups on goal contagion. *Journal of Experimental Social Psychology* 44: 1,555–1,558.

73 Competitive bird watching regularly seeks out rarities and higher species tallies at superfund sites, landfills, and sewage ponds. Though often environmental in use of physical space, the pastime is decidedly not environmental*ist*. "Sport aspects of birding tend to minimize the dire significance of toxicity and mask the significance of the polluted sites.... The obvious contradiction within big-year birding is that it has typically been practiced in ways that involve large expenditures of fossil fuels ... and in

this ... adding to greenhouse gases.... In the case of competitive birding, the relationship between environmental sporting and environmentalism is such that environmental problems are either masked or made light of." Consequently, "among competitive birders, competition obviates engagement with environmental dangers." Schaffner, S. 2009. Environmental sporting birding at Superfund sites, landfills, and sewage ponds. *Journal of Sport & Social Issues* 33: 206–229.

74 Some of the divergence between the identities of birding and conservation arises from a peculiar ***framing effect*** exemplified in field guides, namely a visual depiction that divorces birds from nature. "Forms of understanding that field guides promote are not necessarily conservationist. Instead field-guide authors represent birds and the environments they live in as strangely detached from and unaffected by a wide variety of currently pressing environmental challenges." Field guides act to portray "birds as living apart from one another and from the consequences of human intervention." Schaffner, S. 2011. *Binocular Vision: The Politics of Representation in Birdwatching Field Guides*. University of Massachusetts Press, Amherst, p. 3. In addition, the "high expectations of many birdwatchers are not always beneficial and the excessive zeal of some birdwatchers to see or photograph certain species may have harmful consequences." Sekercioglu, Ç.H. 2002. Impacts of birdwatching on human and avian communities. *Environmental Conservation* 29: 282–289. Finally, although ecotourism from bird-watching can bolster local economies, no mechanisms yet exist to direct those expenditures back into conservation, as is the case with hunters and anglers. Birding thus does not in any fundamental manner "pay its way" for financing the hefty costs associated even with routine upkeep on the nation's conservation infrastructure.

75 Dunn, J.L., and J.K. Alderfer. 2011. *National Geographic Field Guide to the Birds of North America*. National Geographic Books, Washington, DC, p. 324.

76 Rediscovery of species once thought extinct can provide valuable insights for conservation, especially if the species is one for which the decline and disappearance has been well studied before its rediscovery. Ladle, R.J., P. Jepson, A.C.M. Malhado, S. Jennings, and M. Barua. 2011. The causes and biogeographical significance of species' rediscovery. *Frontiers of Biogeography* 3: 111–118.

77 Gauchat, G. 2011. The cultural authority of science: public trust and acceptance of organized science. *Public Understanding of Science* 20: 751–770; Master, Z., and D.B. Resnik. 2013. Hype and public trust in science. *Science and Engineering Ethics* 19: 321–335.

78 Ladle, R.J. 2009. Forecasting extinctions: uncertainties and limitations. *Diversity* 1: 133–150. Misuse of the term "extinction," which accelerated notably after the 1950s, is one source of potential distrust in the messaging from conservation. See: Smith-Patten, B.D., E.S. Bridge, P.H.C. Crawford, D.J. Hough, J.F. Kelly, and M.A. Patten. 2015. Is extinction forever? *Public Understanding of Science* 24: 481–495.

79 If ivory-bill was highly nomadic or reliant on exceedingly large landscape mosaics, however, it is doubtful that any single park or other set-aside would have been sufficient to provide protection anyway.

80 Wilhere, G.F. 2008. The how-much-is-enough myth. *Conservation Biology* 22: 514–517; Hilderbrand, R.H., A.C. Watts, and A.M. Randle. 2005. The myths of restoration ecology. *Ecology and Society* 10: 19. http://www.ecologyandsociety.org/vol10/iss1/art19/ (accessed 27 February 2016).

81 An eventuality in which the ivory-bill might have persisted for a substantial time after the 1940s fits the definition of a ***black swan event*** because it would have colossal ramifications for status quo beliefs held by science, conservation, birding, environmental history, and media. According to the theory of ***black swan events***, rather than being governed by the "normal" and the average, the real world is instead steered by the random, the unpredictable, and at times even the unknowable. Taleb, N.N. 2007. *The Black*

Swan: The Impact of the Highly Improbable. Random House, New York.

82 For the ivory-billed woodpecker, the ***black swan*** phenomenon might play out as follows: "If ivory-bills are ever documented, the frustrated cry arising from the ornithology community will be, 'How could this possibly be? HOW could we have missed them along the way!?' But some relatively small assumptions (about behavior and habitat requirements) can account for it.... Nature is subtle, but persistent; humans on-the-other-hand are rarely as competent as we think we are." Cyberthrush. 2016. Explaining the inexplicable...http://ivorybills.blogspot.com/ (accessed 1 May 2016).

83 "American exceptionalism has always had two sides: the one eager to set the world to rights, the other ready to turn its back with contempt if its message should be ignored." MacMillan, M. 2002. *Peacemakers: The Paris Conference of 1919 and Its Attempt to End War*. John Murray, London, p. 22.

84 "Last of its kind" reverberates cogently in American historical memory. Bill Fought, who accompanied Eckelberry and warden Jessie Laird to the site of the "last" ivory-bill in the Singer Tract, remarked: "I just thought it would be a sad thing to be the last of anything on earth." Severson, F.J. 2007. Memories from the Singer Tract. *Birding* March/April: 42–46. "Last of its kind" also relates to a broader national awareness inclined toward (and a nostalgia about) the abrupt ending to the American frontier. Ishi, a last survivor of California's Yahi people, did not step out of what little persisted of the Sierra Nevada wilderness until 1911. Ishi's remaining life and his eventual death in 1916 were widely chronicled, and kindled sensational interest across early twentieth-century American society. Ishi's story overlapped exactly the timelines for two other last of their kinds, the passenger pigeon and Carolina parakeet. Like Ishi's, these deaths transpired largely in public view. Heizer, R.F. 1979. *Ishi, the Last Yahi: A Documentary History*. University of California Press, Oakland. Kroeber, T. 2004. *Ishi in Two Worlds: A*

Biography of the Last Wild Indian in North America. University of California Press, Oakland.

85 The American Birding Association's Checklist Committee recapped faulty logic with ***argument from ignorance***: "The Committee is waiting for unequivocal proof that the species still exists." Hoose, P. 2004. *The Race to Save the Lord God Bird*. Farrar, Straus, and Giroux, New York, p. 169. Similarly, and given that the evidence gathered had "not been proven to have been made by this woodpecker [ivory-bill]," the Florida Ornithological Society Records Committee fell to the same fallacy. Kratter, A.W. 2008. Seventeenth report of the Florida Ornithological Society Records Committee: 2007. *Florida Field-Naturalist* 36: 94–111. Bird-watching compounds its decision errors by linking ***neglect of probability*** (showing contempt for scientific uncertainty) to ***arguments from ignorance*** and ***authority***.

86 I here evoke shameless parallels to such iconic portrayals as those by Alan Ladd in *Shane* or Gary Cooper in *High Noon*, i.e., the largely unconscious associations that we make because in the end our "myth expresses ideology." See Springhall, J. 2011. Have gun, will travel: the myth of the frontier in the Hollywood Western. *The Historian* Winter 2011: 20–24.

87 "there can be no doubt that the ivory-billed woodpecker persists ... in the hearts, the memories, and the cultural conscience of those of us who will never see the bird." Wright, R. 2007. Taking it personal: where the ivory-bill survives. *Birding* March/April 2007: 48–52, p. 52. And from where comes the impetus for such dogged persistence? "The mythology of a nation is the intelligible mask of that enigma called the 'national character.' Through myths the psychology and world view of our cultural ancestors are transmitted to modern descendants, in such a way and with such power that our perception of contemporary reality and our ability to function in the world are directly, often tragically affected." Slotkin, R. 1973. *Regeneration Through Violence: The Mythology of the American Frontier*. University of Oklahoma Press, Norman, p. 3.

88 Ivory-bill did not succumb to an easy, quick conquest either from our continent's indigenous peoples, who long killed it for its presumed totemic and ceremonial power, or from later European colonizers, who also assigned various commercial, aesthetic, and symbolic values on the big woodpecker. "Savage liberty is a prerequisite of [the ivory-bill's] existence and its home is the depth of the woods, remotest from the activities of civilized man." Thompson, M. 1885. A red-headed family. *The Elzevir Library* 4: 5–21.

89 "He accepts no gifts from man, and asks no favors." *Ibid.*, p. 16.

90 And so art imitates life. Towards the end of sweeping epic *Once Upon a Time in the West*, completion of the railroad symbolized that the American frontier was gone forever. Two protagonists, Cheyenne (Jason Robards) and Harmonica (Charles Bronson), had once lived as nomads, but can never be confined to this newly domesticated world. Now just relics of an "ancient race" (Appendix B), they flee a burgeoning civilization, each drifter realizing a life-or-death destiny in some ever-shrinking corner of freedom.

Afterword

Faith Too Far

Doubt is a pain too lonely to know that faith is his twin brother.
—Khalil Gibran

If reason were a siren, it is a devious creature luring us to greatly overprize our thinking skills. Conservation practices of tomorrow will need to devote as much attention to the alternating currents darting around inside our heads as to the unswerving sentiments tugging at our hearts. It has sobered me, someone tutored in empiricism, to behold the immense powers of history, myth, psychology, and sociology so mangle our perceptions about the material ivory-billed woodpecker. A journey to perceive this feathered icon through the lens of cognitive framing has roused me to remain vigilant over the hidden power of unconscious memes in my own life.

Neutrality on ivory-bill borders on the hopeless. We are dreadful at managing the vast obscurity that surrounds this bird's real fate. Nowhere close to the rational confidence we project, we still persist in assuming that we are the only logical ones left in the room. Our minds are hard-wired for colorful story-telling, not for weighing how considerably the randomness and uncertainty of the natural world come to rule how much we can truly know.[1] Despite our own species' faculty for thoughtful deliberation, the ivory-bill has made us look foolish and feeble. When all is said and done, it was (is) after all just a really large woodpecker.

We ceded a stunning amount of clout to this denizen of the natural world. In the main I do not think that the ivory-billed woodpecker owned biological traits that were so greatly different from other woodpeckers, and was certainly not so unusual if measured against qualities in all members of the avian tribe. No, what makes this bird most distinctive is its amazing facility to prejudice our thinking and manipulate so many of our social conventions, all the while we stay mostly clueless to the bird's inadvertent trickery.

Like the Korean Armistice, stalemate over the ivory-bill's living existence seems endless. Birders will never wade deep into lonely swamps to search[2] for a bird they cannot count on their precious lists, only to dodge scorn from haughty peers[3] ready to keep order in the house – *their* house, as they see it now. Modern science cannot afford the time, money, and other foregone options gambling on sample sizes that hover between zero and one.[4] Given this entirely rational avoidance of ***sunk cost fallacy***,[5] conservation groups might render help, but not so long (nor as well) if no one agrees whether there is a live species out there to still protect.[6] So, if more tangible evidence is ever unearthed, it will likely come through serendipity by volunteer searchers or rural denizens. The former toil with conviction under internet aliases, the latter just go about their normal routines, each crowd seemingly less jaded than the rest of us.

I wonder, too, if deniers would agree on *any* verification for a still-living ivory-bill anyway. The evidentiary goalpost has been shifted so high that this contest looks already called. My hunch is that a decent photograph will be struck with accusations of bogus fraud intended by credibility police to muzzle us into silence. An excellent image will be denounced as counterfeit because the quality is *too* good, an inferior one disparaged as too insulting to qualify even as hoax. Video will come with its own conspiracy: Pixar Animation Studios was in on the scam. Feathers? Planted after being plucked from museum specimens[7] for a nefarious purpose of restricting civil liberties, doubtless for onerous land grabs and attempts to restrict our right to bear arms. We can only hope that "proof" exacted by these rigid cynics never becomes a demand for the fresh kill.[8]

Even if the ivory-bill were incontestably verified as still extant, how soon would it be before we buried the bird alive all over again.[9] So sure of our conceited selves when it comes to divining nature's course, we again insist: "No, this one, this one really, really is the very last of its kind." And then turn right around and exploit the bird's next epithet to achieve our insentient aims. Imprisoned by clashing narratives that no longer serve us at all, we remain incapable of contemplating that the bird really does not need us one bit.

Still, there is room for some optimism. It requires rather a lot from sensibly inclined persons to disbelieve every single report since the mid 1940s, from each observer who saw and wrote down their direct experience. Most were not out to make a wider claim, to score points with foes or impress friends. Many weren't looking for an avian legend at all.[10] Imagine what it must have been like, though, for the mindful ones to wrestle with seeing a large blackish woodpecker with white plumage or bill color where it didn't belong, the mental gymnastics required to deal with this unbidden surprise, the mind swirling into an endless loop of "this can't be what I see ... but, yes, this is what I see."

A faith too far is exacted if we are bidden to close our eyes to all of the ivory-bill incidents from the last 70 years. Such special pleading mocks basic canons of probability. Mistakes made in identification? Certainly some, perhaps many, even most. But 100 or more consecutive hoaxes,[11] guesses, and good-faith errors? Even with more than one witness? Proponents of this stance insist that we convert to their peculiar dogma, endorsing some strange ideology that makes flat earth certitude look passably reasonable.

Neither can I buy the notion that it is our hopeful, optimistic nature (really?) that deludes us into seeing ivory-bills that were (or are) not there. Some speculate that the more dramatic and flamboyant animals, once gone forever, cast such psychic holes from their absence that "the less likely we are to accept [their] loss, and the more apt we are to keep hunting and hoping, even when the evidence is pretty grim."[12] This sort of wild conjecture about everybody's unseen intentions just forces us to commit cognitive transgressions of ***ultimate attribution error***. Anyway, our vulnerability to the ***negativity bias*** means that the doom of extinction, not any happy optimism for survival, will instinctually rule our mental outlooks.[13]

More than I had anticipated at the outset of this exploration I came to esteem those who stood up to harsh verdicts levied on persons who distrusted the ivory-bill myths. We should thank profusely those who saw, came forward and said cleanly, with no hedging, that they knew the ivory-bill was a still-living creature. Whitney Eastman, Herbert Stoddard, George Lowery, John Dennis, the Cornell Lab folks, Geoff Hill, Mike Collins, among all the various and many others.

Unexpectedly, this journey alleviated the unease over my own contradictions. The entrenched nature of cognitive error in all of us means that I need not make any special claim to impartiality nor defend an impossible objectivity. As to that particular clash that gives my daughter such fiendish delight, apparently I am not entirely alone in these discrepancies. Those who engage in both birding and hunting are more active in conservation than those who practice either pursuit singly.[14] Inconsistencies are pervasive in the human species. Who knows, they might even be vital ingredients to enrich a life with meaning.

Much better now I understand why the ivory-bill so seized our attention. Sure, its ecological form and function must have been spectacular to see in living display. Whether justified or not, this species stood in for what was once "a more environmentally pristine South."[15] Ivory-billed woodpecker also summons some fine aspirations that recount our national identity. Not solely is this because of the woodpecker's historical (and perhaps wholly symbolic) link with the vanished frontier, or some anthropomorphic semblance of its cryptic fate to the stoic heroes of our legendary American Westerns.

Democratic ideals are embodied by figurative ivory-bill, too. In its revelations to our citizenry, the bird has been no respecter of tutelage, profession, or social standing. It might appear to those who look for it, but also to those who don't. A squirrel hunter in Louisiana is arguably more apt to see the real woodpecker[16] than a plodding researcher or well-heeled bird-watcher who jets into the swamplands for a quick addition to a life list. Our big woodpecker possesses uncanny flair for making a grand appearance to each successive generation of post-frontier America, reprising its role as historical touchstone and social agitator. Abiding along some uncharted border of antiquity, each and every ivory-bill manifestation has, since the mid 1940s, at once demythologized and then re-mythologized all over again the legitimate bird.

For an extinct or a near-extinct species, the ivory-bill has also delivered to us an astonishing level of entertainment. The big woodpecker is a laudable opponent, one that I cannot quite come to visualize now as a frail, vulnerable species that surrendered meekly to our relentless persecutions. And *Campephilus principalis* is a match worthy to our own species' galling hubris. Merely for its talent in taunting us, the ivory-bill should merit our admiration.

Ivory-billed woodpecker never divulged all of its life history secrets. The bird's extreme rarity led us to draw correlations, make linkages, and establish associations that were never passably verified. We made conclusions on wholly inadequate evidence for the hypotheses that we did choose. And then we succumbed to unrestrained ***congruence bias*** in never bothering to specify alternative theories that also fit the extraordinarily sparse data that were gathered. If the bird is really gone now, it carried some of its ecological and behavioral confidences to the grave.

Now, what about that bewildering Arkansas video from 2004?

I frittered away too many hours peering at it, freezing its frames. I tried diligently to make it conform neatly to either large native woodpecker. At the end of that test I held with uncertainty, but it was not a wavering of equal prospect. To those who insist it was just a normal pileated woodpecker, my first retort would be rather impolite. My more diplomatic response would assign those odds at 5%, tops. On the other hand, I could guess at the probability of that mystery bird fitting the identity of an ivory-bill as somewhere a bit north of 50%. If caution gets thrown to the gale,[17] maybe nudged all the way to 65%.[18]

At the time of this coda, I have no clue as to whether the American ivory-billed woodpecker still lives on. But I will grant that odds for it having survived past the various extinction dates that we blindly imposed upon it were, and still could be, tolerably far from zero. For the time being, my wager gets placed on this mathematical axiom rather than arbitrary edicts pronounced by birding orthodoxy, in the end a pick for me no closer than a long Tennessee mile.

Endnotes

1 "There is a growing mismatch between our psychology ... and the increasingly technological and globalized world we inhabit.... The uncertainty inherent in predicting environmental change makes ... problems particularly vexing.... It seems, perhaps, that we are facing not so much a tragedy of the commons, but a tragedy of cognition." Johnson, D., and S. Levin. 2009. The tragedy of cognition: psychological biases and environmental inaction. *Current Science* 97: 1,593–1,603.

2 "With a few exceptions amateur ornithologists were of little help because few of them were familiar with the wilder parts of the country." Tanner, J.T. 1942. *The Ivory-billed Woodpecker*. Dover Publications, Inc., Mineola, NY, p. 20. "Birders typically watch from the road, a bridge, or a canoe on open water. They rarely penetrate more than a hundred yards into a forest unless there is

a road or a trail, making chances of an ivory-bill sighting slim." Jackson, J.A. 2006. *In Search of the Ivory-billed Woodpecker,* 2nd ed. HarperCollins, New York, p. 241. "Perhaps the greatest myth regarding ivory-billed woodpeckers is that birders would have found them if they still existed." Hill, G.E. 2007. *Ivorybill Hunters: The Search for Proof in a Flooded Wilderness.* Oxford University Press, London, p. 70. "Slogging through bottomland forest to find this reclusive bird is difficult, and the time, money, and manpower for intensive field expeditions are in short supply." Steinberg, M.K. 2008. *Stalking the Ghost Bird: The Elusive Ivory-billed Woodpecker in Louisiana.* Louisiana State University Press, Baton Rouge, p. 15.

This historical consistency makes irrelevant the argument that since the modern birding community has not documented the ivory-bill to its own or anyone's satisfaction, the species must be extinct. If this point is argued solely from a birding ***frame***, the claim makes for exemplary validation of the ***Dunning-Kruger effect***.

3 Although referring to night parrot, the following portrays accurately North American birders' take on the ivory-bill: "I've never gone looking...and doubt that I'd ever have enough time, money or commitment...to look for a bird that no-one will believe I have seen..." Gosford, B. 2010. Bird of the week – the night parrot resurfaces...again...maybe. *Crikey.* http://blogs.crikey.com.au/northern/2010/06/09/bird-of-the-week-the-night-parrot-resurfaces-again-maybe/ (accessed 22 September 2015).

4 The overall *process of science* is cognitively sound, i.e., a rigorous, iterative testing and then retesting of hypotheses followed by revisions that incrementally lead towards greater accuracy. Nevertheless, the *behavior of scientists* within that process requires scrutiny and, if necessary, judgment and intervention. To whit: some scientists displayed serious cognitive error during peer reviews, then abused this responsibility with ***ad hominem*** attacks and use of ***fallacy of division*** that prevented and/or delayed meaningful ivory-bill content from otherwise reaching print.

See Supplement in: Collins, M.D. 2020. Application of image processing to evidence for the persistence of the ivory-billed woodpecker (*Campephilus principalis*). *Scientific Reports* 10: 14616.

5 After two years of surveying central Arkansas, the Cornell search team capitulated to rational injunctions of ***sunk cost fallacy***. The searchers noted: "Systematic transect searching and intensive field monitoring of 12% of the area of the Big Woods has cost roughly 1.6 million dollars.... To cover the entire remaining Big Woods with similar intensity would take significant resources. No federal, state, or private funding sources are available for such an expense, and the resources required for this hypothetical approach should be weighed against other national and global conservation priorities." See: Rorhbaugh, R., K. Rosenberg, M. Lammertink, E. Swarthout, R. Charif, S. Barker, and M. Powers. 2006. Summary and conclusions of the 2005-6 ivory-billed woodpecker search in Arkansas. Final Report, Grant Agreement #401816G060, submitted by Cornell Laboratory of Ornithology to U.S. Fish & Wildlife Service, http://www.birds.cornell.edu/ivory/pastsearches/0607season/0607stories/FinalReportIBWOtext.pdf (accessed 16 May 2015), p. 23.

6 "The requirement of ideal evidence as a prerequisite for establishing such a program could have the effect of closing the door to an opportunity to save this iconic species from extinction." Collins, M.D. 2020. Application of image processing to evidence for the persistence of the ivory-billed woodpecker (*Campephilus principalis*). *Scientific Reports* 10: 14616.

7 Conceivability of this very hoax was raised over the 2013 find of night parrot in Australia (Chapter 13). "The DNA and feathers don't_necessarily_help, unless there is some way to date the feathers and make sure they aren't recent (or find signs of taxidermy or curation, which wouldn't necessarily be present, but if they are [sic] are obviously problematic)." Blogging skeptic spookyparadigm, commenting on John Young's find of the night parrot. http://doubtfulnews.com/2013/07/has-the-extremely-rare-

night-parrot-been-rediscovered-alive/ (accessed 15 September 2015).

8 Yet it already has, regrettably. "You've got no evidence at all unless you bring us a live or dead specimen. Put up or shut up." An incitement from apparent ivory-bill denier Mark Gelbart, August, GA, January 21, 2015, available at: http://www.birdforum.net/showthread.php?t=297876&page=2 (accessed 19 April 2015).

9 Such intemperate ***lack of object constancy*** in our historical dealings with the ivory-bill seems profoundly immature, especially contrasted with our cognitive facility in handling other intermittently recurring events that occur routinely in the natural world.

10 Assertions by ivory-bill deniers notwithstanding, the ***observer-expectancy bias*** works two ways. Due to formation of strong prior expectations, field observers can be primed unconsciously either to misinterpret seeing an object as an ivory-bill that really isn't, or be swayed to disbelieve the veracity of an object that really is this species. During the Arkansas and Florida controversies in 2004–2008, prominent members of the birding hierarchy were conspicuously out of touch with this ***bias blind spot***.

11 Hoaxes that involve fabricated reports (and even doctored photographs) of the big woodpecker have certainly transpired. Bloggers Bill Pulliam and "cyberthrush" furnish some background on several bogus claims made as recently as 2009-2011. See: http://bbill.blogspot.com/2010/01/recent-history.html, and http://ivorybills.blogspot.com/2011/12/unknown.html (both accessed 18 May 2016).

12 Weidensaul, S. 2002. *The Ghost with Trembling Wings: Science, Wishful Thinking and the Search for Lost Species*. Farrar, Straus and Giroux, New York, p. 39.

13 Rozin, P., and E.B. Royzman. 2001. Negativity bias, negativity dominance, and contagion. *Personality and Social Psychology Review* 5: 296–320. A potent characteristic of ***negativity bias*** is contagion, i.e., a dilute trace of some negative essence is enough to entirely

contaminate a much larger domain. With respect to the ivory-bill, this negativity is death, a ***meme*** that carries more powerful emotive weight in us than does life, thus tending to outcompete and overwhelm any facile prospect of the bird's continued survival, at least in our unconscious awareness. Given the strength of ***negativity bias***, only through diligent redirection of our cognitive instincts might we be able to view survival as a logically and/or probabilistically feasible option for the woodpecker.

14 Cooper, C, L. Larson, A. Dayer, R. Stedman, and D. Decker. 2015. Are wildlife recreationists conservationists? Linking hunting, bird watching, and pro-environmental behavior. *Journal of Wildlife Management* 79: 446–457.

15 Steinberg, p. 9.

16 Approximately 8 months after I first penned these words, Joseph Tyler Saucier, pastor, avid outdoorsman, and Louisiana native reported one and possibly two ivory-billed woodpeckers while out squirrel hunting during fall 2017; see https://projectcoyoteibwo.com/2017/11/28/change-of-pace-change-of-place-trip-report-and-sighting-follow-up-november-16-21-2017/ (accessed 13 December 2017).

17 My tribute to Bill Hardy's ultimate verdict for the acoustic evidence once given him to evaluate: "throwing scientific conservatism to the wind … I … lean toward ivory-bill." Hardy, J.W. 1975. A tape recording of a possible ivory-billed woodpecker call. *American Birds* 19: 647–651.

18 Out of deference to the ***congruence bias***, I am obliged to assign the remaining balance of probability to all other options that might explain the identity of that odd Arkansas bird.

Appendix A

Did anthropogenic Allee effect kill off ivory-bill?

Might the escalating "value"[1] placed on ivory-bill while it was becoming so scarce have gone on to trigger mounting demand and even more takings of the woodpecker at the end of the nineteenth century? "In a dangerous downward spiral repeated for a disturbing number of bird species, a decrease in the number ... served to increase their value to collectors, whose predations further decreased the remaining population."[2] Does that kind of negative feedback loop, today known as the anthropogenic Allee effect,[3] explain the mystery behind the (putative) extinction of this charismatic species?

> **Box A1**. The ***anthropogenic Allee effect*** ensues when human exploitation intensifies as population size or density falls, leading to the "rarity paradox," wherein persecution accelerates the population decline due to ever increasing demand for consumption, harvest, subsistence take, or other human uses.

Humans undeniably place higher value on scarce items. This communal taste extends to the worth we set on rare species as well.[4] Whether our aim is for consumptive ends like trophy hunting,[5] or renewable pursuits like bird-watching,[6] species rarity is habitually associated with greater value. In the context of wildlife conservation, however, this rarity-fueled demand can have two adverse outcomes. First, a species may be stuck at low numbers by levels of hunting effort that depress population size well below what it could be otherwise. Second, hunting of sufficient intensity may eventually force a species into outright extinction.[7]

Anthropogenic Allee effect is a fairly new addendum to the theoretical and applied underpinnings of wildlife conservation.[8] Nevertheless, the fundamental setback caused by rarity itself as a culprit for disproportionate harm to birds was recognized a century earlier. Gifford, writing a brief commentary in the 1917 issue of *The Condor*, summarized the problem thusly:

> *In the case of a rare bird, the collector becomes a relatively far more important influence in the process of extermination; and where a declining species is undergoing a sectional re-adjustment to changed conditions, he may well be the factor that turns the scale toward extinction.*[9]

Gifford's observations were stunningly prescient, not just about the anthropogenic Allee effect (AAE) generally, but also for revealing how the AAE might reinforce other threats faced by America's imperiled birds. Early in the new twentieth century, Gifford had articulated a link between heightened risk of extinction and the problem of a contracting range size.

Conventional economic theory predicts that as the over-exploitation of a population causes it to become rarer, escalating costs of finding the last few remnants serve as a financial incentive to abandon the exploitation.[10] If the species' unit value increases as its numbers decline, the anticipation is that the item eventually becomes so over-priced that demand for it wanes. Profit margin disappears past this tipping point. Human utilization either stops entirely or switches to a cheaper alternative (e.g., pileated woodpecker) after a species has become too scarce to readily acquire. Consequently, an economic extinction[11] first acts to depress the population size well before biological extinction should be able to wipe out the entire species.

Unfortunately, AAE does not obey the tidy rules of standard economic theory. Failure of AAE to conform to expectation arises when any of several human preferences for rarity *per se* instead strengthen our cultural or social demands for ever more exploitation. Six different human activities may be elicited by such rarity, including: 1) hobby-motivated collections,[12] 2) trophy hunting,[13] 3) luxury items,[14] 4) exotic pets,[15] 5) ecotourism,[16] and 6) traditional medicine[17] (or similar aesthetic utilization). At one time or another, and bearing in mind the cultural practices of Euro- and Native Americans alike, at least four categories above (numbers 1, 2, 3, and 6) were relevant to the historical values we placed on the ivory-billed woodpecker.

Box A.2. Our impulse for valuing rarity and over-using natural resources stems from Tragedy of the Commons, a situation denoted by an exploitation contrary to the common good. Consumer preferences for luxury items linked to ***anthropogenic Allee effect*** can depress such species through obsession with wildlife artifacts. Species affected may include parrots (pets), rhinos (horns), bluefin tuna, sharks (fins), crocodiles (skins), sturgeons (caviar), Asian pangolin (traditional medicine), and African ebony (wood).

So, instead of a positive correlation between population size and some biological trait(s) seen in a typical expression of the Allee effect,[18] the AAE embodies a viscous spiral leading to smaller and smaller numbers having an ever-greater price that just perpetuates the exploitation all the more. Human enthrallment with scarcity leads to this depletion. Our irrational temperament is what leads to a "human attraction for rarity [that] fuels a disproportionate exploitation."[19]

But did a rarity-fueled lust for ivory-bill cause its (near) extinction? This is not a straightforward question to answer. Wanton killing of ivory-bill is a valid proposal for the species' decline. Snyder et al.'s treatise on direct killing as the primary threat behind the ivory-bill's demise is a notable exception to the orthodox emphasis placed on habitat loss.[20] Significantly, the killing hypothesis depends largely on whether one accepts the premise that "hunters were not mistaking pileateds for ivory-bills."[21] That assumption is highly suspect. In fact, it defies common sense given how much the two large woodpeckers evidently so resemble(d) each other that even skilled ornithologists and avid birders have failed since the 1940s to reliably distinguish them (Chapter 11).

Regardless of the key assumption about identity, woodpecker persecutions were certainly undertaken. Largest in scale and most worrisome for its potential to harm ivory-bill populations was the financially-motivated enterprise of commercial taxidermy. Specimen dealers during late nineteenth century America[22] could be quite ruthless in their pursuit of the representative, the unusual, and especially the rare. Our country's natural history mania took off in the years right after the Civil War. An obsession for acquisition can be traced back several centuries earlier to an upper-class European taste for display cabinets filled with various novelties brought back by imperialist adventurers as souvenirs from the natural world.[23]

Beginning in the 1880s, commercial specimen dealers were a key stimulus behind much (but not all) of the prominent acceleration seen in lethal takings of the ivory-billed woodpecker (Chapter 3). Although early ornithology is often blamed for this surge (***Image A.1***), "most were not collected by scientists but rather by hobbyists."[24] In an advertisement featured prominently in the frontispiece of a 1910 edition of *The Oölogist*, one hobbyist's passion to get the rare bird heralded: "Wanted.—For cash in mounting condition: skins of ivory-billed pileated woodpecker. L.A. PARRE, Batavia, Ill."[25]

Wholesale pursuit of the woodpecker on the part of private dealers was motivated by a retail market demand from these avid hobbyists. Given orders from buyers, the list prices for ivory-bill specimens and skins could range widely. "FOR SALE.—Three finely mounted ivory-billed woodpeckers at $10 each...," read a 1905 add placed in *The Oölogist* from Pennsylvania hobbyist James Neal, who was also vending a California condor and 400 other birds.[26] Prices for the ivory-bill started as low as $2–$4 (in the 1880s),[27] but went on to garner at least $40–$50 ("collected rather late"),[28] with those taken towards the end of our collection mania fetching up to $75.[29]

Image A.1. Early American ornithologists Frank M. Chapman (on right, holding shotgun) and William Brewster (on left, holding a skin of the ivory-billed woodpecker) during an expedition to collect birds along the Suwannee River, Florida, in 1890. From the Ernst Mayr Library and Archives of the Museum of Comparative Zoology, Harvard University (used with permission)

Numbers of ivory-bill specimens acquired in the 1880s, 1890s, and 1900s were more than two, four, and three times, respectively, the number collected during the 1870s (Chapter 2). This trend of course cannot be equated to a population increase in the woodpecker (a signal of market availability or *supply*), but it does provide us one metric (albeit of unknown accuracy) for tracking the economic *demand* placed on the ivory-bill. Without question, a ready trading and auction market existed for ivory-billed woodpeckers during the late nineteenth century.

Just as abruptly, though, the trend in acquisition was cut short. The number of ivory-bill specimens plummeted enduringly in all later decades. Reasons for this sudden drop were several: the Financial Panic of 1893, changing tastes in interior design,[30] the power of bird protection movements (Chapter 4), much stricter legal prohibitions on take, and shifting prospects for naturalists who were pushed out of specimen dealing altogether, either pressed into bankruptcy or forced to seek other occupations.[31] This same plunge may or may not have been associated with a corresponding decline in the bird itself.

A strong likelihood exists that some early ornithologists were driven by the bird's very scarcity to get this "last of its kind" in order to complete or expand a collection, giving in to the AAE's irresistible allure. Indeed, their own words betray this inducement:

> *I could write a volume on the ivory-bill, and I believe that I have seen and taken more birds of this species than any man. I have been offered large sums of money to go to Florida to get these birds and their eggs....*[32]

On the other hand, despite widespread "sanctioning of collection of ivory-bills a century ago, the numbers of ivory-bills in collections are relatively low."[33] Only about 413 ivory-bill specimens are archived today across 90 institutions in North America and Europe.[34]

In addition to specimen dealers, hobbyists, and museum collectors, two other social identities were linked to a deadly pursuit of the ivory-bill. One group consisted of those who shot the bird as a curio. In this rendering, the bird was perceived as something so peculiar that it roused human visitors in the bird's native haunts to shoot the woodpecker just to see it up close, to examine its very oddity.[35] Another group allegedly went after ivory-bill as a subsistence food, or for some other everyday purpose, such as to cut up for use as fishing or trapping bait.[36]

Rarity-fueled pursuit of American wildlife certainly played a role in diminishing our native bird populations. "By 1880 some observers were commenting on the decrease in the species, and this served to attract more collectors."[37] Perhaps the most wrenching account of AAE as motivating this very kind of persecution was penned by Frank M. Chapman. Looking to acquire representatives of the showy Carolina parakeet, and thwarted at first by inclement weather, Chapman was at last successful in his pursuit of the rarity in Florida. After the first round of shooting he was:

> *illumined by a serene joy which casts a radiance over the darkest landscape, for I have met* Cornurus *and he is mine, three adults and a young one.*[38]

But then, after shooting five *more* specimens, Chapman's pleasure morphed into something akin to regret, a reaction that at first seemed to temper his enthusiasm for any further killings of the parakeet. His initial reflections even seemed to hint at a nascent outlook towards the bird's conservation:

> *I admired them to my heart's content ... for now we have nine specimens and I shall make no further attempt to secure others, for we have almost exterminated two of the three small flocks which occur here, and far be it from me to deal the final blows. Good luck to you poor doomed creatures....*[39]

Chapman's attempt at judicious restraint did not last long. Only days later, a temptation to add to his series of the parakeet proved too powerful to resist:

> *Good resolutions like many other things are much easier to plan than to practice. [T]he parakeets tempted me and I fell; they also fell, six more of them making our total fifteen.*[40]

What made AAE so lethal was the parakeet's fateful habit of maintaining tight social structure. Upon seeing initial members of the clustered (and easily targeted) flock drop to the gun, surviving parakeets would just return again and again to gather around fellow members killed or injured earlier, setting up a mini-spiral of annihilation. That sort of defenselessness is textbook validation of how deadly AAE can be.

Carolina parakeet was not the only American bird for which AAE aggravated the extinction risk. In passenger pigeon, it was not direct killing in isolation that made AAE so devastating. "Social facilitation in food finding may have become increasingly ineffective with decreasing population size ... loss of critical breeding habitat and lack of social facilitation at low densities would have been enough to lead the passenger pigeon to extinction even without killing a single bird and despite the existence of considerable remaining forest."[41] The pigeon was clobbered from two sides: a conventional Allee effect pressed the species' numbers ever lower, while the AAE also made killing off survivors easier due to the bird's strongly colonial social system.

Some traditional views hold that ivory-bill, too, was eradicated by over-zealous collectors. This outlook may have started with Tanner's mentors, Arthur Allen and Paul Kellogg. "Once a community has been located by collectors,... it [was] possible in the past to exterminate local groups of ivorybills, and this may well have happened even in the name of science."[42] Tanner further endorsed this viewpoint: "ivory-bills were wiped out of the Suwannee River region of Florida by the collecting of A.T. Wayne in 1892 and 1893."[43] Although holding fast to his hypothesis on logging as primary reason for the woodpecker's imperilment, Tanner admitted: "now that there are so few ivory-bills living, the shooting of a few birds might become the final cause for their extinction."[44]

A catastrophic effect from Wayne's zest for the ivory-bill, however, falters somewhat at closer inspection. Wayne himself stated more than 10 years later that "almost any cracker" could take Chapman to Suwannee swamps which Wayne claimed then still "must contain innumerable ivory-bills."[45] Wayne also reported he had spared almost half of the ivory-bills that he had detected during one expedition along the Suwannee River.[46] Tanner's monograph mentions additional records of ivory-bill along the upper Suwannee between 1910 and 1915, and from the lower portion of this river in 1917 and 1925.[47] So, not just Tanner's, but *all* accounts ever contending that ivory-bills were "wiped out" failed to deliberate suitably how nomadic habits (see Chapter 1) might render only a false *appearance* of local eradications at a given site.[48]

Credence that specimen dealers, scientist collectors, and subsistence hunters could readily exterminate the ivory-bill locally is repudiated by first-hand testimony on the bird's cagy reactions at being pursued. The woodpecker greatly frustrated its trackers: "in disposition [ivory-bill] is wild and wary ... they are a wild and suspicious bird."[49] Bearing in mind that those who could have shot out the bird entirely for some profitable end were likely to have done just that, repeated avowals of the difficulty in shooting even one bird are not only credible, the wider implications of these trials are crystal clear.

Witnesses to that challenge were explicit: "we saw a single pair of the rare ivory-bills.... They were very shy, restlessly swinging from tree to tree, and taking good care to keep beyond gun-range."[50] And: "I saw and heard four ivory-bills ... but could not get a shot as they were too wild, and couldn't be approached nearer than 300 or 400 yards."[51] Then: "very rare and hard to get a shot at.... I tried to get her but failed as she was exceedingly wary."[52] And this: "I saw six in one day ... two of these I secured, but the others were so wild that I got only a long distance shot at one flying above the pine tops."[53] Scott got only three of 11 ivory-bills he saw in March 1887, and these were of a pair that were closely attending the third, a very young bird still in the nest cavity.[54]

Writers were also quite plain about not getting most, never mind all, of the pairs or groups that they so much wanted on expeditions to bag the elusive ivory-bill.[55] One collector admitted that "the specimens which I value most, however, are the skins of three fine ivory-billed woodpeckers." But he didn't (or couldn't) wipe out the entire group. Indeed, as he elaborated, "I saw eight of the birds and also saw one of them leave a hole, high up in a hickory tree – How I wanted to get up there! But it was utterly impossible."[56]

Mere gunshots provoked caution in ivory-bill. On 29 March, 1890, William Brewster shot at (but missed or wounded) a yellow-crowned night-heron *Nyctanassa violacea* in Florida. "The report of the gun started an ivory-billed woodpecker which uttered its trumpet note a dozen times or more. It was several hundred yards off apparently. I turned back and paddled hard but it stopped calling and I failed to find it."[57] One cannot help but wonder if the guesswork about a particular vulnerability of ivory-bill groups to getting easily wiped out instead came from accounts that do substantiate this vulnerability in the closely-related imperial woodpecker.[58]

Guns were not even necessary to provoke the ivory-bill's keen vigilance. Egg collectors and idle watchers of the big bird also found ivory-bill to be unusually guarded. Hoyt wrote with frustration of the toil he expended in trying to find a nest in Florida:

> *If among all the feathered tribe there is a bird whose nest is more difficult to locate than that of the ivory-billed woodpecker it belongs to some species with which I am unacquainted.... I have had some of the most expert woodsmen and trappers in my employ and withal, after three seasons' [calendar years] hard work I have obtained but two sets of eggs, and these from the same pair of birds.... The collector who goes for a set of eggs will find he has to work for them, and if he succeeds in getting a set the first one or two seasons he will do what I could not.*[59]

Hoyt's quest began in 1902, three decades before Tanner's study. Hoyt marveled how this pair of wily ivory-bills would try to hoodwink his woodsman by going "in and out of every old nest in the swamp, but steer wide of the new [active] one."[60] The account gives us indubitable evidence that finding a nest was no easy task even if the woodpecker was already exhibiting site fidelity while breeding.

Adding nests or eggs to a hobbyist's cabinet proved as challenging as shooting the birds. "It is almost impossible to get specimens of its eggs."[61] As late as 1895, ivory-bill "eggs ... [were] still quite rare in collections."[62] One final triumph on acquiring nest contents was achieved only after three years with paid experts familiar with local terrain. This level of toil ought to disabuse anyone of that pernicious myth of "tame ivory-bill" so beloved by today's arm-chair, snark-loving cynics.[63]

In the late nineteenth century, Thompson ran into the ivory-bill's caginess in the Okefenokee Swamp of southern Georgia, an incident that furnishes crucial glimpses as well into historical outlooks of some woodsmen towards the woodpecker. After days watching a pair from a makeshift blind 50 feet away from a nest cavity, the birds at last spotted Thompson, and:

> *Thenceforth my observations were few and at a long distance. No amount of cunning could serve me any turn. Go as early as I might, and hide as securely as I could, those great yellow eyes quickly espied me, and then there would be a rapid and long flight away into the thickest and most difficult part of the swamp.*[64]

Thompson's host had escorted him out to a roost and nest tree that had been found by the woodsman a year before. That man expressed no particular interest in collecting or shooting (never mind eating) the birds. And as for Thompson: "I could have killed her easily with the little sixteen-gauge breech-loader at my side, but I would not have done the act for all the stuffed birds in the country."[65]

Later in the same account, Thompson described an encounter with a local deer-hunter who explained why he, too, did not bother shooting the woodpecker. This hunter ignored the bird despite aggravation of being roused abruptly with a bloody nose caused by an ivory-bill's energetic scaling of a heavy wood chip that fell directly upon the man while he was sleeping below. His reason for restraint? "I cudn't stan' the expense er the thing. Powder'n' lead air mighty costive."[66] Thompson's chronicle was published in 1885,[67] so the ivory-bill's cunning habits were apparently established already. His story also calls into question inevitable interest of rural people in always shooting every ivory-bill.

An inordinate vigilance in the ivory-billed woodpecker was depicted by naturalists who pursued it between 1877 and 1905,[68] during but also prior to the zenith of market hunting. Challenges faced by hunters do not portray a species that made an easy mark out of a lopsided vulnerability from the AAE. Rather, the ivory-bill's evasion tactics point to buffering from the very worst harms brought on by the

AAE. As accounts came from practiced collectors, woodsmen, or field naturalists who archived their efforts in writing, there is no reason to regard subsistence hunters and curiosity seekers across the South as any more proficient (or motivated) in their pursuit and killing of the ivory-bill.

Even high hunter success with ivory-bill was not necessarily what it seemed.[69] Arthur T. Wayne's field catalogue reveals he was responsible for 44 ivory-bill specimens,[70] more than 10% of specimens known today from anywhere in the world! But Tanner clarified that "many ... [of these] were collected by local hunters."[71] A mere $2–$5 was all that it took to induce locals to "keep their eyes open" for an ivory-bill.[72] William Brewster had already convinced Wayne that he could make money by providing bird skins to private hobbyists and museums.[73] "Sub-contracting" out wholesale the woodpecker's acquisition weakens certitude in the ease of large numbers of ivory-bills readily shot out by solo hunters.[74]

Moreover, Wayne's large collection was apparently procured in its entirety between 1892 and 1894, in a single (but rather extensive) region of northern Florida, and from a very narrow temporal window matching the irruptive habits expected from a nomadic species prone to wander and concentrate into local hot spots created by recent forest disturbances (see Chapter 1). Indeed, Wayne clearly emphasized that the birds he was shooting were aggregated in areas that had been subject to recent forest fires or "burn outs."[75]

Other hunters' success with ivory-bill was rather spread out in time, too.[76] A hunter from Hawkinsville, Florida, revealed that he and his brothers had shot 20 to 25 ivory-bills in one region during the previous 10 years. Assuming that their identity was correct, i.e., no pileated woodpeckers were involved, and depending on the land area from which these birds were collected, their age classes, and a total population size, two birds per year is neither a particularly remarkable nor ineludibly unsustainable rate. That level of take might represent no more than about the annual fecundity of one or two pairs. Indeed, demographic impairment to the species caused by taking 400 or more individuals might hinge substantially on how much of that mortality came from year-one (presumably more naïve) birds versus breeding adults.

Heavy, widespread subsistence taking of the ivory-bill pivots, too, on how much confidence we place in anecdotes about the woodpecker being eaten, used as bait, or shot for idle amusement. If our two big woodpeckers were so alike as to lodge total confusion about their identity among professional ornithologists and keen bird-watchers, we cannot exempt the identification skills of rural residents from such mix-up. Indeed, good reasons exist for distrusting all assertions the ivory-bill was hunted for food, at least widely. One report that ivory-billed woodpecker was "served up as quail on toast" and "frequently to be seen hanging at the doors of restaurants in New York" referred instead to the pileated woodpecker.[77] Confirming the smaller woodpecker's gustatory appeal, Jackson attested to the flavor of the pileated as "a bit gamy, maybe almost nutty, but not bad."[78]

Like so much else about this species (Chapter 7), the ivory-bill's palatability is quite open to dispute. In the Wacissa and Aucilla rivers area of Florida, Wayne did in fact exclaim: "Every one is shot by being systematically followed up. They are shot for food, and the people – the crackers – consider them 'better than ducks'!"[79] Yet in his wide-ranging review Tanner "found no other instance of this occurring. In some localities of the South, woodpeckers are shot by fishermen and the meat used to bait the hooks of set or trot lines, and trappers occasionally use woodpecker meat for baiting traps. The much commoner pileated woodpecker is usually shot for such purposes."[80] Alexander Wilson did not regard the ivory-bill as large enough to merit a place at the table.[81] Laurent, too, described hunters' reactions to ivory-bill as "rather poor eating."[82]

Skepticism intensifies about AAE propelling the ivory-bill past a point of no return considering how difficult it is to detect and deliberately eradicate remnant members from a closed population. Removal of alien species makes for a good test of the AAE's efficiency under conditions similar to imperilment. While removing feral pigs from Santa Cruz Island, California, investigators found it enormously difficult to detect even these large animals at very low numbers. As populations get pushed down to ever fewer individuals, the pursued species has "a refuge below that threshold, and the population could be sustained despite continued allocation of [hunting] effort."[83] I submit that this very outcome likely occurred with ivory-bill, the woodpecker's numbers finally shrinking to a point where the AAE was no longer efficient. Rarity put brakes on further depletion because the bird was so wary, and because it was difficult to detect at low density past the early 1900s.[84]

Arguably the strongest challenge to the AAE as having finished off the ivory-bill pivots on the timing of heightened persecutions from market-based hunting. No one disputes that ivory-bills were alive and still breeding in at least Louisiana during the mid-1930s, almost a half century *after* the apex of the collection epoch. If instead the mid-1940s are benchmarked as the ivory-bill's time of extinction, and the bird's lifespan was about 15 years, the bird reproduced for at least two generations past our *fin de siècle* frenzy over bird collection. Failure of these dates to match up is strong evidence that other factors also must have contributed to the bird's fate.

Some regional hotspots for ivory-bill remained largely free of persecution for decades after the end of the collection rush. In his monograph Tanner regarded Gulf Hammock of Florida as among the more promising locales that may have still harbored the ivory-bill into the middle 1930s. Laurent was adamant that birds there had not been pursued around the turn of the century: "between the years 1886 and 1916, and from what I have seen and heard, I doubt if there has been an ivory-billed woodpecker shot in that part of Florida in twenty-five years."[85]

Table A.1. Traits alleged in one recent synthesis to have predisposed the ivory-billed woodpecker to amplified vulnerability from human persecution contrasted with observations from direct witnesses to the woodpecker's habits as reported during the late nineteenth and early twentieth centuries

Contended vulnerabilities of ivory-billed woodpecker to human depredation	**Actual first-hand accounts of ivory-bill habits**
Steady calling behavior for easy detection, location, and stalking?	**No**. "Very silent at all times so far as their voices are concerned.[86] The call lacks carrying capacity and can rarely be heard over 100 yards on a still morning.[87] A man would probably have to come... close enough to disturb them, before he would be likely to hear them.[88] The ensuing noise scarcely equaled in volume the work of the downy woodpecker [*Dryobates pubescens*]."[89]
Apparent palatability?	**No**. Ivory-bill was considered "rather poor eating."[90]
Frequent approachability in early times?	**No**. "Very shy and not easy to approach.[91] These birds are the shyest and most cunning of anything that wore feathers."[92]
Large size making for a massive target and thus a worthwhile game item?	**No**. "Few persons were found who thought they knew the bird.[93] I [couldn't stan[d] the expense [of shooting] the thing. Powder'n' lead [are] mighty costive, plus: I had come as a visitor … not as an assassin.[94] The difficulty of securing one even with a gun...."[95]
Spatial proximity of pairs and family groups for multiple kills per hunter?	**No**. "The male shot … the female escaping … another pair.... I shot the male … but [the female] was too sagacious to appear.[96] I saw six … two of these I secured, but the others were so wild.[97] I saw eight … [but obtained at most] the skins of three fine ivory-billed woodpeckers.[98] The same day that the nest was found eleven were counted … the three spoken of were all that were obtained."[99]
High value for anthropogenic use?	**Yes**, sometimes, but not invariably. "I [couldn't stan[d] the expense [of shooting] the thing. Powder'n' lead [are] mighty costive.[100] If you will kill [ivory-bills] and we catch you at it, we will prosecute you to the fullest extent of the law.[101] Collectors who secured ivory-bills usually paid local hunters a small price for fresh specimens."[102]
Low reproductive rate?	**No**. Variable fecundity; clutch size and fledgling numbers ranging as high as 4 to 6. Reproductive rate substantially higher relative to other *Campephilus*, e.g., *magellanicus*.

Snyder and colleague's modern premise that ivory-bill could not survive persecution because the woodpecker was "conspicuous and relatively unwary"[103] is blatantly contradicted by Bendire's nineteenth-century insistence on the bird as "exceedingly wild and suspicious."[104] Another assumption, that "time was evidently too short ... to evolve truly effective protection,"[105] ignores key findings from modern behavioral ecology that demonstrate how rapidly avian threat reactions are attained in wild birds (synthesized in Chapter 1), even if the centuries to millennia of earlier predations on the bird by Native Americans get discounted altogether. Of seven features Snyder et al. listed as predisposing the ivory-bill to exceptional vulnerability,[106] only one (high value) is not utterly refuted by the first-hand reports from early naturalists who described exactly opposite characteristics associated with this species (***Table A.1***).

Ivory-billed woodpecker was certainly placed in greater danger of extinction from human persecution and by market forces that steer the anthropogenic Allee effect. But unlike several other now- or near-extinct American birds, the ivory-billed woodpecker owned some vital life history traits that gave it a level of resistance to detrimental Allee effects at small population size (Chapter 1). Moreover, some of those same traits, including low density, nomadism, and a supreme wariness around humans, fortified the woodpecker against the very worst penalties that can arise from the anthropogenic Allee effect.

Therefore, I conclude that the specimen collection and market hunting eras (1880s to 1910) did *not* drive the ivory-billed woodpecker into an inexorable demise. After exiting a period of heighted risk posed by this historical bottleneck, the scope of human persecution never again spanned a duration nor encompassed a geographic extent large enough capable of finishing off the species. Anthropogenic Allee effect may have been *one* contribution to the woodpecker's protracted history of decline, but it was not the sole factor pushing it into an ultimate or certain extinction. Yet again, the ivory-bill resists our facile yearning to pin down a single, clear-cut explanation for its bizarre disappearance.[107]

Endnotes

1 "Value" here signifies any economic, financial, scientific, recreational, cultural, or aesthetic worth we might place on the ivory-bill, regardless of whether or not such worth can be monetized. For example, see: Martín-López, B., C. Montes, and J. Benayas. 2007. The non-economic motives behind the willingness to pay for biodiversity conservation. *Biological Conservation* 139: 67–82.

2 Barrow, M.V., Jr. 1998. *A Passion for Birds: American Ornithology after Audubon*. Princeton University Press, Princeton, p. 102, referring here to pursuit of the Carolina parakeet *Cornuropsis carolinensis* by its collectors.

3 Courchamp, F., E. Angulo, P. Rivalan, R. Hall, L. Signoret, L. Bull, and Y. Meinard. 2006. Rarity, value and species extinction: the anthropogenic Allee effect. *PLoS Biology* 4: e415.

4 Angulo, E., and F. Courchamp. 2009. Rare species are valued big time. *PLoS One* 4: e5215.

5 Mysterud, A. 2012. Trophy hunting with uncertain role for population dynamics and extinction of ungulates. *Animal Conservation* 15: 14–15.

6 Species rarity can predict birdwatcher numbers at a site, and birdwatchers also travel further to see rarer species. Booth, J.E., K.J. Gaston, K.L. Evans, and P.R. Armsworth. 2011. The value of species rarity in biodiversity recreation: a birdwatching example. *Biological Conservation* 144: 2,728–2,732.

7 Hall, R.J., E.J. Milner-Gulland, and F. Courchamp. 2008. Endangering the endangered: the effects of perceived rarity on species conservation. *Conservation Letters* 1: 75–81.

8 Apparently first coined by Courchamp et al., the term anthropogenic Allee effect may also be known as the *paradox of value* within the discipline of economics. Cox, A. 2004. *Win-win?: The Paradox of Value and Interests in Business Relationships*. Earlsgate Press, London.

9 Gifford, H. 1917. Communication: to the editors of *The Condor*.

The Condor 19: 73. Gifford here gave a perfectly modern definition of the anthropogenic Allee effect. He also reveals how much American social values had changed (or at least diversified) when it came to bird killing. "So it is hoped that in the near future the man who collects bird-skins or eggs for private gratification or gain will be classed with the plume-hunter and be banished from respectable ornithological society." *Ibid.*

10 For example, see Clark, C.W. 1990. *Mathematical Bioeconomics: Optimal Management of Renewable Resources,* 2nd edition. John Wiley & Sons, New York, NY, USA. When a selling price always exceeds the unit cost, however, and if the discount rate is sufficiently large, then maximization of present value will *always* result in extermination of the species. Clark, C.W. 1973. Profit maximization and extinction of animal species. *Journal of Political Economy* 81: 950–961.

11 Lobo, A.S., A. Balmford, R. Arthur, and A. Manica. 2010. Commercializing bycatch can push a fishery beyond economic extinction. *Conservation Letters* 3: 277–285.

12 Slone, T.H., L.J. Orsak, and O. Malver. 1997. A comparison of price, rarity and cost of butterfly specimens: implications for the insect trade and for habitat conservation. *Ecological Economics* 21: 77–85. In the late nineteenth and early twentieth centuries, specimens, nests, and eggs of rare species like ivory-bill were highly sought after by amateur collectors, individuals who were neither science-motivated ornithologists nor associates of any professional institution. Largely from this direction came the substantial *market demand* for ivory-bills.

13 Harris, R.B., R. Cooney, and N. Leader-Williams. 2013. Application of the anthropogenic Allee effect model to trophy hunting as a conservation tool. *Conservation Biology* 27: 945–951.

14 Gault, A., Y. Meinard, and F. Courchamp. 2008. Consumers' taste for rarity drives sturgeons to extinction. *Conservation Letters* 1: 199–207. Despite international protection through CITES, quota-based management, and a well-publicized threat status, an accelerated

commercial demand feeds a thriving illegal trade in sturgeon-derived caviar. Perceived rarity of sturgeon caviar positively influences consumer attraction, confirming that rarity itself signifies an exaggerated (and irrational) consumer preference. *Ibid.*

15 Tella, J.L., and F. Hiraldo. 2014. Illegal and legal parrot trade shows a long-term, cross-cultural preference for the most attractive species increasing their risk of extinction. *PLoS One* 9: e107546.

16 Booth et al.

17 Lever, C. 2004. The impact of traditional Chinese medicine on threatened species. *Oryx* 38: 13–14.

18 Stephens, P.A., and W.J. Sutherland. 1999. Consequences of the Allee effect for behaviour, ecology and conservation. *Trends in Ecology & Evolution* 14: 401–405.

19 Gault et al., p. 204.

20 Snyder, N., D.E. Brown, and K.B. Clark. 2009. *The Travails of Two Woodpeckers*. University of New Mexico Press, Albuquerque.

21 *Ibid.*, p. 45. Moreover, "many early pileated woodpecker populations were also heavily impacted by hunting." *Ibid.*, p. 47.

22 Barrow, Jr., M.V. 2000. The specimen dealer: entrepreneurial natural history in America's gilded age. *Journal of the History of Biology* 33: 493–534.

23 Poliquin, R. 2012. *The Breathless Zoo: Taxidermy and the Cultures of Longing*. Pennsylvania State University Press, University Park.

24 Jackson, J. A. 2006. *In Search of the Ivory-billed Woodpecker*, 2nd ed. HarperCollins, New York, p. 74.

25 *The Oölogist* 27. A juxtaposition here of the names for each of the two large American woodpeckers further darkens our confidence in historical reliability for the identification of these species.

26 *The Oölogist* 22: 2.

27 Laurent, P. 1917. My ivory-billed woodpeckers. *The Oölogist* 34: 65–67.

28 Jackson, p. 74.

29 Laurent. A later addendum corrected Laurent's error of $5 to $75.

See *The Oölogist* 35: 81.

30 "Tasteless as some of these things now seem, they remain an interesting and characteristic feature of late nineteenth century household decoration." Morris, P.A. 1993. An historical review of bird taxidermy in Britain. *Archives of Natural History* 20: 241–255.

31 Barrow, 2000.

32 Letter of Arthur T. Wayne to Frank M. Chapman, sent from Mount Pleasant, South Carolina, October 12, 1905, as reproduced in Snyder et al., pp. 136–137. Others were more poetic in their obsession: "How long and how diligently I had sought the home of *Campephilus principalis,* the great king of the red-headed family..." Thompson, M. 1885. A red-headed family. *The Elzevir Library* 4: 5–21.

33 Jackson, *Ibid.*

34 Capainolo, P. 2007. Extended-wing preparation made from a 117-year-old ivory-billed woodpecker (*Campephilus principalis*) specimen. *The Auk* 124: 705–709.

35 Tanner, J.T. 1942. *The Ivory-billed Woodpecker*. Dover Publications, Inc., Mineola, NY, p. 56. In this section of his monograph, Tanner placed quite an emphasis on killing from curiosity.

36 "I have been told that in southern Georgia and the Florida Panhandle, everyone used to know that if you wanted something for the stew pot, you went to the swamp, banged twice on a bucket or a log, and shot the woodpecker when it came in. I was also told that ivorybill flesh was the favored mink bait used by trappers on the Pascagoula River in Mississippi...There was a time when ivory-billed woodpeckers were regular targets of hunters." Hill, G.E. 2008. Book review: An alternative hypothesis for the cause of the ivory-billed woodpecker's decline. *The Condor* 110: 808–810.

37 Hall, G.A. 1988. Ornithological literature. [review of *The Carolina Parakeets in Florida,* by D. McKinley, 1985, Florida Ornithological Society, Gainesville]. *The Wilson Bulletin* 100: 342–343.

38 As taken from Chapman's *Parakeet Journal,* quoted in Barrow, 1998, pp. 105–106.

39 *Ibid.*

40 *Ibid.*

41 Bucher, E.H. 1992. The causes of extinction of the passenger pigeon. *Current Ornithology* 9: 1–36. Plenum Press, New York.

42 Allen, A.A., and P.P. Kellogg. 1937. Recent observations of the ivory-billed woodpecker. *The Auk* 54: 164–184.

43 Tanner, p. 19.

44 *Ibid.*, p. 56.

45 1905 letter of Arthur T. Wayne to Frank M. Chapman, as given in Snyder et al., p. 136. Wayne also informed Chapman that he "left more than 100 birds in a radius of 20 square miles." *Ibid.* Whether this figure was an enticement seeking funds to shoot even more ivory-bills, or mere boasting about his search skills, the veracity of Wayne's claims are today unclear.

46 Wayne, A.T. 1893. Notes on the birds of the Suwanee River. *The Auk* 10: 336–338.

47 Tanner, pp. 5–6. Despite inclusion of these records from the Suwannee River in his own monograph, Tanner made the bizarre claim that "after Wayne's work there have been no reliable records of ivory-bills having been seen in that region." *Ibid.*, p. 19.

48 Failure to consider this alternative hypothesis leads to the ***congruence bias***. An original fixation on the woodpecker's absence as inevitably stemming from being shot out or eradicated from habitat loss sets us up for ***illusory correlation***.

49 Higley, W.K. 1906. *Birds and Nature*, Volume III, A.W. Mumford and Company, Chicago, p. 123.

50 Brewster, W. 1881. With the birds on a Florida river. *Bulletin of the Nuttall Ornithological Club* 6: 38–44.

51 Field notes of Arthur T. Wayne from April 22, 1892, as given in Tanner, p. 63.

52 D., W.A. 1885. The great woodpecker. *Field & Stream* 24: 427.

53 Account by Vernon Bailey, field naturalist with the U.S. Biological Survey, in describing his experience with collecting the species in Texas; see Oberholser, H. C. 1974. *The Bird Life of Texas*. University

of Texas Press, Austin.

54 Scott, W.E.D. 1888. Supplementary notes from the Gulf coast of Florida, with a description of a new species of marsh wren. *The Auk* 5: 183–188.

55 These historical accounts of missed shots and lack of success convincingly refute a false assertion that hunters slaughtered "from about 1880 until about 1920 … every known individual." Hill, p. 7.

56 Whitehead, R.B. 1892. A trip through "Wa-hoo-Hammock." *The Oölogist* 9: 71–72.

57 *Journals of William Brewster, 1871–1919 Inclusive,* untitled p. 72. http://transcribebhl.mobot.org/display/display_page?ol=w_rw_p_pl&page_id=2104#page/n0/mode/1up (accessed 22 August 2016). Brewster's observation here also suggests a wariness about human dangers.

58 Pursuers of the imperial noted that "during the next few days this entire party [five birds] fell victims to our guns," apparently because they "showed [a] strange persistence in returning to their haunt." Nelson, E.W. 1898. The imperial ivory-billed woodpecker, *Campephilus imperialis* (Gould). *The Auk* 15: 217–223.

59 Hoyt, R.D. 1905. Nesting of the ivory-billed woodpecker in Florida. *The Warbler* 1: 52–55.

60 *Ibid.*, p. 52.

61 Thompson, p. 12.

62 Bendire, C. 1895. Life histories of North American birds, from the parrots to the grackles. *Special Bulletin of the U.S. National Museum* 3: 44.

63 "I see no reason why the ivorybill would stay half-tame through over 200 years of hunting pressure, then make a quantum leap to ultra-wariness during 60+ years of no hunting pressure." Tom Nelson, Minnesota birder, blogger, climate change and ivory-bill skeptic, writing in 2005. (http://tomnelson.blogspot.com/2005/08/just-how-wary-was-ibwo.html) (accessed 27 May 2016). Nelson managed to fall to ***Dunning-Kruger effect*** twice, once clueless

over the bird's wary habits prior to the historical era marked by Tanner's study, and as documented in more than 30 accounts (Chapter 7), and then again through ignorance of findings from behavioral ecology on how contextual wariness is acquired or expressed rapidly in bird populations, whether from natural selection, cultural and social learning, or threat acclimation.

64 Thompson, pp. 13–14.

65 *Ibid.*, p. 9.

66 *Ibid.*, pp. 18–19.

67 Thompson's sojourn to the Okefenokee took place no later than 1884, and possibly even earlier. In the same account, he relates he shot a male ivory-bill "since writing the foregoing" in January 1885. *Ibid.*, p. 21.

68 Audubon witnessed the ivory-bill as exceedingly cautious as early as the 1830s. "I observed that in two instances, when the woodpeckers saw me thus at the foot of the tree in which they were digging their nest, they abandoned it for ever." Audubon, J.J. 1832. *Ornithological Biography, Or an Account of the Habits of the Birds of the United States of America*. Volume 1. James Kay, Jun. & Co., Philadelphia, p. 344.

69 Few records exist of high numbers of ivory-bills collected all at once in lone locale. Apparently the most was 10 specimens near Punta Rassa, south Florida, all collected between 22 and 29 February 1904, but few details are available (C. Hunter, *pers. comm.*). Although barely more than a large family group, the next highest was six (one male, five females) from Orange County, Florida, by J.T. Mason in 1906.

70 As many as 37 "ivory-bills had been shot from ... near St. Marks, Florida, alone." Jackson, p. 74.

71 Tanner, p. 32.

72 Laurent, p. 65.

73 Chamberlain, W.D. 1986. Arthur T. Wayne. *The Chat* 50: 105–110.

74 Ease of acquiring multiple ivory-bills may have also been substantially exaggerated by the lax reporting standards used

for specimen labeling in earlier historical eras. It was not unusual for specimens to have missing data. If a particular collector was responsible for, say, eight specimens arriving at some museum, that number may have been aggregated in a manner that disguised whether the birds had been acquired on the same day, at the same location, or even by that particular collector given the common practice of "sub-contracting" out specimen acquisitions.

75 Wayne, 1893. Wayne was never successful in seeing, never mind collecting, a single ivory-bill in his home state of South Carolina.

76 W.A.D., *ibid.*

77 Shanly, C.D. 1874. Winter in Canadian Forests. *The Aldine* 7: 61–63. Shanly refuted the gastronomic appeal of woodpeckers in general, and then described the woodpecker in question as "black in plumage, with a crimson crest upon his head, and a tinge of sulphur-yellow under his wings. The cry of this bird is very loud, resembling a series of short, angry yelps, as he goes dipping through the forest in his undulating flight. He is a very shy, vigilant bird, difficult to get a shot at, as he usually hunts for his living in the tops of tall, decayed trees, round the stems of which he dodges the marksman with provoking adroitness. But the woodsmen do not often waste powder upon members of the woodpecker family, because they consider them unfit for food." Key details (light color on underwings, undulating flight, calls) in this account instead better describe pileated woodpecker.

78 Jackson, p. 69.

79 Wayne, A.T. 1895. Notes on the birds of the Wacissa and Aucilla River regions of Florida. *The Auk* 12: 362–367.

80 Tanner, p. 56.

81 Wilson, A. 1811. *American Ornithology, Volume 4*. Bradford and Inskeep, Philadelphia, PA.

82 Laurent, *ibid.*

83 Morrison, S.A., N. Macdonald, K. Walker, L. Lozier, and M.R. Shaw. 2007. Facing the dilemma at eradication's end: uncertainty of absence and the Lazarus effect. *Frontiers in Ecology and the*

Environment 5: 271–276.

84 After sustained persecution "the remaining animals will probably possess traits that render them even more difficult to detect, as a result of selection and/or learning." *Ibid.*

85 Laurent, p. 67.

86 Ridgway, R. 1898. The home of the ivory-bill. *Osprey* 3: 35–36.

87 Hoyt, R.D. 1905. Nesting of the ivory-billed woodpecker in Florida. *The Warbler* 1: 53–55.

88 Tanner, p. 22.

89 Dennis, J.V. 1948. A last remnant of ivory-billed woodpeckers in Cuba. *The Auk* 65: 497–507.

90 Laurent, p. 65.

91 Nehrling, p. 170.

92 Hoyt, p. 52.

93 Ridgway, p. 35.

94 Thompson.

95 Tanner, p. 55.

96 Ridgway, p. 36.

97 Bailey quoted in Oberholser, *ibid.*

98 Whitehead, p. 72. This writer makes clear he saw a total of eight birds, but it is not evident from the text that he actually took even the three in his possession from this group; they may have been collected or traded from elsewhere.

99 Scott, p. 186. On March 17, 1887, and despite having as many as four or five ivory-bills in sight simultaneously, Scott only shot two adults and took a down-covered young bird they were attending at a cypress tree nest cavity.

100 Thompson, pp. 18–19.

101 Ellis, J.B. 1918. Ivory-billed woodpecker not yet extinct. *The Oölogist* 35: 11–12. Ellis is here warning would-be collectors coming down from the north not to try to find the bird in his home state of Florida.

102 Tanner, *ibid.*

103 Snyder et al., p. 9.

104 Bendire, p. 45.

105 Snyder et al., *ibid.*

106 *Ibid.*, p. 46.

107 Regard for ***congruence bias*** requires our remaining open to more than one explanation for the woodpecker's imperilment or its putative extinction. For example, one of several possible explanations might combine *both* the anthropogenic Allee effect *and* demographic impairment from a very low population density.

Appendix B

Was the ivory-bill a Pleistocene relict?

Overview

American ivory-billed woodpecker has long been perceived as vanishing from habitat loss aggravated by dependency on old-growth bottomland hardwoods. Some historical reports, however, originated from younger, upland, and heavily logged forest. Biomes, plant composition, and forest continuity varied momentously during the protracted evolution of northern *Campephilus*. American ivory-billed woodpecker could not have specialized on late Holocene forest conditions because this woodpecker lineage diverged almost 1 million years BP when starkly different vegetation conditions prevailed. Given narrow wing profiles, plus larger body and clutch sizes, traits that set northern *Campephilus* apart from tropical congeners, I propose this lineage was adapted originally to live in open, dynamic, and park-like temperate woodlands. Northern *Campephilus* used scattered large trees and/or more isolated stands embedded in arid landscapes that prevailed during much of the Pleistocene. Forest structure (not composition) more broadly dictated lifestyles in the large *Campephilus*. Given its ostensibly uncommon status in historical times, the ivory-bill may represent an evolutionary relict[1] of ancient ecosystems.

Ivory-bill during ecological time (1000 years BP to present)

Forbidding, densely-canopied forest, draped in long strands of waving Spanish moss. Tea-colored waters camouflaging unseen hazards. Human travel fraught by a monotonous, disorienting terrain. Since our first naturalists, the American ivory-billed woodpecker has been tied to our vast southeastern swamp forests. Calling it the bird's favorite resort, Audubon regarded the bird as most at home in "those deep morasses, overshadowed by millions of gigantic dark cypresses, spreading their sturdy moss-covered branches, as if to admonish intruding man to pause and reflect on the many difficulties which he must encounter, should he persist in venturing farther into their almost inaccessible recesses."[2]

Narrow habitat specialization furnished an intuitively appealing pretext for why the ivory-bill was regarded as doomed to an inevitable extinction. Tanner was explicit about this theory: "If we are to preserve the ivory-billed woodpecker from extinction, we must maintain for the remaining birds and their offspring a habitat that will supply them with food."[3] Tanner construed these habitat requirements as bottomland hardwoods dominated by "oak, gum, and other big trees."[4] Not only was the tree composition thought to be crucial, but forest age dictated the ivory-bill's ultimate living needs. "Ivory-bills ... can find enough food only in ... old, virgin forests."[5]

Alluring as this portrayal may be, wider evidence refutes a firm conviction that the ivory-billed woodpecker was a limited habitat specialist. Indeed, it could not have been restricted to contemporary vegetation conditions. Evidence that countermands the narrow habitat utilization in ivory-bill comes from two primary directions. One stems from the first-hand reports gathered at the same time or even before Tanner's Singer Tract study, observations that dispute any one forest composition type or stand age for the woodpecker. More potent evidence concerns how the woodpecker's habitat affinities could not have remained constant across the centuries and millennia before its putative extinction. Biomes during just the late Pleistocene were not only exceedingly variable, these vegetation configurations did not conform to modern-day species arrangements.[6]

Virgin, old-growth, or mature stands of timber were not invariably used by the ivory-bill during historical times. In the decade after Tanner's study, Richard Pough conducted a reconnaissance of the Singer Tract, a locale "frequently ... referred to as entirely or largely covered by virgin forest." But Pough was adamant: "This is not true."[7] Pough instead highlighted that the trees found in such fertile alluvial conditions could "grow with...amazing speed," with the large 40-inch diameter trees attaining that size in just 30 years. Ivory-bills certainly foraged in parts of the Singer Tract having large, old trees, but they did not use *only* such areas.

Another of Tanner's contemporaries, Herbert L. Stoddard, also took discreet issue with Tanner's narrow views. Stoddard based his evaluation in part on an opposing report "that a taxidermist had shot a pair of the birds in 1924 in a tract that had been cut over in 1904."[8] Pough underscored how several stands that made up the Singer Tract actually consisted of trees that largely post-dated the Civil War. Pough detected that the lone female later so venerated by others as the very last ivory-bill (Chapter 9) was feeding "almost entirely" on Nuttall oaks *Quercus nuttallii* just 12 to 20 inches in diameter.[9]

When Dennis studied the closely related Cuban ivory-bill *C. bairdii* in the late 1940s, he relied on observations only slightly sparser than Tanner's (again one pair) to reach a precisely opposite conclusion about the woodpecker's habitat use:

> *From observations in Cuba it would seem that the ivory-bill can find such a food supply [larvae of cerambycid and other beetles] more advantageously in cut-over pine forests where millions of trees were already dead or were in the process of being killed by fire and, presumably, by the attacks of insects.*[10]

Dennis' annotations were remarkable for several reasons. They conformed to a forest type (pines), disturbance history (logged and burned within the previous seven years), and tree size (small diameter) that directly contradicted those emphasized by Tanner. Indeed, Dennis pointed out that dead pines near a nest cavity used by Cuban ivory-bill ranged from only 10 to 13 inches in diameter.

Despite "time searching through the most likely areas of the remaining deciduous forest," which Dennis contrasted as "very luxuriant with huge trees towering above the forest floor," in this old-growth zone they "saw no sign whatsoever of the ivory-bill."[11] At a small scale, then, Cuban ivory-bill was avoiding older hardwood forest and instead using the cut-over "scrub growth" pine found nearby. Moreover, Cuban ivory-bill "for nearly a half century occupied [this] region that had been devastated by logging."[12]

A major reason for Tanner's strenuous objections to Dennis' later reports of the ivory-bill occurring in the Big Thicket of Texas[13] came from the comparatively scrubby, logged-over nature of woods found in that region during the 1960s. Despite the Big Thicket's stark differences from the Singer Tract, however, a sound recording that had been secured there by Dennis did in fact in the long run furnish evidence of at least one bird identified as "almost certainly an ivory-bill."[14]

Tanner routinely demoted the prospects of ivory-bill where he deemed bottomland hardwoods not sufficiently old or extensive. He discounted recent reports of ivory-bill in the mid-1930s from the Savannah River Waterfowl Refuge in South Carolina as "unsuitable" because "the only swamp nearby is of small cypress trees."[15] Conversely, he played up prospects for the ivory-bill in virgin sweet gum and oak swamp forest of the Wee Tee Lake section of Santee Bottom in the same state, even though he "found no sign of the birds there."[16] Tanner derided almost all of the then-recent reports of ivory-bill from southeastern Texas, in line with his unwavering conviction that the region was "greatly over-rated as a wilderness area."[17]

Box B.1. The ***myth of the pristine*** stems from misguided belief that in 1492 the Americas were a sparsely populated wilderness. Quite to the contrary, the continent was made up of human-modified landscapes. Some indigenous populations could also be quite dense. "Forest composition had been modified, grasslands had been created, wildlife disrupted, and erosion was severe in places. Earthworks, roads, fields, and settlements were ubiquitous."[17]

Closed-canopy, mesic climax forests witnessed by early European immigrants in the 1700s (when the ivory-bill was first described) were unlike conditions for the species in ecological time just prior. Denser stocking and greater contiguity in the eastern forest between 1700 and 1900 was in part a historical artifact created by the relaxed burning frequencies after the depopulation of native peoples from diseases transmitted by the first Europeans. Once estimated to number more than 5 million inhabitants, Native American populations in the United States fell to only 600,000 by 1800.[18] Native peoples were arguably less visible in 1750 than they were in 1492.[19]

Native Americans exploited fire to clear brush and understories in order to facilitate travel, improve game habitat, and make it easier to hunt and grow food. This burning regime promoted heterogeneous mosaics of vegetation types, some fire-adapted, heightening the contrasts across vegetational boundaries and elevating the landscape diversity overall.[20] After burning was curtailed, however, those once fire-maintained open lands converted to closed-canopy forest made up of more shade-tolerant, mesophytic trees and plants.[21] First impressions of the ivory-bill's living space, then, were skewed by a set of environmental conditions that had had no real precedent for the previous several thousands of years.

Ivory-bill used various open-canopy woodlands. The bird readily came to clearings when trees were girdled to clear fields for agriculture.[22] Bendire linked the ivory-bill to southern live oak *Quercus virginiana* woodlands on the South Atlantic and Gulf barrier islands where it seasonally fed on and even cached acorns.[23] Ivory-bill also used open pine woodland to nest and feed in Florida, the Carolinas, Alabama, and Texas.[24] In the Okefenokee Swamp, it nested on a small tussock of "enormous pine trees" rather than in the nearby cypresses.[25] One cannot help but conjecture what might otherwise have been had our first views on the woodpecker's living space been shaped more by Allen and Kellogg's photograph of an ivory-bill foraging in an open parkland of pines[26] instead of being so colored by Tanner's unyielding advocacy for bottomland hardwoods as the primary habitat.

The ivory-bill in geological time (pre-500 years BP)

No habitat conditions over the last 10,000 years could be presumed to faithfully represent what the ivory-billed woodpecker encountered over its protracted evolution. The Holocene epoch, a period represented by the roughly 11,700 years since the last major ice age ended, was branded by unprecedented changes from anthropogenic and climatic influences alike. Not only did humans immigrate to and radiate across the New World during this epoch, the Holocene was distinguished by a turn to a substantially warmer and more humid climate.

Besides the start to direct persecution of ivory-bill by early Native Americans, other environmental conditions changed strikingly in the last ~10,000 years. Today's southeastern forests, including forest wetlands on the lower coastal plains and interior riverine systems, attained modern characteristics only about 5,000 years BP as water tables stabilized from melting of the Laurentide Ice Sheet. Sedimentary geology of the Tensas River Basin where Tanner conducted his study of ivory-bill consists entirely of Holocene alluvial deposits, signifying a floodplain but a few thousand years old.[27] Indeed, much of the original, interior distribution for ivory-billed woodpecker as mapped by Hasbrouck in the late 1800s[28] matches closely the geographic extent of the Holocene deposits that today delimit the Lower Mississippi Valley.[29]

Box B.2. Without an unambiguous benchmark for how a natural situation truly used to be, we can be misled by the ***shifting baseline syndrome***. This limitation leads us to redefine "natural" based on some artifact from human history, or an arbitrary selection from geological timescales. The American ivory-billed woodpecker evolved during the early- to mid-Pleistocene, under environmental conditions very unlike those found today. Consequently, what we tend to regard today as its "natural" habitat, old-growth bottomland hardwood forest, may distort our conception of its optimal living space in the past.

Pleistocene landforms and vegetation in the southeastern United States were so different from contemporary analogues, however, that what we think of now as the forest type(s) so tightly fastened to ivory-bill could not possibly have represented the primary (much less singular) habitat(s) used by this species over long geological time spans.[30] Too much instability in vegetation composition, spatial extent, and even geographic location of the major vegetation biomes took place for such constancy throughout the ivory-bill's long ecological and evolutionary history. This temporal volatility affected bottomland hardwood and upland pine forests alike.

Coastal plains of the southeastern U.S. support bottomland hardwoods today due to the region's frequently-inundated hydrology and low-gradient fluvial landforms. But extensive stands of bottomland hardwoods would not be an invariable resource for a large-bodied, ecologically-specialized woodpecker throughout the Pliocene[31] and Pleistocene. River valleys during these periods were narrower and shallower, with braided, scoured floodplains at higher elevations relative to the wider alluvial valley surfaces found in the region today.[32] "Modern conditions" in the Mississippi's alluvial valley, where a significant fraction of bottomland forests occurred prior to logging, arose only 2,000 years BP.[33] Given their high fertility for cultivation, floodplains also were partially cleared of forest and managed for agriculture by early Native peoples.[34]

For 2.8 million years, the Lower Mississippi served as an enormous sluiceway that carried vast amounts of glacial meltwater and outwash from the vast interior of the continent to the Gulf of Mexico. Just 14,000 years ago the Mississippi's volume was five times greater than today,[35] a force that scoured adjoining landscapes but made for poor development of mature riparian forests. The Mississippi adopted its current meandering regime only after meltwater flow from continental glaciers ceased and discharge levels fell. One of the oldest and larger back-swamps associated with the big river, the St. Francis Basin of eastern Arkansas, therefore dates from only ~9,800 years BP.[36]

As Gulf of Mexico shorelines regressed (moved inland) during the various interglacials, coastal plain forests were destroyed from erosion while riverine floodplain forests were killed by saltwater intrusion.[37] During periods of major glacial advance and dropping sea levels, the same shorelines instead transgressed (moved seaward). Falling water tables then caused greater fluvial incisions along the older alluvial valleys, narrowing (if not eliminating) the geomorphological conditions most conducive to development of extensive bottomland hardwoods. Influences of these fluvial incisions could extend 600 km inland.[38]

Over geological time, then, areas optimal for today's bottomland hardwoods could be left high and dry. At other times, these same bottomland sites were drowned. Given a generally more arid climate affecting the flows in all southeastern rivers, the lowered water tables, and the fluvial incisions linked to shoreline transgression, bottomland hardwood forest likely faced critical bottlenecks in both its location and extent. Mature floodplains wide enough for the multiple meander belts and shallow gradients necessary for development of bottomland hardwood forest may not have arisen in the Lower Mississippi River until 135,000 years BP.[39]

Tree species typical of bottomland hardwoods survived, but they probably did so at isolated sites contracted greatly from extents attained by the late Holocene. (Long-lasting but small wetlands with forested margins served to collect the plant pollens that furnish glimpses into these environments past.).[40] Extensive wetland forests on the coastal plain could only migrate in one dimension – toward or away from the coast. In contrast, upland forest types could adapt and migrate in three dimensions – inland/seaward, east/west, and upslope/downslope. Nevertheless, upland forests across the ivory-bill's range in the southeastern U.S. also changed dramatically throughout the Pleistocene.[41]

West of the Appalachians, fossil and pollen data reveal that boreal plants advanced and retreated distances of 600–1,000 km during the last glaciation.[42] As recently as the late Pleistocene and early Holocene, 12,700 to 3,500 BP, white spruce *Picea glauca*, tamarack *Larix laricina*, and northern white cedar *Thuja occidentalis* grew south to the Gulf Coastal

Plain of southeastern Louisiana and southwestern Mississippi.[43] At the peak of Wisconsin glaciation, 18,000 years BP, oaks grew far out on what was then the exposed continental shelf of the Gulf of Mexico.[44] Southern pines in that era became restricted to a small north-south band along the Atlantic Coastal Plain.[45] Not until 8,000 years BP did southern pines expand throughout Florida, and only by 6,000 years BP did southern pines approach the outer limits that circumscribe this biome today.[46]

During the long Pleistocene epoch, then, American ivory-billed woodpecker encountered vegetation conditions totally unlike those found today. "Due to the extremely arid climate, vegetation consisted of trees clumped in favorable locations or scattered over the landscape in open park-like savannahs."[47] At the apex of Pleistocene glaciation, annual precipitation in the southeastern United States fell by more than 50%, with a droughty, arid climate fostering a mosaic of grasslands, savannahs, and woodlands. Closed forest was restricted to the most mesic of sites.[48] Vegetation was "ecologically restive" even as late as 16,500 to 12,500 years BP,[49] such that trees and other plants of that era organized into assemblages having no modern analogues.[50]

Box B.3. The colorful notion that eastern North America consisted of vast stretches of climax, closed canopy forest where a squirrel might travel through tree-tops from the Atlantic coastline to the Mississippi River without ever touching the ground was an ***environmental myth.*** It arose out of an illiteracy about roles of disturbance, especially fire and Native cultivation, in promoting open woodland during the times prior to European arrival.

During much of geological time across the ivory-bill's range, woodland rather than dense, closed-canopy forest prevailed.[51] Reconstruction of ancient plant assemblages using fossil pollen and other sources reveal that "a broad *woodland of open-grown vegetation* existed south of the ice sheet" (emphasis supplied).[52] Water tables in Florida were 15 meters lower,[53] leading to mere scrub growing on sandy terrain in the middle of the huge peninsula. Even "forests along the rivers were probably open."[54] Sandy coastal plain soils were well-drained and droughty, however, facilitating pine woodlands. Ancient climates enabled pines and some broadleaved trees to migrate south into the Sierra Madre of northeastern Mexico across the Texas "barrier,"[55] providing at times more contiguous woodland across the western Gulf than occur there today.

Besides climate and geology, open, park-like vegetation on Pleistocene landscapes was promoted by browsing and grazing from a diverse megafauna community. Large herbivores transformed closed woodland or thicket into open savanna, and converted some woody vegetation to grasslands.[56] Declines in these mega-herbivores at the end of the Pleistocene era then promoted reverse changes in the vegetation.[57] Increases in fire activity also took place at intervals of rapid climate change, including the Pleistocene/Holocene boundary.[58] Prior to man's arrival to the New World, lightning fires maintained a park-like structure to the southern pine biome, as exemplified in woodlands of wiregrass and long-leaf pine *Pinus palustris*.

Hypothesized evolution in northern *Campephilus*

Closure of the Isthmus of Panama is thought to have enabled *Campephilus* to radiate out of South America northwards into Middle America.[59] By the late Pliocene, the genus was found north to the south-central United States.[60] At the latitude of the Great Plains where the fossil record confirms presence of *Campephilus*, Pliocene climate was wetter (by 4 to 8 mm per day), far less seasonal, and lacking freezing winters.[61] The woodpecker found here in this epoch, *C. dalquesti*, was not nearly as large as *C. principalis*.[62] In keeping with subtropical and tropical affinities for most of the genus, *C. dalquesti* had body dimensions more like the smaller pale-billed woodpecker *C. guatemalensis*.

Vegetation in Middle America (northern Mexico) and southeastern North America changed markedly during the Pleistocene. Mild, subtropical conditions of the Pliocene disappeared. New environments arose from climate instability and changing sea levels. "Migration of tree species, high erosion due to dying vegetation, and droughty growing conditions would not have favored closed-canopy forest."[63] Natural selection on resident biota included these vast stretches of open woodland, including some that were likely made up of scattered big trees. Stands of woodland were separated by zones of prairie or sagebrush,[64] a feature that made travel distances between tree patches longer. At the same time, past climates in northern Mexico were wetter and cooler during glacial maxima,[65] conditions that favored woodland instead of the desert and shrubland vegetation found there now. Consequently, northern *Campephilus* likely encountered more vegetation continuity than occurs in that region today.

Avian speciation accelerated from the new niches opened by climate change.[66] Woodpeckers responded to fragmentation from changing environments by diversifying world-wide.[67] *Campephilus* at the northern limit of what was once a subtropical zone had to adapt to newly emerging conditions, retreat southward back into subtropical biomes, or die out. On the one hand, newly emerging types of woodland offered a unique and vacant niche for a large woodpecker. Conversely, the same conditions were not optimal for a small or weakly-flying woodpecker possessed of low fecundity living in what could be such an unstable environment. Climate had turned notably cooler, too, conditions that favored a larger body size.

I propose that this northern, ancestral *Campephilus* was an ecological pioneer to the new frontiers presented by open woodlands,[68] going on to evolve for a niche created by novel geology, climate, and vegetation. Adapted primarily to open pine and/or oak woodlands,[69] this ancestral *Campephilus* only incrementally acclimated to the warm or cool temperate zones then arising at more northern latitudes or higher elevations. These novel environments favored acquisition of traits like larger body size, variable and potentially high fecundity, and a wing morphology that expedited more efficient flight over vast distances. Secondarily, such traits facilitated a demography that was more nomadic and/or more irruptive than is the case for *Campephilus* in the stable tropics.

This ancestral line of northern *Campephilus* eventually split into three derived species, the imperial *C. imperialis,* Cuban ivory-billed *C. bairdii,* and American ivory-billed woodpeckers *C. principalis.* Divergence in northern *Campephilus* occurred sometime between 0.6 to 1.4 million years BP.[70] The split was broadly (but not necessarily precisely) coincident with the onset of the mid-Pleistocene revolution, a time when global climate switched from oscillating at 40,000- to 100,000-year cycles that delimited the glacials from the interglacials. Longer cycling may have prolonged a time of separation among disjunctive *Campephilus,* thereby enhancing the processes that promote speciation.

Vicariance led to taxa separation along a broad west-to-east axis. Vicariant barriers to gene flow may have included discontinuities from isolation of montane woodland upslope towards the west (*imperialis*), non-woodland vegetation intervening across the vast western Gulf of Mexico coastal plain (i.e., the Texas barrier), and epochal restrictions of pine woodland to the Atlantic Coastal Plain of the far southeastern corner of the United States[71] (*principalis*). Extensive pine woodlands covered the northern Bahamas between Florida and Cuba during the Pleistocene.[72] Cross-water distances between Florida, the Bahama Banks, and Cuba were as narrow as 20 km during times of greatly lowered sea levels.[73] A route for what would become *bairdii* reaching Cuba first via the Bahamas Bank is more plausible on the grounds of dispersal[74] and corresponding ecology[75] than would be the longer over-water route across the Yucatan Channel.[76]

Implications for the fate of American ivory-billed woodpecker

Timing of origins for the ivory-bill infers it could not have steadily associated with late Holocene vegetation conditions. American ivory-billed woodpecker was certainly a Pleistocene-adapted bird. Although I propose that ivory-bill exploited mosaics of open, park-like woodland rather than dense closed-canopy forest, the full characteristics of this vegetation cannot be known with exactitude. Archaic ecology is, by definition, speculative.[77] Especially hidden to us today are the disturbance regimes for which the ivory-bill was originally adapted. But the precise conditions are less vital than the fact that forest vegetation dynamics were so different from those with which the species was associated during recent historical times.

The very traits that had enabled the ivory-bill to flourish in environmental settings so long ago, however, might have become less adaptive or even turned maladaptive during later geological eras, particularly by the Holocene. To whit, some Pleistocene vertebrates dwindled in body size just prior to going utterly extinct at the end of that epoch.[78] The Pleistocene's end was marked by numerous vertebrate extinctions,[79] as well as by extensive reorganizations to the plant assemblages and biomes making up the continent's woody vegetation.[80]

Ostensibly strong philopatry and conservative movement patterns by the ivory-bill at logged forest sites[81] might be explained by Pleistocene-epoch adaptations for open forest. At the same time, such an affinity may have been at least partially maladaptive, especially if human-caused mortality exacerbated effects from habitat alterations like logging that mimicked open woodland. Yet because the ivory-bill did in fact survive well into the Holocene, that long tenure vouches for at least *some* flexibility in its basic life history.

As a Pleistocene throwback, the ivory-bill might share traits with other evolutionary relicts. Such birds became ever more restricted to remnants of once broader plant communities,[82] or they earlier ranged more widely in biotic associations that have no real analogues to what we see today.[83] Avian relicts also exhibited striking dietary changes over geological time.[84] If American ivory-billed woodpecker was also a Pleistocene relict, that evolutionary shadow could account for much of the lingering paradox about its abundance, ecology, and behavior. Indubitably a descendent from pioneer *Campephilus* ancestors, offspring of a notably ancient race,[85] some of the ivory-bill's most enigmatic riddles are likely to remain forever shrouded.

Endnotes

1 "Relict" in the sense of a taxon that has survived as a trace from the past. For the ivory-bill, I propose that a status of biogeographic (rather than a taxonomic) relict better describes this putative condition. Habel, J.C., T. Assmann, T. Schmitt, and J.C. Avise. 2010. Relict species: from past to future. Pp. 1–5 in *Relict Species: Phylogeography and Conservation Biology* (J.C. Habel and T. Assmann, eds.) Springer-Verlag, Berlin.

2 Audubon, J.J. 1832. *Ornithological Biography, Or an Account of the Habits of the Birds of the United States of America.* Volume 4. James Kay, Jun. & Co., Philadelphia, p. 215.

3 Tanner, J.T. 1942. *The Ivory-billed Woodpecker.* Dover Publications, Inc., Mineola, NY, p. 88.

4 *Ibid.*, p. 87.

5 *Ibid.*, p. 88.

6 Williams, J.W., B.N. Shuman, T. Webb, P.J. Bartlein, and P.L. Leduc. 2004. Late-quaternary vegetation dynamics in North America: scaling from taxa to biomes. *Ecological Monographs* 74: 309–334.

7 Pough, R.H. 1944. Report to the executive director, National Audubon Society: present condition of the Tensas River forests of Madison Parish, Louisiana and the status of the ivory-billed woodpecker in this area as of January, 1944. 9 pp.

8 Gallagher, T. *The Grail Bird: The Rediscovery of the Ivory-billed Woodpecker.* Houghton Mifflin Harcourt, Boston, MA, p. 19.

9 Pough, p. 3.

10 Dennis, J.V. 1948. A last remnant of ivory-billed woodpeckers in Cuba. *The Auk* 65: 497–507. Dennis would go on to regard the American ivory-bill as possibly a "disaster" species, i.e., one that specialized on stand disturbance. That emphasis placed Dennis at odds with Tanner's theory on old-growth, and may explain the antagonism with which Tanner met Dennis' later reports of ivory-

bill from disturbed forests.

11 *Ibid.*, p. 505.

12 Snyder, N., D.E. Brown, and K.B. Clark. 2009. *The Travails of Two Woodpeckers*. University of New Mexico Press, Albuquerque, p. 7. "How are we to rationalize the surprisingly long persistence of the last known Cuban population in degraded habitat?" *Ibid.*, p. 57.

13 The Big Thicket region, however, lies squarely inside the Piney Woods region of Texas, an area once notable for its vast woodlands of large, open-canopied pines.

14 Greg Budney, audio curator at the Macaulay Library at Cornell, drew this conclusion about Dennis' Texas recording, one that had been mislaid and/or ignored for several decades: "I have no doubt that it's an ivory-billed woodpecker ... and the tape is not the one that Allen and Kellogg made." Gallagher, p. 228.

15 Tanner, p. 24.

16 *Ibid.*, p. 23.

17 *Ibid.*, pp. 25–26. In searches he conducted away from the Singer Tract in Louisiana, Tanner was prone to ***observer expectancy bias*** about what he deemed suitable forest habitat for the ivory-bill. Given his strong fixation on wilderness as a necessary condition for the woodpecker, his bias was likely influenced by both ***myth of the pristine*** and ***tainted nature delusion*** (see Chapter 5).

18 Thornton, R. 1987. *American Indian Holocaust and Survival: A Population History since 1492*. University of Oklahoma Press, Norman, p. 43.

19 Denevan, W.M. 1992. The pristine myth: the landscape of the Americas in 1492. *Annals of the Association of American Geographers* 82: 369–385.

20 Delcourt, H.R., and P.A. Delcourt. 1997. Pre-Columbian Native American use of fire on southern Appalachian landscapes. *Conservation Biology* 11: 1,010–1,014.

21 Nowacki, G.J., and M.D. Abrams. 2008. The demise of fire and "mesophication" of forests in the eastern United States. *BioScience* 58: 123–138.

22 Tanner, p. 96.

23 Bendire, C. 1895. Life histories of North American birds, from the parrots to the grackles. *Special Bulletin of the U.S. National Museum* 3: 42–45.

24 Jackson, J. A. 2006. *In Search of the Ivory-billed Woodpecker*, 2nd ed. HarperCollins, New York, pp. 56–57; Nehrling, H. 1882. List of birds observed at Houston, Harris Co., Texas, and in the counties Montgomery, Galveston and Ford Bend. *Bulletin of the Nuttall Ornithological Club* 7: 166–175.

25 Thompson, M. 1885. A red-headed family. *The Elzevir Library* 4: 5–21.

26 Allen, A.A., and P.P. Kellogg. 1937. Recent observations of the ivory-billed woodpecker. *The Auk* 54: 164–184.

27 Saucier, R.T. 1994. Geomorphology and Quaternary geologic history of the Lower Mississippi Valley. U.S. Army Corps of Engineers, Mississippi River Commission, Vicksburg, MS. 364 pp.

28 Hasbrouck, E.M. 1891. The present status of the ivory-billed woodpecker (*Campephilus principalis*). *The Auk* 8: 174–186.

29 Saucier, p. 8.

30 Fleischer, R.C., J.J. Kirchman, J.P. Dumbacher, L. Bevier, C. Dove, N.C. Rotzel, S.V. Edwards, M. Lammertink, K.J. Miglia, and W.S. Moore. 2006. Mid-Pleistocene divergence of Cuban and North American ivory-billed woodpeckers. *Biology Letters* 11: 466–469.

31 The earliest *Campephilus* known from the United States is Dalquest woodpecker, *C. dalquesti*. Its holotype is from Upper Pliocene deposits in Scurry County, north-central Texas. Brodkorb, P. 1971. The paleospecies of woodpeckers. *Quarterly Journal of the Florida Academy of Sciences* 33: 132–136.

32 Saucier, p. 216.

33 *Ibid.*, p. 329.

34 Doolittle, W.E. 1992. Agriculture in North America on the eve of contact: a reassessment. *Annals of the Association of American Geographers* 82: 386–401.

35 Teller, J.T. 1990. Volume and routing of late-glacial runoff from

the southern Laurentide Ice Sheet. *Quaternary Research* 34: 12–23.

36 Saucier, p. 247.

37 *Ibid.*, p. 248.

38 Shen, Z., T.E. Törnqvist, W.J. Autin, Z.R.P. Mateo, K.M. Straub, and B. Mauz. 2012. Rapid and widespread response of the Lower Mississippi River to eustatic forcing during the last glacial-interglacial cycle. *Geological Society of America Bulletin* 124: 690–704.

39 Saucier, p. 326.

40 Delcourt, P.A., H.R. Delcourt, and J.L. Davidson. 1983. Mapping and calibration of modern pollen-vegetation relationships in the southeastern United States. *Review of Palaeobotany and Palynology* 39: 1–45.

41 Area of pine-dominated vegetation across the southeastern U.S. expanded and contracted, at times shrinking to less than 25% of its modern-day extent. Williams, J.W., B.N. Shuman, and T. Webb, III. 2001. Dissimilarity analyses of Late-Quaternary vegetation and climate in eastern North America. *Ecology* 82: 3,346–3,362.

42 Whitehead, D.R. 1973. Late-Wisconsin vegetational changes in unglaciated eastern North America. *Quaternary Research* 3: 621–631.

43 Delcourt, P.A., and H.R. Delcourt. 1977. The Tunica Hills, Louisiana–Mississippi: late glacial locality for spruce and deciduous forest species. *Quaternary Research* 7: 218–237.

44 Watts, W.A. 1980. The late Quaternary vegetation history of the southeastern United States. *Annual Review of Ecology and Systematics* 11: 387–409.

45 Jacobson, G.L., Jr., T. Webb III, and E.C. Grimm. 1987. Patterns and rates of vegetation change during the deglaciation of eastern North America. Pp. 277–288 in *North America and Adjacent Oceans during the Last Deglaciation* (Ruddiman, W.F., and H.E. Wright, Jr., Eds.). Geological Society of America, Boulder, CO.

46 *Ibid.*, p. 280.

47 Carroll, W.D., P.R. Kapeluck, R.A. Harper, and D.H. Van Lear.

2002. Background paper: historical overview of the Southern Forest landscape and associated resources. Pp. 583–605 in *Southern Forest Resource Assessment*, General Technical Report SRS-53. U.S. Department of Agriculture, Forest Station, Southern Research Station, Asheville, NC.

48 *Ibid.*, p. 586.

49 Edwards, R.L., and A.S. Merrill. 1977. A reconstruction of the continental shelf areas of Eastern North America for the times 9,500 BP and 12,000 BP. *Archaeology of Eastern North America* 5: 1–43.

50 Davis, M.B. 1983. Holocene vegetational history of the Eastern United States. Pp. 166–181 in *Late-Quaternary Environments of the United States* (Wright, H.E., Jr., Ed.). University of Minnesota Press, Minneapolis.

51 Hamel, P.B., and Buckner, E.R. 1998. How far could a squirrel travel in the treetops? A prehistory of the southern forest. *Transactions of the North American Wildlife and Natural Resources Conference* 63: 309–315.

52 Jacobson et al., p. 281.

53 Watts, W.A. 1971. Vegetation history of Georgia and central Florida. *Ecology* 52: 676–690.

54 Carroll et al., p. 590.

55 Perry, J.P., Jr., A. Graham, and D.M. Richardson. 1998. The history of pines in Mexico and Central America. Pp. 137–152 in *Ecology and Biogeography of* Pinus (D.M. Richardson, Ed.). Cambridge University Press, New York.

56 Gill, J.L., J.W. Williams, S.T. Jackson, K.B. Lininger, and G.S. Robinson. 2009. Pleistocene megafaunal collapse, novel plant communities, and enhanced fire regimes in North America. *Science* 326: 1,100–1,103.

57 Owen-Smith, N. 1987. Pleistocene extinctions: the pivotal role of megaherbivores. *Paleobiology* 13: 351–362.

58 Marlon, J.R., P.J. Bartlein, M.K. Walsh, S.P. Harrison, K.J. Brown, M.E. Edwards, P.E. Higuera, M.J. Power, R.S. Anderson, C. Briles,

A. Brunelle, C. Carcaillet, M. Daniels, F.S. Hu, M. Lavoie, C. Long, T. Minckley, P.J.H. Richard, A.C. Scott, D. S. Shafer, W. Tinner, C.E. Umbanhowar, and C. Whitlock. 2009. Wildfire responses to abrupt climate change in North America. *Proceedings of the National Academy of Sciences* 106: 2,519–2,524.

59 A majority (9) of *Campephilus* species reside only in South America. Five to six *Campephilus* occur to the north of that continent, two of these in Panama only. Diversification in other South America picids coincides broadly with the emergence of the Isthmus of Panama. Moore, W.S., L.C. Overton, and K.J Miglia. 2011. Mitochondrial DNA based phylogeny of the woodpecker genera *Colaptes* and *Piculus,* and implications for the history of woodpecker diversification in South America. *Molecular Phylogenetics and Evolution* 58: 76–84.

60 Casto, S.D. 2002. Extinct and extirpated birds of Texas. *Bulletin of the Texas Ornithological Society* 35: 17–32.

61 Haywood, A.M., P.J. Valdes, B.W. Sellwood, J.O. Kaplan, and H.J. Dowsett. 2001. Modelling Middle Pliocene warm climates of the USA. *Palaeontologia Electronica* 4: 5.

62 Brodkorb, p. 135.

63 Carroll et al., p. 595.

64 Watts 1980.

65 Metcalfe, S.E., S.L. O'Hara, M. Caballero, and S.J. Davies. 2000. Records of Late Pleistocene–Holocene climatic change in Mexico–a review. *Quaternary Science Reviews* 19: 699–721.

66 Drovetski, S.V. 2003. Plio–Pleistocene climatic oscillations, Holarctic biogeography and speciation in an avian subfamily. *Journal of Biogeography* 30: 1,173–1,181.

67 Fuchs, J., J.I. Ohlson, P.G.P. Ericson, and E. Pasquet. 2006. Synchronous intercontinental splits between assemblages of woodpeckers suggested by molecular data. *Zoologica Scripta* 36: 11–25.

68 "A dense understory or midstory would limit the flight of a bird as large as the ivory-bill." Jackson, p. 56.

69 Several common denominators broadly unite the habitat conditions in which *imperialis, principalis,* and *bairdii* were all reported historically. These include geographic ranges that encompass and habitat use of: 1) *Pinus,* 2) *Quercus,* and 3) open-canopy stand structures. In the case of *bairdii,* however, just one species of oak, *Q. sagraeana,* occurred in its Cuban range.

70 Fleisher et al., p. 467.

71 For example, see Williams et al. 2001, p. 3,353.

72 Steadman, D.W., N.A. Albury, B. Kakuk, J.I. Mead, J.A. Soto-Centeno, H.M. Singleton, and J. Franklin. 2015. Vertebrate community on an ice-age Caribbean island. *Proceedings of the National Academy of Sciences* 112: E5,963–E5,971.

73 "The Bahamian Archipelago, although never higher than 200 m elevation even at the lowest late-Pleistocene sea levels, was essentially Greater Antillean in its land area, location and proximity to other islands during glacial times." Steadman, D.W., and J. Franklin. 2014. Changes in a West Indian bird community since the late Pleistocene. *Journal of Biogeography* 42: 426–438. This huge expanse of pine woodland offered a stepping-stone for continental *Campephilus* to Cuba, although conditions on the Bahamas Banks may have deteriorated such that the genus died out or never permanently established there.

74 A supposed low proclivity of woodpeckers to cross open water fails scrutiny because: 1) northern *Campephilus* had especially powerful flight, and 2) yellow-bellied sapsucker *Sphyrapicus varius,* a much smaller species, regularly migrates across large water gaps from continental North America to wintering grounds in the Greater Antilles and Bahamas. Walters, E.L., E.H. Miller, and P.E. Lowther. 2002. Yellow-bellied sapsucker (*Sphyrapicus varius*), *The Birds of North America Online* (A. Poole, Ed.). Cornell Lab of Ornithology, Ithaca, NY. http://bna.birds.cornell.edu/bna/species/662 (accessed 7 July 2016).

75 Fossil evidence reveals that bird communities of the Bahama Banks had strong affinities for pine woodland, grassland, or

savanna that dominated during the Pleistocene. Steadman et al. 2015, p. E5,968.

76 Dispersal of ancestral *C. bairdii* to Cuba via the Yucatan seems less likely on both ecological and evolutionary grounds. *Campephilus guatamelensis* in the Yucatan is allopatric in tropical and subtropical forest of southeastern Mexico. Yet *bairdii*'s strong morphological resemblance to *principalis* infers an origin derived from that more northern line rather than through some tropically distributed *Campephilus*.

77 "Understanding the behavior of a species that is extinct or so rare that it can no longer be studied is a challenge." Jackson, p. 13.

78 Guthrie, R.D. 2003. Rapid body size decline in Alaskan Pleistocene horses before extinction. *Nature* 426: 169–171.

79 Gill et al. 2009.

80 Williams et al. 2001.

81 Snyder, N., D.E. Brown, and K.B. Clark. 2009. *The Travails of Two Woodpeckers*. University of New Mexico Press, Albuquerque, p. 42.

82 Forest encroachment in the late Pleistocene caused major reductions to the once wider range of New Zealand's takahe *Notornis mantelli*. Holocene conditions left only isolated remnants of alpine and sub-alpine grassland where this bird is confined today. Mills, J.A., R.B. Lavers, and W.G. Lee. 1984. The takahe – a relict of the Pleistocene grassland avifauna of New Zealand. *New Zealand Journal of Ecology* 7: 57–70.

83 California condor, *Gymnogyps californianus*, once occurred in spruce-jack pine woodland as far east and north as New York. This range, climate, and biotic association are nowhere apparent from the restricted settings in which the species was or is recently known to be extant. Steadman, D.W., and N.G. Miller. 1987. California condor associated with spruce-jack pine woodland in the late Pleistocene of New York. *Quaternary Research* 28: 415–426.

84 Chamberlain, C.P., J.R. Waldbauer, K. Fox-Dobbs, S.D. Newsome, P.L. Koch, D.R. Smith, M.E. Church, S.D. Chamberlain, K.J. Sorenson, and R. Risebrough. 2005. Pleistocene to recent dietary

shifts in California condors. *Proceedings of the National Academy of Sciences* 102: 16,707–16,711.

85 *Campephilus* bears molecular similarities to woodpecker genera of southeast Asia, suggesting an origin at or near the Miocene-Pliocene boundary. See Benz, B.W., M.B. Robbins, and A.T. Peterson. 2006. Evolutionary history of woodpeckers and allies (Aves: Picidae): placing key taxa on the phylogenetic tree. *Molecular Phylogenetics and Evolution* 40: 389–399. More recent evidence points to an age of 8.2 million years BP for the *Campephilus* lineage. Fuchs, J., J.-M. Pons, L. Liu, P.G.P. Ericson, A. Couloux, and E. Pasquet. 2013. A multi-locus phylogeny suggest an ancient hybridization event between *Campephilus* and melanerpine woodpeckers (Aves: Picidae). *Molecular Phylogenetics and Evolution* 67: 578–588.

Appendix C

Dividing to conquer: cognitive gaffes in repudiating the Luneau video

History of the video controversy

Given the raging firestorms that it provoked, one could be excused for imagining that the identity and persistence of ivory-billed woodpecker as an extant species in the 2000s hinged entirely on dogged convictions argued by rivals over short, blurry images caught by random in one amateur video. Pardoned for thinking so, yes, but still sorely mistaken.

Cognitive derailment factored markedly into the group-refereed interpretations placed upon a short video clip taken of a large woodpecker during 2004. Social perceptions abridged this clip into the only material evidence that mattered for assessing the ivory-bill's identity *and* persistence. ***Framing*** prejudices across social identities then steered antagonistic evaluation of key content revealed in that accidental image. Establishing the species' ongoing persistence was held hostage to, rather than being viewed merely as a logical precedent for, a correct species assignment. In this instance our evidentiary yardsticks for bird documentation were haphazard, bordering on the nonsensical, and thus compromised the application of much rigorous interpretation.

On April 25, 2004, Robert Henderson and David Luneau, an electrical engineer from Little Rock, were canoeing in Bayou de View, part of the Cache River National Wildlife Refuge in Arkansas. Using a constantly running video camera fixed to their craft, the two canoeists serendipitously captured a grainy, 4.5-second video clip of what appeared to be a big woodpecker as it took off from behind the lower trunk of a large water tupelo *Nyssa aquatica*. The video went on to be fiercely debated and vigorously defended as a primary justification for persistence of ivory-billed woodpecker into the twenty-first century.

The Big Woods of Arkansas, however, was neither the first nor only location in the southeastern United States from which ivory-billed woodpecker was reported between just the years 2000 and 2009 (***Table C.1***). Prior reports of the woodpecker from that decade came from South Carolina, south Florida, and the Pearl River region of Louisiana. By the time the decade had ended, an astonishing *26* separate incidents of the woodpecker were tallied across seven states and from every season. Some incidents involved several observers, more than one ivory-bill, direct comparison with the pileated woodpecker, and supplemental physical evidence.[1] Although some of these incidents arose from dedicated surveys intended to confirm whether the bird was still extant after the Arkansas incident, other reports of ivory-bill in this decade were entirely coincidental.

Logical breakdowns over the Luneau video started from failing to place the entire Arkansas case inside a larger context of these near-concurrent ivory-bill incidents, revealing our penchant to flaunt ***Bayesian caution*** and capitulate far too readily to error from ***Bayesian conservatism***.[2] Reports detecting the ivory-bill during aerial waterbird surveys in North Carolina and Florida (***Table C.1***), for example, stand out as notable departures from a narrative perforated with ***ultimate attribution error*** whereby, primed by a quixotic hope, observers purportedly go find just what they set out to see. A search purpose aimed at something else entirely contradicts a simplistic notion that ivory-bill reports are generated invariably by some skewed agenda on the part of expectant novices.

Table C.1. Reports of the ivory-billed woodpecker grouped into geographically linked incidents compiled for the first decade of the twenty-first century, 2000–2009.[3] Any given incident may have included multiple observers, dates, and lines of evidence, e.g., audio detections, sightings, poor-quality video, and so on.

Region	*State, County/Parish*	*Date(s)*	*Description, Observer(s)*
Carolinas	NC, Brunswick	2005	Pair in Green Swamp, J. Condrey
	SC, Marion	Sep 2004, Sep 2005	Female seen twice, F. Ervin
	SC, Marion	Mar 2007	Female, J. Godbold
	NC, Robeson	Apr 2008	During aerial waterbird survey, A. Houston
	SC, Berkeley	Sep 2007	Pair with pair of pileateds, J. Cork
	SC, Berkeley	Jan 2005	Lower Santee River, L. Riney
	SC, Richland	2005–2009	Multiple visuals, audio recordings
	SC, Sumter & Calhoun	2001, 2002, 2005	Multiple reports, observers
	SC, Dorchester	Dec 2006	Four Holes Swamp, B. Teagardin
	SC, Jasper	Jun 2005	Three birds, D. Hamilton
	SC, Beaufort	Oct 2006	Heard and seen, D. Dunlap
Georgia–n. Florida	FL, Franklin	Jan 2003	Possible *kent* calls, T. Spahr
	FL, Madison	Apr 2007	Female fly-over highway near Aucilla River
S. Florida	FL, Collier	Unknown	Three independent reports, *fide* Jackson
	FL, Monroe	Feb 2003, Jan 2007	During aerial waterbird survey, L. Oberhofer
Panhandle	FL, Washington	2005–2007	"Many observers," as well as audio recordings and poor video, from Choctawhatchee River
	LA/MS, St. Tammany	2000, 2005–2009	Multiple sightings, poor video, Pearl River
Lower Mississippi Delta	LA, St. Mary	2005–2006	Multiple sightings and auditory encounters, Bayou Sorrel near Patterson

Upper Mississippi Delta	AR, Desha, Phillips, and Arkansas	Mar 2003, 2004, Jan–Dec 2005	Multiple sightings and auditory encounters, southern White River National Wildlife Refuge
	AR, Monroe	Feb 2004–Dec 2005	Multiple sightings of male, possible vocalizations and double-knocks, plus Luneau video, Bayou de View, Cache River
	AR, Prairie	Dec 2006–May 2007	Multiple *kent* calls and double knocks recorded, plus two sightings by separate observers, R. Everrett and A. Mueller, Wattensaw Wildlife Management Area
	TN, Lauderdale and Tipton	Jan 2006	Numerous auditory encounters, *kent* calls and double knocks, one visual, R. Ford
	TN, Dyer	2007–2009	Several sightings, multiple observers, numerous double-knocks, one series recorded possibly a response to simulation
	IL, Pulaski	2005, 2008	Several independent reports, J. White, A. Albores, G. Erdy
Arkansas–Oklahoma	AR, Ashley	Aug 2007	Several visual reports by refuge staff, Felsenthal National Wildlife Refuge, including two birds at once at close range
	AR, Sevier	Oct 2007	Visual report while deer hunting, B. Petersen

Even more capitulation to ***Bayesian conservatism*** came from divorcing the Luneau video from other support that was germane to the identity or persistence of ivory-bill just in the Big Woods of Arkansas. The video was not the only bit of evidence collected, not the first obtained for the woodpecker in that region, and not necessarily the most dispositive.[4] Cornell claimed modestly that the Arkansas incident for putative ivory-bill offered "more data for public scrutiny, than any previous case ... since 1944."[5] But because the 1940's reports of the big woodpecker had delivered no physical evidence at all for any dispassionate verification (see Chapter 9), the Arkansas incident actually had furnished the superior evidence.[6] This disparity makes even more obvious such irrational contentions as: "the Luneau video may loom as one of the most unfortunate things to ever happen to the Laboratory of Ornithology or to the conservation of the ivory-billed woodpecker."[7]

Luneau's video was an *outcome of* (not the instigation for) sight reports of ivory-billed woodpecker in Arkansas. Strident critics failed to weigh the logical prominence of *when* and *how* this video was obtained. Tangible evidence appeared *after* less verifiable sight observations had been made already. The original trigger came months earlier on February 11, 2004, when Gene Sparling saw a huge, odd woodpecker with a red crest while kayaking in Cache River National Wildlife Refuge. Sparling watched this bird land on a tree about 20 meters in front of his kayak. Cornell staff and others learned of Sparling's encounter by reading his description on an internet site devoted to boating. By February 14, 2005, Cornell's team had deployed searchers to the Big Woods, then went on to garner at least 15 more sightings of ivory-bill. Seven of these were deemed to have sufficient details or quality to be included as part of their original *Science* report.[8]

Sightings of this apparently solo ivory-bill waned later during 2005, but Cornell acquired other evidence in the Big Woods for a bit longer. That support included recordings that were close acoustic matches to the woodpecker's known sounds.[9] Passive, digital autonomous recording units (ARUs), physical devices disconnected from the biased or inflexible expectations of human observers, captured sounds having various of the acoustic properties consistent with or constituting best fits to the ivory-bill. These sound recordings even originated from two adjacent regions, the Cache River (the location of the initial sighting reports) and the White River National Wildlife Refuges.[10]

Sound recordings included the bird's characteristic *kent*-like calls, vocalizations lasting from 8 to 41 seconds, some of which extended for 11 and 26 notes, as well as double-knocks like those made by other *Campephilus* woodpeckers. One of these double-knocks contained a sequence and cadence that matched a potential exchange between *two* birds. Cornell eventually secured about 60 double-knock sound recordings. In a very rare exception to the thinking gaffes from ***Bayesian conservatism*** that were epidemic elsewhere in the Arkansas incident, three critics withdrew a scholarly rebuttal of Cornell's original *Science* paper once they had learned of this supplemental acoustic evidence.[11] Such logical reactions were the exception, however, never the norm.

How was it, then, given so much complementary material, that the Luneau video came to so narrowly define social perceptions about accuracy in the identity and persistence of living ivory-billed woodpecker in Arkansas (and elsewhere)? Ambiguities in the video tempted its critics to apply logical fallacy and capitulate (however unwittingly) to cognitive biases when discrediting Cornell's identification of the bird. By cleaving off the video from everything else consistent with ivory-bill, the blurry image could be assaulted on its weakest flanks. Arguing for inadequate evidence was made easier by juxtaposing an allegedly "extraordinary" claim that ivory-bill still lived alongside tangible evidence that itself was demonstrably fuzzy in its literal appearance.

Out of convention, not all situations for documenting bird presence or identity require the same force of physical evidence (***Table C.2***). During the Arkansas incident, critics and detractors tried constantly to drive the standard of evidence ever higher towards an "extraordinary" criterion, that is, a situation wherein the presence or identification of ivory-billed woodpecker demanded greater thresholds of physical support than routinely adopted for other bird species in parallel situations. Because humans so often guess wrongly about the true rate of extinction (***Romeo error***; see Chapters 13, 14), this tendency to exaggerate a necessity of proving to everyone's satisfaction an ivory-bill "rediscovery" also was inherently susceptible to miscalculation from ***base rate fallacy***.

Table C.2. Relative (and thus inherently subjective) levels of "extraordinariness" that might encompass various contexts for judging the evidentiary merits of a blurry video as sufficient documentation for a bird species' mere presence. Greatest levels of "extraordinariness" are generally presented at the top of the table, with declining levels depicted towards the bottom of the table.

Context	**Usual evidentiary standard**	**Sufficiency of Luneau-quality video?**
Species new to science	Specimen; maybe multiple clear images along with in-hand measurements	Never
1st rediscovery of species long lost to science	Specimen; probably multiple clear images along with in-hand measurements	Highly unlikely or never
2nd rediscovery of species long lost to science	Probably multiple clear images along with in-hand measurements	Unlikely but possibly
nth rediscovery of species temporarily lost to science	Images	Unlikely to possibly[A]
A new, long (>1,000's km) range extension	Multiple and clear images	Highly unlikely
A new, short (100-1,000 km) range extension	Diagnostic image(s) or published account	Possibly to likely
nth range extension	Written and/or published account	Possibly
Less than a few hundred sighting reports inside original range	Diagnostic image; written and/or published account	Possibly
500th to 1,000th sighting report inside original range	Written and/or published account	Image not necessary[B]

A By the time of the 10th to 20th "rediscovery" of a lost species, however, an evidentiary standard that demands never-ending image documentation starts to appear arbitrary and extreme. For example, once the status of a bird species is revealed to be merely uncommon, written documentation is usually adequate.

B Given hundreds and hundreds of specimens, plus all of the pre-1944 written sighting reports, this context best fits the status of ivory-billed woodpecker in modern times. No adequate reasoning has been offered for why the ivory-bill continues to require far higher levels of evidentiary standards, than do other birds, or indeed other wildlife species and natural objects that display a similarly sparse, erratic, or recurrent pattern in their occurrence.[12]

Moreover, throughout social discourse over the Arkansas rediscovery, it was common to hear even professional researchers (who really should have known better) refer to evidentiary prospects of a living ivory-bill as aligned to those of Bigfoot, the Loch Ness Monster, and similar phenomenon never substantiated by genuine science.[13] This illogical broadcasting of a ***false equivalence*** by "experts" was then seized upon as gripping theater by public journalism. This in turn led the media to engage in a rampant ***equivalency bias*** during its extensive reporting of the Arkansas rediscovery incident (Chapters 9, 12).

The Luneau video was also widely condemned as falling short of decisive *proof* for a living ivory-billed woodpecker. Though this proclamation is self-evident and essentially true, it is also irrelevant. Science does not deal in proof. Pronouncing a piece of evidence as falling short of that standard, mirrors a deep miscalculation of what science truly is: the "psychology of science is deeply connected to uncertainty."[14] Science is an iterative practice geared towards reducing the uncertainty about the natural world through repeated hypothesis-testing. A ***red herring*** device of demanding ongoing and never-ending proof for a living ivory-bill (or any other species already amply documented by science) echoes a deep cognitive illiteracy about ***neglect of probability***.

For more than 100 years, ornithological science and bird-watching placed their respective trust in incongruent evidentiary standards. After the two broke apart in the very early twentieth century (Chapter 4), specimens were the only acceptable standard then in use by professional ornithology. But as that profession expanded into subject areas beyond nomenclature and distribution (mostly leaving these topics to bird-watching and natural history museums), the need for specimens in many lines of avian research diminished. At the same time, bird-watching was consigned to its amateur standing in substantial part from its reliance on sight reports, information that was deemed suspect by professional ornithology given the inaccuracies that can arise from ***observer expectancy bias***.

Elbowed aside by professionals over documentation, then, bird watching struggled to gain its own autonomous credibility. So the past-time fashioned ***argument from authority*** by setting up review committees to scrutinize each other's sight reports. Over time bird-watching adopted its own eccentric standard of "proof," one achieved via non-lethal and non-destructive methods such as photography and sound recording. Today, the social identity inside bird watching consistently employs terms like "definitive," "conclusive," and "irrefutable" during its group discourse as constituting the proper standard for gauging reliable information about bird presence. Although this standard is easy to fathom from the standpoint of history, it is wholly inadequate for unanimous application. Standards of bird watching are *for that hobby*, not for universal implementation during mankind's complex search for insight and accuracy about the natural world.

Skeptics and detractors, especially those enmeshed in a birding perspective, were quite frank about their inclination to use divide-and-conquer tactics as a means to discredit the Arkansas ivory-bill reports as sturdy evidence. "Focus on the misidentified video, the crappy cluster of sight records, and the driving force of RennFitzenPatrick."[15] Bailing out entirely from ***Bayesian caution*** to endorse such partitioning, this strategy enabled its adherents to easily select from a broad menu of cognitive breakdown: ***argument from authority***, ***ad hominem***, ***ultimate*** and ***fundamental attribution errors***, ***fallacy of division*** and ***composition***, and so forth (see Chapter 12).

Others divulged a more cryptic prejudice that made it difficult to trace exactly where the breakdown had begun or which bias was the proximate trigger. Kaufman lectured that because searchers "have yet to get any kind of real documentation ... it becomes extremely implausible that even one individual [ivory-bill] could be ... evading the cameras of the entire Cornell search team for the last two winters."[16] But whatever it was that searchers pursued in Arkansas, it had not evaded the Luneau camera. In his synthesis of the ivory-bill, Jackson revealed a predilection towards ***argument from authority, outcome reporting bias***, or ***attribution error*** by choosing to interview only "the few who I truly believe have watched these birds."[17] Sibley protested that "experience with other rare birds, especially resident species, suggests that any valid sighting should very quickly lead to more sightings."[18] Aside from the qualifier about pace, this is exactly what transpired – an initial sighting report led to more visual reports, then the Luneau video, and still later the acoustic matches from sound recordings! So what was really going on here?

Divergent social contexts had hardened a deep, unresolved friction across two group identities. ***Framing*** diverged between professional ornithology and birding, insuring contrary expectations for evidence used to evaluate the identity and ongoing presence of living ivory-bill. These ***framing*** differences, however, are little more than a shadow cast by the caprice of history. By insisting on its own evidentiary standard, birding capitulated to ***false consensus effect***, too, much like Cornell had when it overstated the degree to which others would agree with the quality or quantity of their evidence about an Arkansas ivory-bill. Birding committed its error by overstating the supposed requirement for "proof." Still, this arbitrariness in evidentiary standards between each identity explains only part of our madcap bungling over the prospect of living ivory-billed woodpecker.

What makes so peculiar the demands for ceaseless authentication for a living ivory-bill, including disputes over the Luneau video, is that neither professional ornithology nor bird-watching apply standards to the big woodpecker that customarily pass muster elsewhere in natural science. Consider the case of Halley's Comet (and certain other celestial bodies). Known to mankind for millennia, recognized for centuries as having a periodicity of 75–76 years, we have no literal (touchable) evidence whatsoever for this object. All evidence is indirect: photographic and spectrographic. Halley's Comet is solely an inferred entity.[19] Halley's Comet cannot be conjured up at will when some individual or group demands it. Reasonable persons accept a factual existence for this indistinct object, including some of its physical properties, despite its very long invisibility to us.

Yet out of some weird ***lack of object constancy***, perhaps attended by other impulse of human thought, we persist in demanding ceaseless proofs for a demonstrably real object like the ivory-bill, something for which ample physical evidence exists already. I attribute that whim to a strong, precise ***anchoring bias***. For nearly a century, the predominant belief has been that ivory-bill is extinct. This belief is the starting point from which human "reason" treats any new, dispositive information received about the woodpecker. Because belief in the species' extinction has been so reinforced through ***availability cascades***, the human mind is forced grudgingly to overcome its rigidity and revise a wrongly-held belief, or fall to increasingly fanatical displays of special pleading to rationalize away so many contradictions. Dogged opposition in those choosing the second response triggers a strong ***backfire effect*** that in turn leads skeptics to double down on their resistance.[20]

In the case of the Luneau video, blurry imagery was extended from a tangible reality into metaphoric certainty. Luneau's canoe-mounted video had been in a constant running mode before it fortuitously captured its renowned avian oddity. Images of the bird were tiny, fleeting, and indistinct, lasting only seconds. Essentially on auto-pilot, the camera was unmanned for purposes of clear focusing or ideal tracking of any one object in which the canoeists might have had a particular interest. Accordingly, the video was plagued by pixilated resolution, limitations on angle and perspective, and other pitfalls in quality. These substantial deficiencies made even its staunchest defenders confess that "the Luneau video ... [was] a challenging identification puzzle."[21]

Table C.3. Nine attributes identified by the Cornell Laboratory of Ornithology in their evaluation and analysis of the image content in the Luneau video. These anomalies were regarded by that search team to represent diagnostic features for ivory-billed woodpecker.

Diagnostic feature of ivory-bill	**Alternative explanation by critics?**	**Alternative explanation debunked by Cornell, others?**
Underwing pattern in flight largely white	Yes. Pileated woodpecker also has extensive white on underwing	Partially. De-interlaced video failed to reveal wide black border on trailing edge expected in pileated
Upperwing pattern in flight consistently white on trailing edge	Yes. Wing-twisting in pileated woodpecker made bottom of wing visible throughout down and up strokes	At least partially. Critics' hypothesis on avian flight mechanics has no support in theory or fact
Wings longer relative to body diameter than pileated	Yes. Measurements difficult or inaccurate in Luneau video	Mostly. Body size ratios are more robust to measurement error than linear dimensions
Dynamically flapping models produced images similar to video, for both woodpecker species	Yes. Models don't properly mimic flexing wings	Not fully. Models from other material perform substantially different than live birds[22]
Wingbeat frequency of 8.6 beats per second virtually identical to the rate audible in an ivory-bill recording	Yes. Pileated woodpeckers fly just as fast on take-off	Yes, utterly. Claims otherwise fraught by technical errors, improbable experimental conditions, and deceptive misuse of statistics[23]
White plumage on back is visible	No	Yes. Placement cannot be accounted for in a normal pileated woodpecker
Dorsal view of right wing shows white triangle as it unfolds behind tree trunk	Yes. White triangle was instead a pileated's underwing on take-off	At least partially. Much later analysis of a movie of closely-related imperial woodpecker showed rotation at takeoff more like bird in Luneau video

Distance between wrist area and tip of tail 32–36 cm, too large for pileated	Yes. Measurements difficult or inaccurate in Luneau video	Not fully. Linear measurements inexact in poor-resolution imagery
20 seconds before Luneau woodpecker flees, a bird with size and color pattern fitting ivory-bill recorded on a tree 3 m away	No. Special pleading for object as a tree stub demonstrably wrong	At least partially. Alternative hypothesis was debunked

Such deficiencies made it easier to implement divide-and-conquer tactics (***subadditivity effect***) as a means to refute key details that could be extracted from the video (***Table C.3***). Critics never succeeded convincingly in refuting every line of evidence Cornell relied on to reach its conclusion that the Luneau bird was an ivory-billed woodpecker. Instead, critics attacked the most technically ambiguous content, frame by individual frame, then built their refutations centered on these subjectively-chosen images, the most equivocal parts-of-the-whole, in order to discredit Cornell's identification in its entirety (***fallacy of composition***).[24]

Critics also ignored or glossed over material that most inconveniently countered the usual alternative proposed for the bird, one that claimed instead that it was a normal pileated woodpecker. The most surprising of these several neglects was why so little attention was paid ultimately to an object that matched the pattern and size of an ivory-bill, an object visible on another tree in several video frames *before* the 4.5 second clip was taken. This feature (among others)[25] never was explained satisfactorily by staunch critics of the Luneau video.

Ensuing commentary over the Luneau video habitually deviated from common sense. Debate showed a level of acrimony hard to fathom about a matter so trivial, dare I say, as bird identification. Individuals in birding and ornithological circles proclaimed that "we all see what want to see in the Luneau video,"[26] but this declaration is neither accurate nor rational. The bird in the video shifts from an upright stance aligned with a vertical tree trunk, then launches into horizontal flight in a manner that, alongside other attributes all visible and measurable from the video imagery, are jointly consistent only with a big woodpecker. Even if the video was "just good enough to show a black and white bird fleeing into the forest," while at the same time "just poor enough to show no field marks clearly," it does not follow cogently that "the image in the video could be pretty much whatever you chose to see."[27]

Despite causal interrogations grounded on the matter of plain identity, the arguments over the species of woodpecker captured in the Luneau video were unusually heated, remarkably voluminous, and exceedingly prolonged. Debates relevant to the image content in this video extended for at least *seven* years.[28] Rather than condense all of the twists and turns of those deliberations, I send readers to the original summaries from the principal searchers,[29] major and minor criticisms that disputed the bird depicted in the Luneau video,[30] responses to those criticisms,[31] and a selection of blogging sites. The latter offered at times more insightful appraisal because the discussions had no limitations on length in that more flexible format.[32]

Social and cultural implications

For myself the Luneau video is evocative but not adequate to insist on a categorical identity of the imaged bird as an ivory-billed woodpecker. But my hesitation is strictly personal: it stems from my lack of familiarity with intricacies of video capture, possibly misleading effects from image distortion, and the reliability of pre- or post-processing techniques used to eliminate image artifacts. My reservation arises, too, from caution over how much we honestly ever knew about the field appearance of this cryptic species, one that largely bolted our grasp before its identification analytics had been fully worked out. But bending selected images from the Luneau video so as to fit a normal pileated woodpecker[33] requires more mental contortion than I care to try.

Cornell's *claims* for evidence were imperfect, but the results presented were probably as thorough as human nature generally allows when confronted with untidy information. Messy data is quite prevalent in the real world. I give deference to Cornell for that search team's allegiance to thoroughly vet their evidence, to follow loose threads amenable to little experiments and tests. Around the time of the Arkansas incident, far too many who pontificated about ivory-bill identification had no experience with the topic at all. Challenges to the video from "top bird identification experts"[34] were riddled with a perniciously skulking brand of ***argument from authority***. Such ungrounded opinion is even less dispositive than the original, raw sighting reports of the bird.

Taking the Luneau video alone, divorced from all other evidence, I tend to regard the identification offered by Cornell as more likely correct than wrong. My judgment is akin to adopting a standard that aligns with *preponderance of evidence* (>50% likelihood) rather than *beyond all reasonable doubt* (close to but not reaching 100%).[35] The total ***neglect of probability*** paraded by some who weighed in for a contrary identity of the Luneau bird is startling to me. Such insistence on total certitude might arise from our own species' heedless rush to attain a ***cognitive closure*** when some matter conflicts with our long-cherished, preconceived beliefs. Whatever else the ivory-bill denotes for us, it seems to trigger some obstinate anxiety in our mental circuitry, one that demands a quick resolve so we can return once again to belief relief.

Although retaining some measure of caution about the identity of what the Luneau imagery portrays, I am less disposed to hold back about the impressions I gleaned from the sociological dimensions to the Arkansas incident. Admitting freely to a ***hindsight bias***, much about human nature in this episode troubles me. "The human side of the story could not be murkier."[36] Disputes over the Luneau bird "soon escalated beyond polite discussion into decidedly unscientific name-calling. The bird world split into Believers and Skeptics, and splintered further into True Believers, Agnostics, and Atheists, with former friends and colleagues at each other's throats."[37] If this correspondence to religious fervor is true, the parallel can only sober us all.

More than the impropriety of this bickering, the primal origin(s) for all this hullabaloo are what I find most mesmerizing. I am hardly the first ever to be held spellbound by the "highly charged emotions and sometimes irrational behavior [that] have figured into the search for the ivory-billed woodpecker."[38] No kidding. Considering the historical literature about the species in its entirety, I find it especially difficult to reconcile the content and tone of accusations levied by certain personalities, especially considering the astounding extent of supportive evidence that has since the 1940s accumulated for a still living ivory-billed woodpecker.[39]

Moreover, I find it astonishing that so much scientific procedure (to also include "extraordinary" evidence) must be levied at a matter of plain bird identification.[40] If not actually unprecedented compared to other re-found animals, this level of cultural and social fixation on a species identification problem surely must be abnormal. In my view, requirements to use that degree of hypothesis-testing with the full force of scientific method should not be binding practice for identifying an already-described animal species. Instead, I propose that such documentary obsession in us arises out of bizarre pathways that control the motherboard of our unconscious.

Some of the original arguments waged over the Luneau video, and the later reactions to allied reports of ivory-bill, have not stood up well to aging. Coming up on two decades now, some of these strike me as cheap shots and careless scholarship. Cornell had to correct at least 10 errors of fact[41] levied in a lengthy critique disseminated by Jackson.[42] The latter criticism was little more than scattershot denunciation, a broadside rife with ***fundamental attribution error***. Cornell even had to stipulate in the lead to its response that clarifying "intentions" was a primary justification for their rejoinder. When guessing at others' intentions occur inside a frame ostensibly reserved for science, we can be prudently inclined to look for cognitive prejudices at work.

Pitfalls in human reason over the Luneau video verged on a veritable standoff between birding and professional ornithology. "Two groups of authors ultimately disagree on its interpretation and the same still video images that are used to argue for the sighting are used to argue against it."[43] For those who were casually tuning in, it might have seemed like each side's arguments were of roughly comparable value. But we generally overlooked how the *meaning* placed in such images constitutes a mere abstraction of nature. Meaning varies subjectively by location, depending on whether it is achieved at the site of image production, the site of the image, or the site of the audience viewing the image.[44] Even objectivity can assume at least seven different, normatively acceptable forms, three of which (convergent, detached, and concordant) were each realized in the case of the Arkansas ivory-bill incident.[45]

At times, identical mental mistakes were perpetrated regardless of the particular social identity involved. Cornell fell to a ***false consensus effect*** when relating evidence in terms that neglected uncertainty. But Sibley and his fellow critics, too, grossly overstated their case: "the evidence firmly supports this hypothesis."[46] Birding critics succumbed to ***false consensus effect*** in part by speculating that everyone ought to adopt uncritically that hobby's say-so (***framing***) about bird identification (***argument from authority***). If prominent field guide authors get so utterly wrong core facts that are accessible in the extant literature about the ivory-bill's basic biology (Chapter 12), the pastime loses credibility as a trustworthy source of information about birds.[47]

For valid reasons the two endeavors split apart so long ago. Each endeavor does some things better than the other, yet the Luneau video incident revealed that whether alone or jointly, avian science and birding do not encapsulate all dimensions to evidentiary truth and accuracy in matters of the ivory-billed woodpecker. Should there ever again be a fortunate occasion when some signs for living ivory-bill comes to light, I submit that our cognitive processing and tendency to perpetrate mental gaffes be fittingly dissected.[48] In addition to all of the image analysts, acoustic technicians, and signal processing engineers,[49] we will need to invite the psychologists and sociologists to weigh in as well. We will certainly need them.

Endnotes

1 Overlapping and ensuing searches in Florida and Louisiana also led to acquisition of physical evidence. This included video of similar or lower quality from each location. Florida searches produced more acoustic evidence (210 *kent* calls, 99 double-knocks) than ever gathered in Arkansas, however, as well as sighting reports of what might have been multiple ivory-bills. Hill, G.E. 2007. *Ivorybill Hunters: The Search for Proof in a Flooded Wilderness*. Oxford University Press, London, p. 207.

2 Had there been no reports of ivory-bill at all in years just prior to the Arkansas incident, a charge of bias from a lack of ***Bayesian caution*** itself could be faulted on grounds of ***hindsight bias*** (because many other incidents came later in the decade). But critics of the Arkansas incident never revised their rigid assertions that the woodpecker was extinct, despite the additional reports both before *and* after the Luneau video. Failure at such hypothesis revision displays rampant ***Bayesian conservatism***, a serious cognitive error.

3 These 26 incidents are condensed out of Hunter, W.C. 2010. *Interpreting historical status of the ivory-billed woodpecker with recent evidence for the species' persistence in the southeastern United States.* Appendix E, Recovery Plan for the Ivory-billed Woodpecker (*Campephilus principalis*), U.S. Fish & Wildlife Service, Atlanta, GA. Additional details for ivory-bill incidents from Florida and Louisiana are available from Hill, and Steinberg, M.K. 2008. *Stalking the Ghost Bird: The Elusive Ivory-billed Woodpecker in Louisiana.* Louisiana State University Press, Baton Rouge, p. 152, respectively.

4 "Taken alone, the evidence provided by the [Arkansas] eyewitness accounts, field notes, and sketches, might be viewed as powerful, and even conclusive. If this were a court of law, the eyewitness testimony would be considered quite strong, if not dispositive." Lynch, M. 2011. Credibility, evidence, and discovery: the case of the ivory-billed woodpecker. *Ethnographic Studies* 12: 78–105. Eyewitness testimony is not invariably accurate, but the convergence in testimony given by multiple observers all reporting the same thing

(i.e., a single male ivory-bill) contradicts the appeal of mere wishful thinking as explaining the phenomenon (the latter epitomizing its own cognitive pitfall, ***ultimate attribution error***).

5 Fitzpatrick, J.W., M. Lammertink, M.D. Luneau, Jr, T.W. Gallagher, B.R. Harrison, G.M. Sparling, K.V. Rosenberg, R.W. Rohrbaugh, E.C. Swarthout, P.H. Wrege, and S.B. Swarthout. 2006. Clarifications about current research on the status of ivory-billed woodpecker (*Campephilus principalis*) in Arkansas. *The Auk* 123: 587–593.

6 Tangible evidence included the Luneau video and acoustic recordings of *kent* calls and double-knocks.

7 Hill, p. 144.

8 Fitzpatrick, J.W., M. Lammertink, M.D. Luneau, T.W. Gallagher, B.R. Harrison, G.M. Sparling, K.V. Rosenberg, R.W. Rohrbaugh, E.C. Swarthout, P.H. Wrege, and S.B. Swarthout. 2005. Ivory-billed woodpecker (*Campephilus principalis*) persists in continental North America. *Science* 308: 1,460–1,462.

9 Known acoustic soundtracks of ivory-bill are, essentially, limited to the single sample of the one pair that was studied at one location, the Singer Tract of Louisiana. Because of this extreme ***insensitivity to sample size***, that one recording is much too sparse to faithfully represent variation likely to occur across regions, populations, individuals, ages, or sexes. Large variations in song and other acoustic repertoires are prevalent in birds, but these variations were never measured for the ivory-bill.

10 Charif, R.A., K.A. Cortopassi, H.K. Figueroa, J.W. Fitzpatrick, K.M. Fristrup, M. Lammertink, M.D. Luneau, Jr., M.E. Powers, and K.V. Rosenberg. 2005. Notes and double knocks from Arkansas. *Science* 309: 1,489.

11 "The new sound recordings provide clear and convincing evidence that the ivory-billed woodpecker is not extinct," announced Richard Prum, one of these critics. Dalton, R. 2005. Ivory-billed woodpecker raps on. *Nature*. Published online 2 August http://www.nature.com/news/2005/050802/full/news050801-4.html (accessed 20 July 2016).

12 "We must ask whether over the years we have set the bar too high

in terms of what constitutes a credible sighting." Steinberg, p. 20. Shifting the baseline for evidence in ivory-bill matters has led science researchers not only into errors from ***neglect of probability,*** but also adopting, however unconsciously, the spurious standards of birdwatching for insisting on the incontestable. "No recent sightings have been convincingly replicated or fully confirmed by unassailable hard evidence, and all as yet have failed to achieve universal acceptance." Snyder, N., D.E. Brown, and K.B. Clark. 2009. *The Travails of Two Woodpeckers*. University of New Mexico Press, Albuquerque, p. 1.

13 "It was like getting a report of Bigfoot or the Loch Ness Monster." Hill, p. 4. See also Chapters 9 and 12 for examples of such inapt comparisons from scientists and the media alike.

14 Schunn, C.D., and J.G. Trafton. 2013. The psychology of uncertainty in scientific data analysis. Pp. 461–483 in *Handbook of the Psychology of Sciences* (Feist, G.J., and M.E. Gorman, eds.). Springer Publishing Company, New York.

15 http://tomnelson.blogspot.com/2006/07/some-humor.html (accessed 17 August 2015). Although listed under humor, evidence of these tactics is amply demonstrated in the thousands of comments that viciously attacked the support for and the researchers involved with finding the ivory-billed woodpecker in the first decade of the twenty-first century. http://www.birdforum.net/showthread.php?t=33968 (accessed 5 June 2015).

16 http://listserv.arizona.edu/cgi-bin/wa?A2=ind0603c&L=birdwg01&T=0&O=D&P=4587 (accessed 21 March 2006).

17 Jackson, J. A. 2006a. *In Search of the Ivory-billed Woodpecker,* 2nd ed. HarperCollins, New York, p. 14. Arbitrary screening of second-hand testimony about ivory-bill based on our "belief" about witness credibility is prone to inaccuracies from both Type I error (concluding that information about the bird was reliable when it was not) and Type II error (concluding that information about the bird was undependable when in fact it was reliable). Our belief(s) about others' credibility are also susceptible to ***attribution error.***

18 Comment by David Sibley, October 23, 2007, at: http://sibleyguides.blogspot.com/2007/10/ivory-billed-woodpecker-status-review.html (accessed 27 May 2015).

19 Even periodic celestial objects like comets are prone to their own displays of eccentric unpredictability. Chirikov, B.V., and V.V. Vecheslavov. 1989. Chaotic dynamics of comet Halley. *Astronomy and Astrophysics* 221: 146–154.

20 Unusually hostile reactions to what are rather innocuous bits of information about a rare bird that may not be extinct are eerily reminiscent of other fact-defying trends from the same era in modern American history. A peculiar Alice-in-Wonderland affiliation with facts was exemplified by some in the Bush White House. Senior presidential advisors schooled writer Ron Suskind that the rest of us mere mortals were trapped "in what we call the reality-based community," defined as poor folk who still "believe that solutions emerge from [our] judicious study of discernible reality.... That's not the way the world really works anymore.... We [the Bush White House] create our own reality." Suskind, R. 2004. Faith, certainty and the presidency of George W. Bush. The *New York Times* Magazine http://www.nytimes.com/2004/10/17/magazine/faith-certainty-and-the-presidency-of-george-w-bush.html (accessed 25 July 2016).

21 http://www.birds.cornell.edu/ivory/evidence/segments/segments/intro (accessed 25 July 2016).

22 "Cornell's attempts at reenactment of the bird's appearance in flight are flawed from the outset, because the wings of real birds twist and turn in flight, unlike the wings of painted cardboard models." Kon Kaufman. http://listserv.arizona.edu/cgi-bin/wa?A2=ind0603c&L=birdwg01&T=0&O=D&P=2992 (accessed 19 March 2006). Kaufman exaggerated over the term "flawed," however, as the Cornell team had built models for reenactment that did have some flexing ability.

23 Mean flap rate for wing-beats detected in putative ivory-billed woodpeckers were as much as *10 standard deviations higher* than mean flap rates recorded for pileated woodpecker. Collins, M.D.

2011. Putative audio recordings of the ivory-billed woodpecker (*Campephilus principalis*). *Journal of the Acoustical Society of America* 129: 1,626–1,630; Collins, M.D. 2020. Application of image processing to evidence for the persistence of the ivory-billed woodpecker (*Campephilus principalis*). *Scientific Reports* 10: 14616.

24 E.g., "Luneau's video does not provide confirmation that the ivory-billed woodpecker still lives among us." Patten, M. 2006. Survey researcher disputes ivory-billed woodpecker claim. *Biosurvey News Spring 2006*: 1–2.

25 Some other features never explained adequately by critics included wing-beats that were sustainably faster than any known for pileated woodpecker, and a peculiar wing-bowing observed during flight.

26 Hill, pp. 144, 146.

27 *Ibid.*, p. 144.

28 For example, see: Lammertink, M., T.W. Gallagher, K.V. Rosenberg, J.W. Fitzpatrick, E. Liner, J. Rojas-Tomé, and P. Escalante. 2011. Film documentation of the probably extinct imperial woodpecker (*Campephilus imperialis*). *The Auk* 128: 671–677. The Arkansas incident for putative ivory-bill still has a way to go in order to match the longevity of the Lowery/Lewis photos of putative ivory-bill. That controversy has been "kept simmering for thirty years." Steinberg, p. 85.

29 Fitzpatrick et al. 2005; Charif et al. 2005; Rosenberg, K.V., R.W. Rohrbaugh, and M. Lammertink. 2005. An overview of ivory-billed woodpecker (*Campephilus principalis*) sightings in eastern Arkansas in 2004–2005. *North American Birds* 59: 198–207; Hill.

30 Sibley, D.A., L.R. Bevier, MA. Patten, and C.S. Elphick. 2006. Comment on "Ivory-billed woodpecker (*Campephilus principalis*) persists in continental North America." *Science* 311: 1,555a; Jackson, J.A. 2006b. Ivory-billed woodpecker (*Campephilus principalis*): Hope, and the interfaces of science, conservation, and politics. *The Auk* 123: 1–15; Collinson, J.M. 2007. Video analysis of the escape flight of pileated woodpecker *Dryocopus pileatus*: does the ivory-billed woodpecker *Campephilus principalis* persist in North America? *BMC*

Biology 5: 8.

31 Fitzpatrick et al. 2006.

32 http://ivorybills.blogspot.com/2011/11/further-imperial-analysis.html; http://bbill.blogspot.com/2011/11/woodpecker-wingbeats-revisited.html (accessed 26 July 2016).

33 A possibility that the bird was an exceedingly abnormal pileated woodpecker was not tested directly. Abnormally plumaged ivory-bills are also known. Wayne, A.T. 1905. A rare plumage of the ivory-billed woodpecker (*Campephilus principalis*). *The Auk* 22: 414. However slim those likelihoods might be, respect for the ***congruence bias*** demands we allow for such chances to be >0%. And along with any other possibility, including some other species of woodpecker for which the evidence was never tested, the joint probabilities of all possible outcomes for a correct species identification must sum to 100%.

34 Hill, p. 137.

35 I am hardly the first to use a convention of probability to judge identity of the bird in the Luneau video. "I would not say absolutely, positively, 100 percent, that the bird in the video is an ivory-bill. But I think that is the best explanation." Douglas Stoltz, conservation ornithologist, Field Museum of Chicago. Crewdson, J. 2006. Woodpecker sighting a flight of fancy? *Chicago Tribune*, May 19, 2006 http://articles.chicagotribune.com/2006-05-19/news/0605190139_1_ivory-billed-woodpecker-cornell-university-bird/3 (accessed 19 May 2005).

36 *Ibid*., p. 235.

37 White, M. 2006. Ivory-billed woodpecker. National Geographic http://ngm.nationalgeographic.com/2006/12/ivory-woodpecker/white-text (accessed 2 August 2016).

38 Steinberg, p. 19.

39 I found the regional and decade-by-decade summary by Hunter (end-note #3) to be especially illuminating in this regard.

40 See Supplement to Collins, M.D. 2020. Application of image processing to evidence for the persistence of the ivory-billed

woodpecker (*Campephilus principalis*). *Scientific Reports* 10: 14616.

41 Fitzpatrick et al. 2006.

42 Jackson 2006b.

43 Winn, W. 2009. "Proof" in pictures: visual evidence and meaning making in the ivory-billed woodpecker controversy. *Journal of Technical Writing and Communication* 39: 351–379.

44 *Ibid.*

45 Douglas, H.E. 2009. *Science, Policy, and the Value-free Ideal.* University of Pittsburgh Press, PA, pp. 129–130.

46 Sibley et al. 2006.

47 "[Sibley's] pileated woodpecker in-flight illustration [in his field guide] is the least accurate I have ever seen in terms of ratio of white:black on the underwing. Anyone with a ruler can figure this out from his illustration." A member of the Cornell search team, responding to a query from Steinberg. "Distortions in [Sibley's] drawing of a pileated woodpecker make his claim that the [Cornell] search team mistakenly identified a pileated woodpecker ring hollow." Steinberg, p. 21.

"To date no video of an actual pileated woodpecker exhibits from frame to frame the same plumage characteristics and flight mechanics exhibited by the woodpecker in the Luneau video ... the alternative interpretations of Sibley et al. (2006) and Collinson (2007) fail to credibly support their assertion that the woodpecker in the Luneau video could reasonably be a pileated woodpecker." Hunter, W.C. 2010. *U.S. Fish and Wildlife Service statement on existing evidence for ivory-billed woodpecker* (Campephilus principalis) *occurrence in the Big Woods of eastern Arkansas and elsewhere in the southeast U.S.* Appendix B, Recovery Plan for the Ivory-billed Woodpecker (*Campephilus principalis*), U.S. Fish & Wildlife Service, Atlanta, GA, pp. 44–45.

48 My fundamental proposition in this work has been that cognitive function, not "political volatility" or "personality conflict,' explains our very strong reactions to the ivory-billed woodpecker. Although certainly transmitted by and reinforced within several of our social

networks and identities, the fundamental instability begins inside the neural circuitry of our own heads.

49 More recent and sophisticated means of enhancement and post-processing applied to the images and/or sounds that were gathered during several searches that found putative ivory-bills throughout the 2000s have, in the main, consistently found evidence that is more consistent with that species rather than pileated woodpecker. Collins 2011, 2020.

Further Reading

Ariely, D. 2008. *Predictably Irrational: The Hidden Forces That Shape Our Decisions*. HarperCollins Canada. 247 pp. ISBN 978-0-06-135323-9. "Our irrational behaviors are neither random nor senseless – they are systematic and predictable. We all make the same types of mistakes over and over, because of the basic wiring of our brains.... But because a stereotype provides us with specific expectations about members of a group, it can also unfavorably influence both our perceptions and our behavior."

Dobelli, R. 2013. *The Art of Thinking Clearly*. 326 pp. ISBN 978-0-06-221968-8. "We make complex decisions by consulting our feelings, not our thoughts.... In the presence of other people we tend to adjust our behavior to theirs, not the opposite.... In other words, the more people who follow a certain idea, the better (truer) we deem the idea to be. And the more people who display a certain behavior, the more appropriate this behavior is judged by others. This is, of course, absurd."

Hand, D.J. 2014. *The Improbability Principle: Why Coincidences, Miracles, and Rare Events Happen Every Day*. Scientific American/Farrar Straus Giroux. 288 pp. ISBN 10: 0374175349. "Extremely improbable events are commonplace.... Our intuitive grasp of probability isn't very good.... It's the very essence of science that its conclusions can change, that is, that its truths are not absolute."

Grant, A.M. 2021. *Think Again: The Power of Knowing What You Don't Know*. Viking. 320 pp. ISBN-10: 1984878107. "We favor the comfort of conviction over the discomfort of doubt.... Scientific thinking favors humility over pride, doubt over certainty, curiosity over closure. When we shift out of scientist mode, the rethinking cycle breaks down, giving way to an overconfidence cycle."

Kahneman, D. 2011. *Thinking, Fast and Slow*. Farrar, Straus and Giroux. 512 pp. ISBN 9780374275631. "A reliable way to make people believe in falsehoods is frequent repetition, because familiarity is not easily distinguished from truth.... This is the essence of intuitive heuristics: when faced with a difficult question, we often answer an easier one instead, usually without noticing the substitution ... when people believe a conclusion is true, they are also very likely to believe arguments that appear to support it, even when these arguments are unsound."

Mlodinow, L. 2008. *The Drunkard's Walk: How Randomness Rules Our Lives*. Pantheon. 252 pp. ISBN 10: 0375424040. "We also use our imagination and take shortcuts to fill gaps in patterns of nonvisual data.... There is therefore a fundamental clash between our need to feel we are in control and our ability to recognize randomness.... Our assessment of the world would be quite different if all our judgments could be insulated from expectation and based only on relevant data."

Tsipursky, G. 2020. *The Blindspots Between Us: How to Overcome Unconscious Cognitive Bias and Build Better Relationships*. New Harbinger Publications. 216 pp. ISBN 9781684035106.

Glossary

Cognitive biases and fallacies linked to ivory-bill

Ad hominem

Attack on the individual or even organization delivering the information as opposed to the nature of the information itself. This fallacy makes no attempt to address the particular argument, instead relying on sarcastic, irreverent, or other derogatory attacks. *Example*: Bloggers' censure and name-calling of Cornell absent data or any other dispositive information as a tactic to contradict original evidence, in some cases also relying upon ***appeal to motive*** to reinforce those attacks (as when implying that conservation organizations had hidden motives for land preservation, actually a routine and quite transparent aspect of their mission).

Ambiguity effect

A tendency that can lead us to choose known over uncertain outcomes, regardless of the strength of evidence or probability weighting between those outcomes.[1] *Example*: A bias that may explain a propensity to select (and then insist) on certain extinction for the ivory-bill because the underlying mechanism of species' death is irreversible. In contrast, survival for the ivory-bill may not only perpetuate indefinitely the fundamental ambiguity, a prospect of survival also requires substantial modifications to prior assumptions and beliefs that we have already made (in error) about the woodpecker.

Anchoring bias

Known also as ***focalism***, this irrational stance comes from an excessive confidence placed in the initial observations that are most available, easily received, or effortlessly retained. This initial information is used then to unduly delimit future outcomes or choices.[2] After the anchor is set, subsequent decisions and judgments can be made only by fine-tuning in small increments away from this anchor; bias arises in large part from repeatedly interpreting other information solely around the anchor. *Example*: Tanner's theory of narrow food and habitat specialization in the ivory-bill greatly limited other perspectives that may be relevant to the woodpecker's fate after the 1930s. Even alternative theories proposed for the bird's demise had to be set near, and then contrasted mostly around, Tanner's views.[3]

Appeal to motive

A logical fallacy characterized by attempts to discredit the thesis of an argument using the supposed intentions or predilections of the proposer. This fallacy often assumes that the mere possibility of motive is an adequate basis to refute the argument, or to confirm a counter-argument. *Example*: The Nature Conservancy promoted an "ivory-bill" in the Big Woods merely to raise funds from the public to purchase more conservation lands.

Argument from authority

An argument wherein conclusions are supported by references to who made them as the means to sway debate. Since the argument is based on an appeal to the authority, the strength of support depends on [perceived] quality of the authority in question. *Example*: The following is attributable to James T. Tanner: "I am the world expert on an extinct bird."[4] Note that this declaration requires the listener to accept two debatable premises as factual, each one largely or entirely on the basis of self-assessment.

Argument from ignorance

The fallacy that arises when an individual reaches a conclusion or presumes a fact based primarily on lack of any strong evidence to the contrary. This error is commonly illustrated with the adage "absence of evidence is not evidence of absence." *Example*: "The ivory-billed woodpecker is extinct because you [they] cannot prove that it still exists." Or its converse: "the ivory-billed woodpecker is not extant because you [they] cannot prove that it is not extinct."

Argument from incredulity

Essentially trying to argue that because something is incredulous, or difficult to believe on the part of the speaker, it cannot be true. In order to avoid changing one's mind, then, the speaker attempts to avoid advancing their (or anyone else's) knowledge of the particular topic at hand. *Example*: The exclamation of faux surprise: "[searchers] have yet to get any kind of real documentation ... it becomes extremely implausible that even one individual could be sneaking around in the woods of Arkansas and evading the cameras of the entire Cornell search team for the last two winters."[5] And another incredulous comment: "the implication that birders have been unmotivated and unobservant, and that a population of giant woodpeckers could remain undetected across several states, is simply not plausible."[6] Tones of incredulity are usually quite obvious in these fallacious arguments, e.g., "I am unaware of any animal on the planet as large as the [ivory-bill], living and flying about in habitat surrounded by a sea of humanity, that can escape detection, especially given the great effort expended in eastern Arkansas and northwestern Florida during 2004–06."[7]

Availability cascade

Socially transmitted cognitive bias wherein we tend to believe information to be accurate merely after being exposed to it numerous times. "Repeat something often enough, and it will be true." This self-reinforcing process of collective belief formation then triggers a chain reaction that gives a perception of increasing plausibility due to the topic acquiring increased availability in our social discourse.[8] *Example*: Donald Eckelberry's portrait of a supposed last ivory-bill coming into a roost in a devastated woodland first anchored a date of putative extinction for the woodpecker. After repetition over many decades, that story also formed an ***availability cascade*** due to its pervasive accessibility in most of the subsequent social and professional dialogue about ivory-bill.

Availability heuristic

Mental rule of thumb based on a shortcut in our judgments for arranging observations, usually consisting of the first thing(s) that come to our mind, and often leading to erroneous or irrational decisions. *Example*: Tanner noted in his study that one pair of ivory-bills relied extensively during the breeding season on large beetle larvae extracted from bark scaled from recently dead trees. This led him to conclude that ivory-bills were feeding specialists that, in part to avoid competition with other woodpeckers, were highly vulnerable to the logging of old forests.

Backfire effect

Our tendencies after being confronted with evidence that conflicts with our beliefs to go on to hold our original position yet more strongly.[9] In this bias, people first construe even unfavorable information as aligned with their own beliefs.[10] *Example*: Both believers and deniers, those who are 100% certain that ivory-bills are or are not extant, respectively, interpret arbitrarily (non-rationally) most new information about the woodpecker so as to maintain consonance with a pre-existing position (***belief perseverance***). For example, in order to discredit 70 years of video, sighting reports, and audio recordings that cumulatively suggests evidence consistent with this species' survival past the 1930s, deniers must invoke a vast array of special pleadings to capriciously discredit each line of evidence individually, they must do so using arbitrarily different criteria, and they neglect all ***Bayesian caution*** when doing so.

Bandwagon fallacy

An argument for a position, idea, or perspective based upon irrelevant appeal to the fact that multiple individuals subscribe to its authority.[11] This fallacy often arises when a threat of rejection by one's peers (or peer pressure) is substituted for real evidence in an argument or position. *Example*: A birdwatcher goes to Arkansas in 2005, sees a large woodpecker that to their satisfaction fits the identity of an ivory-bill, counts it on their life list, but later cancels this action entirely after fellow birders gang up and deride their decision as one that violates the sport's established rules.[12]

Base rate fallacy

Our tendency to ignore the background (base) rates of probability in some phenomenon in favor of individuating information (when such is available). A fallacy arising from not logically integrating two kinds of information.[13] This often occurs if we rely upon intuition instead of calculation. *Example*: A number of animal species thought to be once extinct for decades or centuries have actually survived, eventually to be "re-discovered." Deniers believe that this cannot be true about ivory-bill, however, because it is certainly extinct. Ignoring any likelihood that ivory-bill might also be a species once thought gone but in truth survived past a putative extinction date constitutes an example of ***base rate fallacy***.

Bayesian caution

Logical avoidance of assigning an extreme (0% or 100%) probability to what is otherwise a still fallible proposition. *Example*: Those who insist on the ivory-bill's certain extinction, or on its equally certain survival, are (on the basis of current knowledge, at least) not using an appropriate level of ***Bayesian caution***. Some competitive birders were fond of claiming that those who argued for extant status of ivory-bill made a "common error [in] the assumption that lots of weak evidence can be combined to form strong evidence."[14] Here, birders' stark ***neglect of probability*** stemmed from a lack of ***Bayesian caution***, as "lots" of weak evidence is inevitably and always further from 0% or 100% due to the injunctions from the additive law of probability.

Bayesian conservatism

In the face of new or additional information, a belief or position is revised not at all or less than it should be.[15] *Example*: Many otherwise knowledgeable and intelligent persons continue to effectively ignore all 100+ incidents of ivory-bill sightings since the late 1930s, up to and including issuing ever more strident claims that the species is certainly extinct. See also ***subadditivity effect.***

Belief bias

Adherence to a particular argument based on plausibility of its conclusion rather than on how the evidence is actually shown to support the particular conclusion. *Example*: Because its food supply was diminished from cutting down old-growth forests, the ivory-billed woodpecker therefore faced (or faces) extinction.

Belief perseverance

A bias driven by the tendency to maintain existing beliefs in the face of evidence that ought on empirical grounds to weaken or even totally reverse those beliefs.[16] *Example*: Ivory-bill deniers who insist with absolute certainty that the woodpecker went extinct decades ago, despite the availability of much evidence to the contrary, and with such information compiled since the putative subsequent date(s) of extinction.

Bias blind spot

Failure to see the impact of biases in one's own judgments, or only seeing these consequences of bias in other persons.[17] *Example*: A position that observers are unconsciously primed or predisposed to see ivory-billed woodpeckers, if and when such suggestions are feasible, but failing utterly to see that the extinction meme that was promulgated so widely and enforced rigorously since the 1940s also primes observers to strongly *discount or disbelieve* their own (and others') senses if the species was/is observed.

Black swan event

A metaphor for some precedent-setting incident or occurrence that at first glance appears to us as an utter surprise, has a substantial impact (or impacts) on spheres of human influence, and is then rationalized without basis after the fact using some newly uncovered or understood bits of evidence (***hindsight bias***).[18] *Example*: The cataclysm that would likely shake the core premises to science, bird-watching, and conservation practice were ivory-bill ever to be incontestably verified as extant, as well as the consequences for inter-relationships among these (and other) human endeavors (e.g., due to media coverage, challenges to our cognition, and so on).

Cognitive bias

Systematic errors in judgment and decision-making prevalent in all human beings. These errors may arise from cognitive limitations, motivational factors, or adaptations to natural environments (both ancient and modern-day).[19] *Example*: Relative to earlier decades, the rate of disputed reports for the ivory-billed woodpecker fell during the 1990s. If this factor contributed to removal of the woodpecker from newer bird field guides, an influence of ***recency bias*** may have been at work, i.e., a tendency to weight the most recent information that we have most heavily.

Cognitive closure

A very strong proclivity in the human species to seize and then freeze on an answer, *any* answer, in order to avoid further confusion, ambiguity, and uncertainty.[20] The need for cognitive closure varies immensely among individuals, or even among professions and social identities. Birding has a far higher need for cognitive closure (i.e., certainty in species identification) than does science. *Example*: The rush by rigid deniers and rigid believers to fix upon certain extinction and certain ongoing existence, respectively, for the ivory-billed woodpecker.

Confirmation bias

A catch-all term that encompasses several particular biases, confirmation bias is the inclination to search for, interpret, select, recall, and/or report information in a manner that confirms preconceived beliefs or assumptions. Confirmation bias has several sub-types, each of which may have its own separate origin, e.g., the ***myside bias***. These subtypes, while subtly different, nevertheless stem from different origins in our thinking outlook and/or they have quite distinct manifestations when they do arise. *Example*: The ivory-billed woodpecker is extinct; statistical analyses using the temporal pattern of specimen records for this bird (falsely) confirm that conclusion.[21]

Congruence bias

Our tendency to overvalue information that has a high probability of a positive result given what we think is the most likely hypothesis.[22] Also, an insufficient level of attention paid to enumerating all possible hypotheses before seeking a confirmation with more information. This bias can be reduced if alternative hypotheses or the probabilities of negative results are explicitly stated. *Example*: A major assumption about the Cornell search for ivory-bill in Arkansas was a belief that the apparently solo bird came from a source population *in situ*, rather than viewing it possibly as "a surplus, or young bird from private land, out exploring for a new territory or mate [on public land]."[23]

Conjunction fallacy

A formal logical fallacy in which a specific, narrower condition is assumed to be more probable than a broader, general one. *Example*: The ivory-billed woodpecker was driven towards extinction due to direct killing, or from loss of bottomland hardwood forest. In order to minimize risk of conjunction fallacy, reasons for the species' demise might be restated more broadly: The ivory-bill was driven towards extinction because of human impacts, including habitat loss and direct killing.

Dunning-Kruger effect

A cognitive bias in which topically incompetent persons do not, and indeed cannot, recognize how inept their knowledge in a subject matter actually is.[24] *Example*: Ivory-bill deniers who first extrapolated Tanner's limited study of a single pair of human-acclimated woodpeckers as evidence of how unwary was the entire species,[25] then doubled down on that ignorance by disregarding prolific historical literature that indicates exactly opposite behavior in the ivory-bill (see Chapter 7). This sort of illiteracy is epitomized, for example, in the following: "you can only rationalize the failure of the current searches if you don't include the words 'active and noisy', 'called frequently' and 'easy to see and follow' in your dataset."[26]

Egocentric attribution bias

Also known as ***false consensus effect***; see definition and example under that entry.

Equality bias

Inclinations to give everyone an equal opportunity in voicing an opinion, then to weight each of those opinions the same, despite clear differences in their fundamental reliability. A dumbing down to the mean. This unjustified weighting of viewpoints occurs even when doing so is at odds with specific feedback or other explicit incentives that should prompt us to weight them otherwise.[27] Equality bias constitutes an illogical avoidance of using differences in authoritative knowledge among individuals at all as a weighting factor (contrast with ***argument from authority***), thereby leading to more faulty group decisions. *Example*: When the U.S. Fish & Wildlife Service arrived at a decision to retain an uncertain status for the ivory-bill, despite noisy criticism from many individuals in the birding frame, it weighted evidence more heavily from individuals in the science frame as being more reliable in substance, leading them to reach (or keep) that decision, and thereby avoid succumbing to ***equality bias***. Conversely, the media often fell routinely to ***equality bias*** in giving contrary individual opinions on the ivory-bill equal weight in its news accounts.

Equivalency bias

A fallacy to render equivalent two or more "sides" of an issue when the strength of the data or information actually provides substantially more support to only one side. A bias exhibited habitually by the media when it reports contentious issues as being more balanced with respect to the evidence than is factually the case, or when it disregards completely the evidence that disproves or severely weakens the alternative perspective. See also ***Okrent's Law***. *Example*: Tendency of media to rate the scope and importance of criticism by competitive birding of the 2004 Luneau ivory-bill video as equal to the original primary evidence illustrates this bias. This shortcoming included a failure to follow up with the rather protracted refutations of criticisms that, over the long term, tended to still support an original premise of the bird's identity as more aligned with ivory-bill, and refute (but not entirely eliminate) a thesis of the bird as an odd pileated woodpecker or some other explanation.

Fallacy of composition

The error of claiming that is what is true of a member of a group is true for the group as a whole. *Example*: Critics attempting to use refutations for one or only a subset of all (and generally only the weakest) individual supports that were offered for the identity of ivory-bill exemplify this fallacy. Targeting only a few out of the nine total supports for an ivory-bill identity as given by Cornell, then using such refutation of just one or two as justification that the entire identification was incorrect.

Fallacy of division

A logical fallacy in which one infers that something is true of one or more of the parts from the fact that it is true of the whole. *Example*: Kaufmann tried to extend a conclusion from this (disputable) original premise: "[searchers] have yet to get any kind of real documentation," to the inference that it "becomes extremely implausible that even one individual could be sneaking around in the woods of Arkansas and evading the cameras of the entire Cornell search team for the last two winters."[28] Even if the latter were true, it would/might be only one contribution (out of many others that could be possible) to the (putative truth of the) original premise.

False consensus effect

A cognitive error, also known as ***egocentric attribution bias***, in which individuals show a conspicuous tendency to over-appreciate the degree to which most if not all others will agree with them.[29] *Example*: Hyperbole of the Cornell Laboratory of Ornithology when announcing in 2005 that the evidence of ivory-bill in Arkansas constituted a "definitive" or "documented" proof of the continued existence for the woodpecker, when in fact aspects of that particular evidence could be (and indeed certainly were) disputed by others.

False equivalence

An unjustified appearance of equivalence in quantity or quality of evidence between a particular set of opposing arguments, contrasts, or comparisons, when instead the quantity or quality is much greater for one side. Whereas false equivalence arises in and is contextualized by a *particular* argument given, the ***equivalency bias*** is a *tendency* for false equivalence to be repeatedly given in a misguided attempt to represent some issue as more balanced than it truly is. *Example*: A statement that an attempt to search for the ivory-billed woodpecker is no different than a quest for Loch Ness Monster would exemplify a case of ***false equivalence***.

Framing effect

An outcome in which an equivalent description of the evidence or the issues can lead to systematically disparate decisions, often leading to heightened polarity and divisiveness, depending on the nature of the filter(s) through which that evidence is given or received.[30] *Example*: Reactions to a 2004 Arkansas video of a putative ivory-billed woodpecker were framed alternatively by two social identities as: 1) evidence suggestive of or consistent with that species (the *science* frame), and 2) inadequate documentation for "counting" the species as still extant (the *birding* frame). Both frames were used to build and then enforce internal cohesion within their respective social orders. The *birding* frame was more oriented towards a concern about a wrong attribution of the bird's identity; the *science* frame was more concerned about a wrong attribution for the bird's presumed extinction.

Fundamental attribution error

Also called ***correspondence bias***, this error refers to the human tendency to over-estimate the influence of disposition, personality, and other internal, unseen traits on a particular situation to explain the actual bases behind a social behavior.[31] *Example*: Allegations that Cornell and The Nature Conservancy were guarding the search for the Arkansas ivory-bill for some nefarious purpose of secrecy, when in fact it was to protect the bird and not drive up the acquisition costs for conservation lands, respectively, in the Big Woods region of Arkansas. ***Fundamental attribution error*** also led reviewers to abuse peer review and claim without basis that research submitted about putative ivory-billed woodpeckers stemmed from the fact that "wealth and power still give sway to some major expenditures—such as in the case of the ivory-billed woodpecker, but science gives us the tool to call them on it."[32]

Hindsight bias

The fallacy of "I knew it all along." This bias reflects our tendency to overestimate our ability to have predicted an outcome that in reality could not possibly have been known to us previously.[33] It stems from a hubris of thinking we ought to know more than we are in fact capable of knowing, and thus undervaluing greatly the role of uncertainty in much of the natural world. *Example*: What likely would take place were ivory-billed woodpeckers ever to be consensually confirmed as surviving into modern times, and the reason turned out to be due to some previously overlooked or under-appreciated life history trait (e.g., high dispersal, greater longevity, flexible habitat preferences).

Illusion of transparency

A tendency for people to overestimate the real extent to which others can discern their internal states, or to believe that our inner state "leaks out" visibly to others more than is truly the case.[34] *Example*:

For purposes of protecting the location, intentions to keep secret an Arkansas woodpecker might have been believed to be transparent to outsiders by the searchers and land managers in the Big Woods of Arkansas. However, such motivation does not possess great visibility in an exterior sense, which may have led others to misattribute this secrecy to some other intention.

Illusion of validity

An erroneous belief that ever more information will generate additional pertinent data that can be used to improve our predictions, even when it is obvious that it will not, such that individuals have continued sureness in their fallible judgment.[35] Thus, if a compelling impression of a particular event clashes with general knowledge, the impression commonly prevails. *Example*: Both ivory-bill deniers and believers continue to cling tightly to a food specialization hypothesis, even though there are no instances anywhere of other woodpeckers ever having had such a narrow dietary preference. (One side uses this ***illusion of validity*** to justify the species' certain extinction, the other to constrain or prioritize search localities based on the feeding signs that are presumed to be diagnostically unique to this species.)

Illusory correlation

False perceptions that a relationship exists between two variables when either a very minor or no true relationship actually exists.[36] *Example*: Ivory-bill woodpeckers required old-growth bottomland forests because this habitat contained many large engraver beetle larvae. In reality, the woodpecker was not restricted originally to bottomland forests, nor was it limited to feeding solely on these kinds of larvae.

Illusory superiority

A self-serving, cognitive bias that leads us to overestimate our positive qualities/abilities, and to in turn underestimate our negative qualities/ abilities.[37] *Example*: Most (but not quite all) blogging narratives about the ivory-bill were filled with commentators who severely over-weighted how much they actually knew, and depreciated how much they didn't know, about the ivory-bill's behavioral and ecological habits as revealed in the complete record of historical and scholarly literature.

Improbability principle

A phenomenon describing situations in which what seem to be exceedingly unlikely incidents are rather to be anticipated, even understood as commonplace. This principle arises from the manifestation of five different elements, any of which can influence, alone or jointly, the outcome of some otherwise unexpected event. These five elements for the ***improbability principle*** include the Laws of Inevitably, of Truly Large Numbers, of Selection, of the Probability Lever, and of Near Enough.[38] *Example*: Under the Law of Inevitably, the ivory-billed woodpecker necessarily must be either extinct or extant. Because ***base rate fallacy*** shows us that we are often wrong in attributing extinction to each and every individual case of missing animals (the ***Romeo error***), based on current knowledge there is some likelihood, however small or large, that at any given moment this woodpecker species may be dead *or* alive. The Law of Inevitably specifies that even if some of the outcomes have exceedingly small probabilities of occurring, it is certain that one (including the least likely) will happen.

In-group bias

A tendency to grant excess favoritism to those who hold similar views, beliefs, or positions, with the usual result that new or contrary information becomes isolated in social space due to this excessive "stove-piping." Also known as homophily. *Example*: Birding's insistence on a belief frame that all rare birds should be repeatedly found and documented resulted in a vast echo chamber inside that hobby, cutting off the pastime from evidence to the contrary, information that was mostly available elsewhere (in science, conservation, statistics, and other fields or disciplines).

Insensitivity to sample size

A tendency when making a probability judgment to neglect to understand that more variance is likely to occur in smaller samples, and that larger samples provide less variance and therefore provide far more reliable, accurate evidence. Unwarranted generalization from samples that are extended to populations. *Example*: Virtually everything we think we knew/know about the ivory-bill's ecological and conservation needs is based on a ludicrously low sample size of $n = 1$ (Chapter 6), i.e., the Tanner study from northeast Louisiana.

Lack of object constancy

Also known as object impermanence or lack of object consistency, this perceptual bias is one in which missing items are experienced as "lost" or "out of sight, out of mind." It can also reflect a cognitive inability to hold the memory of an extant object. *Example*: In the collective realm, at least some of us seem to be especially (and strangely) challenged in the ability to hold as still possibly extant the ivory-billed woodpecker during those intervals in which it was or is not reported, a phenomenon that dates back at least to the early 1900s.

Loss aversion

Our inclination to view out-of-pocket costs, or deteriorations from the status quo, as worse than the opportunity costs, i.e., the benefits that are lost as a result of continuing the status quo (see also ***sunk cost fallacy***). In the context of risk, people often tend to focus on the losses that are associated with some activity or hazard, and to disregard the gains that might be associated with that same activity or hazard. *Example*: As measured by birders' social currency (loss of peer approval), it is perceived as better to be wrong about the ivory-bill's putative extinction than to be correct about the woodpecker's alleged survival.

Motivational bias

A general class of errors that we commit when trying to evaluate the true rational for either our own behavior or the behavior of others. Misperceptions of what is driving others' behavior leads commonly to communication breakdowns and social conflicts. Motivational bias is a broad category that encompasses, e.g., ***fundamental attribution error*** and the ***illusion of transparency***. *Example*: Jerome A. Jackson's publishing of an essay[39] that criticized the supposed intentions and assumed personal dispositions inside the Cornell Laboratory of Ornithology, and his consequent under-valuing of that rediscovery's situational nature, illustrate ***fundamental attribution error***.

Myside bias

A tendency to evaluate propositions strictly from within one's own perspective when given no instructions or other strong promptings to do otherwise.[40] *Example*: Hill's perceptive acknowledgment that he was being hypocritical in judging that one ivory-bill searcher overstated the evidence yet turned around and was just as emphatic about his own.[41]

Naïve cynicism

An expectation that others are driven by motivational biases, but that we are not nor would we ever be influenced in similar faahion.[42] *Example*: Jackson perceived that Cornell had political or other motivational reasons to time its Arkansas announcement of the ivory-bill. At the same time, Jackson, who was excluded from that search effort despite being an acknowledged authority on this woodpecker, was (or would be perceived as) free of any ulterior motivations that lay behind his own criticisms.

Naïve realism

Inability to appreciate the subjective nature of one's own version of events, and thus not make a sufficient allowance for uncertainties when called on to make behavioral attributions and predictions about others.[43] *Example*: The apparent belief by some birders and bloggers that Cornell's conclusions about the identity of a putative ivory-bill in

Arkansas in 2004 and 2005 were less influenced by evidence available, and/or more influenced than anyone else's by some other, tangential agenda such as alleged funding opportunities.

Negativity bias

Tendencies for negative events, occurrences, objects, and traits to have far greater weight and influence on us than do positive ones, even if of equal intensity. Due to greater salience and potency, these negative contamination effects are much stronger within the realm of human experience than their positive counterparts.[44] *Example*: In social discourse, the finality of extinction, a negative event, is far more potently reactive in us than the mere ongoing existence of a living species. Consequently, even were the species' survival and extinction to have equal numerical weights (i.e., the same probability), the prospect of extinction in the ivory-billed woodpecker would hold much greater unconscious mental sway over us than any outlook of survival.

Neglect of probability

When an outcome is uncertain, the tendency to lose any sense of probabilities in making decisions; a serious inclination to misjudge how likely or unlikely something is to actually happen.[45] Or, when intense emotions are engaged, reaching a decision in which the actual probability matters very little for assessing the true level of risk.[46] *Example*: Those who are absolutely certain in the ivory-bill's extinction depreciate the probability of failure on this decision, in large part because they seriously underestimate the prevalence of species having been re-found after decades or even centuries of being earlier lost to human detection (Chapter 12; see also ***base rate fallacy***).

Observer expectancy effect (bias)

A cognitive bias wherein unconscious tendencies in a researcher's belief systems or expectations cause them to inadvertently influence the outcome of an experiment, trial, or observation. (To safeguard against this effect in some circumstances, researchers can use a double-blind experimental design.) Observer expectancy bias is prevalent in behavioral research with birds.[47] *Example:* Ivory-billed woodpeckers feed or fed in the tree canopy; ivory-bills flushed from on or near the ground instead had to have been cases of mistaken identity.

Okrent's Law

"The pursuit of balance can create imbalance because sometimes something is true."—Daniel Okrent, *New York Times* editor.

Omission bias

An irrational stance to judge as more harmful acts of commission rather than acts of omission.[48] *Example*: The insistence of many ivory-bill deniers who asserted that resources spent on ivory-bill searches were wasteful under the [untested] assumption that higher priorities existed elsewhere in conservation,[49] and despite the potentially irreversible consequence of doing nothing for a charismatic species perched at the edge of extinction.

Omniscience bias

The illusion that we already know everything we need to know in order to make well-informed decisions. A bias frequently promulgated by instantaneous access to high speed internet, 24/7 news availability, satellite coverage of planetary environments, and portable drones outfitted with high resolution cameras. *Example*: Essentially all of us when speculating how much we really knew about the ivory-bill; a tendency to grossly exaggerate our certainty over the woodpecker's true history, its real ecology, and its ultimate fate.

Outcome reporting bias

A specific category of confirmation bias in which the availability of key information conveyed through publication, research, or any other means is skewed via factors that entirely block or severely limit the dissemination from otherwise taking place.[50] This bias can occur often in the field of public health because individual subjects interviewed do not wish to truthfully or fully disclose their actual health risk(s). *Example*: Observers of ivory-bill who *do not* report their sightings, whether due to a desire to protect the bird, avoid the caustic scorn from their peers, and so on.

Overconfidence effect

A fundamental cause of defective decision making that arises from excessive, bordering-on-arrogant faith that one knows the truth. Unwarranted precision in one's belief.[51] "Often wrong but never in doubt." *Example*: "the few reported sightings [of ivory-bill] involve simple, everyday mistakes in perception, and the reason the bird cannot be confirmed is because it is not there.... We need to accept the tragic loss of the ivory-billed woodpecker, and move beyond it."[52]

Presentism

A tendency to interpret past history using mostly current norms and modern-day sorts of understanding. At its worst, ***presentism*** facilitates a type of moral complacency and inclination towards self-congratulation when interpreting past events using contemporary kinds of awareness.[53] *Example*: A ***presentist*** frame embedded in competitive birding holds that given all of our modern technology and innovation, any valid observation of the ivory-bill should lead inevitably and quickly to more sightings, and that these should lend themselves to supporting digital evidence (at least) that is also unassailable.

Primacy effect

One of several manifestations possible from the ***serial position effect,*** this cognitive bias reflects our tendency to have a greater recall for (and thus place greater emphasis on) the things that we see or hear first in a sequence of data, information string, and so on. *Example*: Just after the Arkansas announcement, a theory was put forth that the single putative ivory-bill was part of an overlooked source population in the Big Woods. That theory became so oft repeated that it drove most ensuing discussion and expectation, such that an alternative theory later proposed (a dispersing individual from some other, more distant location) got substantially less attention in social discourse. Indeed, once the ***primacy effect*** for a "population" of ivory-bills in the Big Woods took hold more widely, it was fixed upon by critics because it was relatively easy to refute, albeit mostly using the fallacy from ***argument from ignorance***.

Probability lever

One of the five possible elements that contribute to what otherwise might be unexpected outcomes based on the ***improbability principle***. The law of the ***probability lever*** tells us that "slight changes can make highly improbable events almost certain."[54] *Example*: Any resistance to Allee effects in the ivory-bill from, e.g., nomadism or long pair bonds, would act as a ***probability lever*** that would reduce substantially (but not eliminate) the likelihood of this species' extinction.

Pro-innovation bias

An inclination to hold an undue optimism towards a novel invention (often technological in origin) as having a degree of utility that it cannot or does not really have, usually while failing to see the extent of its true limitations.[55] *Example*: Use of RECONYX™ remote camera traps in the belief that these can easily capture proof with high-resolution, acceptable photographic image of an ivory-billed woodpecker. Some acknowledged this kind of bias: "Listening stations or autonomous recording units are also an ineffective means to search new areas for ivory-bills."[56] ***Pro-innovation bias*** is the erroneous notion that in a high-tech society with instantaneous communication abilities, Ivory-bills cannot readily escape our detection,[57] and thus should be subject to our verification nearly at-will.

Projection bias

A tendency to believe that other people (and groups) think like us (and our group). Also, an inclination to believe that our views will not change over time. *Example*. Professional ornithology was caught off-guard by the strong resistance of others' reactions to the ivory-bill's putative existence in 2004. In this case, a ***projection bias*** arose from failure to see the respective framing effect being relied upon by the other(s). Moreover, by making a profound shift in 2004 in offering a venue in which to air evidence of the ivory-bill (thus greatly changing its own perspective, because it was notoriously resistant to do so in earlier decades), professional ornithology also overcame (at least temporarily) an earlier ***projection bias*** that it had once held.

Reactance

Also known as reverse psychology, it is a tendency when our choices or alternatives are curtailed to select a position that is contrary to what was originally intended or offered. Behavioral reactance reaches a peak when freedom of choice is eliminated completely.[58] *Example*: The Cornell Laboratory of Ornithology's initial announcement of the Arkansas video as constituting definitive evidence or proof of the ivory-bill likely triggered this response because it greatly curtailed any rejoinder by a potential listener. In contrast, Geoff Hill's use of more subtle, nuanced phrasing like "consistent with ivory-bill" in describing evidence from that study did not provoke the same level of ***reactance***.

Recency effect

Another bias caused by ***serial position effect***, this one is prompted by our tendency to more readily recall or otherwise place undue emphasis on the last item(s) found in a list or other series. *Example*: The widespread (but erroneous) belief that the ivory-bill was not particularly wary around humans was likely prompted by Tanner's descriptions from the 1930s, and then used to disregard the substantial weight of evidence from the many and varied sources from earlier historical eras that instead pointed to a very guarded nature in this woodpecker (Chapter 8).[59]

Red herring

A statement intended to mislead or distract from the real issue, often intended to distort the other party's actual position, or to divert them to defend an extraneous statement. *Example*: In matters related to the ivory-bill, the ***red herring*** was widely employed as an attempt to draw false parallels between some fictional creature, like Bigfoot, and the factual woodpecker. Unless we reject the ***red herring*** outright, most of us will feel trapped and then compelled to defend against that extraneous statement.

Romeo error

An erroneous decision in which extinction is assumed wrongly from a lack of detection, often due to insufficient sampling or too few attempts at collecting relevant observations that might otherwise confirm a species' presence. *Example*: Bermuda petrel was thought extinct as early as 1620, went undetected for almost three centuries, but was eventually "rediscovered" in the early part of the twentieth century. The seabird's absence was not due to extinction, of course, so shortcomings in human judgment about the true rates of extinction were caused by this ***Romeo error***.

Semmelweis reflex

A quick predisposition to deny any new information that challenges our already established views. A reflex-like rejection of new knowledge because it contradicts entrenched norms, beliefs, or paradigms. *Example*: Individuals in both birding and ornithological communities who uncritically accept or reject any and all evidence for the ivory-bill's present existence. The ***Semmelweis reflex*** may have influenced the authors who exempted ivory-bill illustrations from their field guides, then later were among those who reacted strongest against any putative rediscovery of the species.

Straw man fallacy

A sham argument put forth and set up to be easily countered, while giving the impression that it refutes an opponent's argument, but instead was not even one advanced by the opponent. A distraction or "bait-and-switch" tactic in formal argument. *Example*: The analogy drawn by a critic who averred that "the body of evidence [for presence of ivory-bill] is only as strong as the single strongest piece – ten cups of weak coffee do not make a pot of strong coffee."[60] This was a ***straw man***, because those who came forth to affirm the identity or the existence of ivory-bill did not always make an express argument that their observations were stronger due to being differentiated along multiple lines of weak evidence.

Subadditivity effect

An irrational stance consisting of a propensity to judge the probability of the whole as less than probabilities of the individual parts.[61] *Example*: This fallacy is most strikingly evident in a conviction that the ivory-bill is extinct (i.e., it has a 0% chance of survival). Holding this stance is irrational in the face of 100 or more separate incidents since the 1940s that suggest otherwise. So, even were the probability exceedingly low, e.g., 1%, that any *single* incident pertaining to possible ivory-bill were actually true, the additive law of probability inevitably requires that the probability of all such incidents be equal to or greater than the probability of any smaller subset of individual incidents.

Sunk cost fallacy

Throwing good money after bad. Sunk cost fallacy is maladaptive behavior that manifests as a greater tendency to continue an endeavor once an investment in money, effort, or time has already been made.[62] *Example*: Rational avoidance of this fallacy generally prevented or minimized new or ongoing searches for the ivory-bill by most scientific researchers and conservation organizations because the return on that investment is perceived to be far too low. Avoidance of sunk cost fallacy ultimately led several organizations, e.g., U.S. Fish & Wildlife Service, Cornell, and so on, to halt searches for ivory-bill in the Big Woods region of Arkansas once any additional sightings had stopped entirely. ***Sunk cost fallacy*** also can be legitimately questioned as a potential driver behind certain quest-like, never-ending searches for the ivory-bill.

System justification

An assemblage of socially-transmitted and -reinforced biases that act to maintain a status quo; resistance to changes in social mores, paradigms, or processes.[63] *Example*: The tendency for both amateur bird watching and professional ornithology to hold tightly to their individual paradigms, procedures, and norms, and also to resist both the influence of each other and any other outside ***frames*** in their deliberations. Ivory-bill deniers alleged a need for prioritizing conservation resources as a sufficient justification for considering the species to be extinct, yet ignored the extensive research on setting conservation priorities, including the requirement to address directly its unavoidably normative, i.e., subjective, dimensions.

Truthiness

An intuitive attachment to a 'truth' that is based on what feels right, a gut- rather than mind-driven characterization of a claim's accuracy. What one sincerely hopes or wants to be true, regardless of the factual support. *Example*: Eckleberry's painting in 1944 of a supposed "last ivory-bill" gave later claims of extinction in the woodpecker a truthy appeal. In this instance, the painting ultimately acted also as "proof" for the claim.

Ultimate (or group) attribution error

An unwarranted extension of stereotypical traits to all individual members of some group or social identity; a gross over-generalization from the specific to the general. *Example*: ***Framing*** insisted upon by birding and casting ivory-bill as extinct was (or is) a prevalent belief held by most birdwatchers. (According to limited research, this was untrue, as most individuals sampled and expressing an interest in birds thought instead that the woodpecker was possibly/probably extinct/extant).[64]

Zero-risk bias

A bias expressed as preference for options that completely eliminate some risk even where alternative, often cheaper or more efficient options will reduce the overall risk by considerably more. It results from our preferences for reaching a cognitive closure as soon as possible through absolute certainty in a smaller benefit compared to a larger benefit having less certainty.[65] *Example*: Given legitimate uncertainty over the identity of a large woodpecker videoed in Arkansas in 2004, many individuals stampeded into a "certain" identity of this bird as a pileated woodpecker instead. In some individuals, the uncertainty over the bird's identity was too difficult to hold cognitively, and an impatience for the protracted technical debates that followed led them to adopt a perceived zero-risk position (even if it was wrong, or less likely).

Endnotes

1 Frisch, D., and J. Baron, J. 1988. Ambiguity and rationality. *Journal of Behavioral Decision Making* 1: 149–157.

2 Fritz, S., and T. Mussweiler. 1997. Explaining the enigmatic anchoring effect: mechanisms of selective accessibility. *Journal of Personality and Social Psychology* 83: 437–446.

3 Snyder, N., D.E. Brown, and K.B. Clark. 2009. *The Travails of Two Woodpeckers*. University of New Mexico Press, Albuquerque, pp. 25–42.

4 As quoted in Hoose, P. 2014. *The Race to Save the Lord God Bird*. Farrar Straus Giroux, New York, p. 150.

5 http://listserv.arizona.edu/cgi-bin/wa?A2=ind0603c&L=birdwg01&T=0&O=D&P=4587 (accessed 21 March 2006). Kaufmann was using a ***fallacy of division***, i.e., he assumed that something that is true of one thing must also be true of all or some of its related parts. He also fell to ***neglect of probability***, as research showed that in fact the likelihood of detecting a single ivory-bill during the Arkansas search given the survey effort actually expended there was quite low, only about 12% (see: Scott, J.M., F.L. Ramsey, M. Lamertink, K.V. Rosenberg, R. Rohrbaugh, J.A. Wiens, and J.M. Reed. 2008. When is an "extinct" species really extinct? Gauging

the search efforts for Hawaiian forest birds and the ivory-billed woodpecker. *Avian Conservation and Ecology* 3: http://www.ace-eco.org/vol3/iss2/art3/)

6 http://www.sibleyguides.com/2007/10/ivory-billed-woodpecker-status-review/ (accessed 30 August 2015).

7 Because the bird(s) in Arkansas and Florida had *not* escaped detection, the statement by this writer is also untrue. Sykes, P.A. 2016. A personal perspective on searching for the ivory-billed woodpecker: a 41-year quest. Pp. 171–182 *in* The History of Patuxent – America's Wildlife Research Story (Perry, M.C., ed.), U.S. Geological Survey Circular 1422, p. 181.

8 Kuran, T., and C.R. Sunstein. 1999. Availability cascades and risk regulation. *Stanford Law Review* 51: 683–768.

9 Nyhan, B., and J. Reifler. 2010. When corrections fail: the persistence of political misperceptions. *Political Behavior* 32: 303–330.

10 Lebo, M., and D. Cassino. 2007. The aggregated consequences of motivated reasoning. *Political Psychology* 28: 719–746.

11 Nadeau, R., E. Cloutier, and J.H. Guay. 1993. New evidence about the existence of a bandwagon effect in the opinion formation process. *International Political Science Review* 14: 203–213.

12 This author was informed of at least two such instances taking place, in each case related to the Arkansas woodpecker.

13 Bar-Hillel, M. 1980. The base-rate fallacy in probability judgments. *Acta Psychologica* 44: 211–233.

14 A frequent reprise by Tom Nelson, exemplified here: http://tomnelson.blogspot.com/2005/09/ivory-bill-skeptic-home.html (accessed 30 August 2015).

15 Corner, A., A.J. Harris, and U. Hahn. 2010. Conservatism in belief revision and participant skepticism. *Proceedings of the 32nd Annual Conference of the Cognitive Science Society* 2010: 1,625–1,630.

16 Tetlock, P.E. 1983. Accountability and perseverance of first impressions. *Social Psychology Quarterly* 46: 285–292.

17 Pronin, E. 2007. Perception and misperception of bias in human

judgment. *Trends in Cognitive Sciences* 11: 37–43.

18 Taleb, N.N. 2007. *The Black Swan: The Impact of the Highly Improbable.* Random House, New York.

19 Wilke, A., and R. Mata. 2012. Cognitive bias. Pp. 531–535 in *The Encyclopedia of Human Behavior* (V.S. Ramachandran, Ed.). Academic Press, New York.

20 Webster, D.M., and A.W. Kruglanski. 1994. Individual differences in need for cognitive closure. *Journal of Personality and Social Psychology* 67: 1,049–1,062.

21 For a notable example, see: Gotelli, N.J., A. Chao, R.K. Colwell, W.H. Hwang, and G.R. Graves. 2012. Specimen-based modeling, stopping rules, and the extinction of the ivory-billed woodpecker. *Conservation Biology* 26: 47–56. This study made the startling assumption that a partial selection of specimen records constituted an unbiased proxy for ivory-bill population size after ~1900 (see Chapters 3, 13). ***Confirmation bias*** is specifically indicated here because the authors are transparent in given us their ostensible motivation for their belief, i.e., "suggest[ing] [that] conservation resources devoted to its rediscovery and recovery could be better allocated to other species."

22 Baron, J., J. Beattie, and J.C. Hershey. 1988. Heuristics and biases in diagnostic reasoning: II. congruence, information, and certainty. *Organizational Behavior and Human Decision Processes* 42: 88–110.

23 Bivings, A.E. 2006. Rediscovery and recovery of the ivory-billed woodpecker. *Journal of Wildlife Managemen*t 70: 1,495–1,496. "This hypothesis is consistent with the observed pattern of events that started with a flurry of sightings at a seemingly unlikely location (fragmented habitat with relatively unlimited public assess at the time) followed by a sharp decline in further encounters." *Ibid.*

24 Kruger, J., and D. Dunning. 1999. Unskilled and unaware of it: how difficulties in recognizing one's own incompetence led to inflated self-assessments. *Journal of Personality and Social Psychology* 77: 1,121–1,134.

25 See, for example, commentary here: (http://www.birdforum.net/

showthread.php?t=33968&page=1-&pp=25)(accessed 19 March 2015).

26 See here: http://proregulus.blogspot.com/2007/12/i-almost-certainly-ignored-this-article.html (accessed 20 April 2015).

27 Mahmoodi, A., et al. 2015. Equality bias impairs collective decision-making across cultures. *Proceedings of the National Academy of Sciences* 112: 3,835–3,840.

28 http://listserv.arizona.edu/cgi-bin/wa?A2=ind0603c&L=birdwg01&T=0&O=D&P=4587 (accessed 21 March 2006). Kaufmann was using a ***fallacy of division*** here, i.e., he assumed that something that is true of one thing must also be true of all or some of its related parts. He also falls to ***neglect of probability***, as research showed that in fact the likelihood of detecting a single ivory-bill during the Arkansas search given the survey effort expended was rather low, only about 12% (see: Scott, J.M., F.L. Ramsey, M. Lamertink, K.V. Rosenberg, R. Rohrbaugh, J.A. Wiens, and J.M. Reed. 2008. When is an "extinct" species really extinct? Gauging the search efforts for Hawaiian forest birds and the ivory-billed woodpecker. *Avian Conservation and Ecology* 3: http://www.ace-eco.org/vol3/iss2/art3/).

29 Mullen, B., J.L. Atkins, D.S Champion, C. Edwards, D. Hardy, J.E. Story, and M. Vanderklok. 1985. The false consensus effect: a meta-analysis of 115 hypothesis tests. *Journal of Experimental Social Psychology* 21: 262–283.

30 Kühberger, A. 1998. The influence of framing on risky decisions: a meta-analysis. *Organizational Behavior and Human Decision Processes* 75: 23–55.

31 Ross, L. 1977. The intuitive psychologist and his shortcomings: distortions in the attribution process. Pp. 173–220 in *Advances in Experimental Social Psychology, Vol. 10* (Berkowitz, L., ed.). Academic Press, New York.

32 See Collins, M.D. 2020. Application of image processing to evidence for the persistence of the ivory-billed woodpecker (*Campephilus principalis*). *Scientific Reports* 10: 14616. The Supplement to this

reference provides other disturbing examples of serious cognitive error by scientists abusing the peer review process, including deployment of ***ad hominem*** and using ***fallacy of division*** to prevent and/or delay content with which they disagreed from reaching a wider audience.

33 Christensen-Szalanski, J.J., and C.F. Willham. 1991. The hindsight bias: a meta-analysis. *Organizational Behavior and Human Decision Processes* 48: 147–168.

34 Gilovich, T., V.H. Medvec, and K. Savitsky. 1998. The illusion of transparency: biased assessments of others' ability to read one's emotional states. *Journal of Personality and Social Psychology* 75: 332–346.

35 Einhorn, H.J., and R.M. Hogarth. 1978. Confidence in judgment: persistence of the illusion of validity. *Psychological Review* 85: 395–416.

36 Hamilton, D.L., and R.K. Gifford. 1976. Illusory correlation in interpersonal perception: a cognitive basis of stereotypic judgments. *Journal of Experimental Social Psychology* 12: 392–407.

37 Hoorens, V., and A.P. Buunk. 1992. Self-serving biases in social-comparison – illusory superiority and unrealistic optimism. *Psychologica Belgica* 32: 169–194.

38 Hand, D.J. 2014. *The Improbability Principle: Why Coincidences, Miracles, and Rare Events Happen Every Day*. Scientific American – Farrar, Straus and Giroux, New York.

39 Jackson, J.A. 2006. Ivory-billed woodpecker (*Campephilus principalis*): hope and the interfaces of science, conservation, and politics. *The Auk* 123: 1–15.

40 Stanovich, K.E., and R.F. West. 2007. Natural myside bias is independent of cognitive ability. *Thinking & Reasoning* 13: 225–247.

41 Hill, G.E. 2007. *Ivorybill Hunters: The Search for Proof in a Flooded Wilderness*. Oxford University Press, London, pp. 94–95, 141, 147.

42 Kruger, J., and T. Gilovich. 1999. "Naïve cynicism" in everyday theories of responsibility assessment: on biased assumptions of bias. *Journal of Personality and Social Psychology* 76: 743–753.

43 Robinson, R.J., D. Keltner, A. Ward, and L. Ross. 1995. Actual versus assumed differences in construal: "naïve realism" in intergroup perception and conflict. *Journal of Personality and Social Psychology* 68: 404–417.

44 Rozin, P., and E.B. Royzman. 2001. Negativity bias, negativity dominance, and contagion. *Personality and Social Psychology Review* 5: 296–320.

45 Fiedler, K., B. Brinkmann, T. Betsch, and B. Wild, B. 2000. A sampling approach to biases in conditional probability judgments: beyond base rate neglect and statistical format. *Journal of Experimental Psychology: General* 129: 399–418.

46 Sunstein, C.R. 2002. Probability neglect: emotions, worst cases, and law. *Yale Law Journal* 112: 61-107.

47 Balph, D.F., and M.H. Balph. 1983. On the psychology of watching birds: the problem of observer-expectancy bias. *The Auk* 100: 755–757.

48 Ritov, I., and J. Baron. 1990. Reluctance to vaccinate: omission bias and ambiguity. *Journal of Behavioral Decision Making* 3: 263–277.

49 ***Omission bias*** was pervasively circulated following a putative 2004 Arkansas rediscovery of the woodpecker. For notable (if unintentional) tendencies to overweight errors of commission, but minimize or even ignore errors of omission, during conservation of the ivory-bill, see: McKelvey, K.S., K.B. Aubry, and M.K. Schwartz. 2008. Using anecdotal occurrence data for rare or elusive species: the illusion of reality and a call for evidentiary standards. *BioScience* 58: 549–555; Roberts, D.L., C.S. Elphick, and J.M. Reed. 2010. Identifying anomalous reports of putatively extinct species and why it matters. *Conservation Biology* 24: 189–196; Rout, T.M., D. Heinze, and M.A. McCarthy. 2010. Optimal allocation of conservation resources to species that may be extinct. *Conservation Biology* 24: 1,111–1,118.

50 Dwan, K., et al. 2008. Systematic review of the empirical evidence of study publication bias and outcome reporting bias. *PLoS ONE* 3(8): e3081.

51 Moore, D.A., and P.J. Healy. 2008. The trouble with overconfidence. *Psychological Review* 115: 502–517.

52 Part of a sustained attack by David Sibley on the hypothesis that the ivory-bill was stll extant. See: http://www.sibleyguides.com/2007/10/ivory-billed-woodpecker-status-review/ (accessed 19 April 2015).

53 Hunt, L. 2002. Against presentism. *Perspectives on history: the news magazine of the American Historical Association*, http://www.historians.org/publications-and-directories/perspectives-on-history/may-2002/against-presentism (accessed 8 April 2015).

54 Hand.

55 Smith, D.W., J.J. Zhang, and B. Colwell, B. 1996. Pro-innovation bias: the case of the Giant Texas SmokeScream. *Journal of School Health* 66: 210–213.

56 Hill, G.E. 2007. *Ivorybill Hunters: The Search for Proof in a Flooded Wilderness*. Oxford University Press, London, p. 245.

57 To increase rates of data acquisition, drones have been tested in attempts to reduce the expected waiting times for documenting this elusive species. Collins. 2018. Using a drone to search for the ivory-billed woodpecker (*Campephilus principalis*). *Drones* 2: 11.

58 Brehm, S.S., and J.W. Brehm. 2013. *Psychological Reactance: A Theory of Freedom and Control*. Academic Press, New York.

59 Collar, N.J. 1998. Extinction by assumption; or, the Romeo error on Cebu. *Oryx*: 239–244.

60 http://www.sibleyguides.com/2007/10/ivory-billed-woodpecker-status-review/ (accessed 1 September 2015).

61 Shanteau, J. 1974. Component processes in risky decision-making. *Journal of Experimental Psychology* 103: 680–691; Tversky, A., and D.J. Koehler. 1994. Support theory: a nonextentional representation of subjective probability. *Psychological Review* 101: 547–567.

62 Arkes, H.R., and P. Ayton. 1999. The sunk cost and Concorde effects: are humans less rational than lower animals? *Psychological Bulletin* 125: 591–600.

63 Jost, J., and O. Hunyady. 2002. The psychology of system

justification and the palliative function of ideology. *European Review of Social Psychology* 13: 111–153.

64 Hayes, F.E., and W.K. Hayes. 2007. The great ivory-billed woodpecker debate: perceptions of the evidence. *Birding* 39: 36–41.

65 Summala, H. 1988. Risk control is not risk adjustment: the zero-risk theory of driver behaviour and its implications. *Ergonomics* 31: 491–506.

TRANSFORMATION

Transform your life, transform your world - Changemakers Books publishes for individuals committed to transforming their lives and transforming the world. Our readers seek to become positive, powerful agents of change. Changemakers Books inform, inspire, and provide practical wisdom and skills to empower us to write the next chapter of humanity's future. If you have enjoyed this book, why not tell other readers by posting a review on your preferred book site.

Recent bestsellers from Changemakers Books are:

Integration
The Power of Being Co-Active in Work and Life
Ann Betz, Karen Kimsey-House
Integration examines how we came to be polarized in our dealing with self and other, and what we can do to move from an either/or state to a more effective and fulfilling way of being.
Paperback: 978-1-78279-865-1 ebook: 978-1-78279-866-8

Bleating Hearts
The Hidden World of Animal Suffering
Mark Hawthorne
An investigation of how animals are exploited for entertainment, apparel, research, military weapons, sport, art, religion, food, and more.
Paperback: 978-1-78099-851-0 ebook: 978-1-78099-850-3

Lead Yourself First!

Indispensable Lessons in Business and in Life

Michelle Ray

Are you ready to become the leader of your own life? Apply simple, powerful strategies to take charge of yourself, your career, your destiny.

Paperback: 978-1-78279-703-6 ebook: 978-1-78279-702-9

Burnout to Brilliance

Strategies for Sustainable Success

Jayne Morris

Routinely running on reserves? This book helps you transform your life from burnout to brilliance with strategies for sustainable success.

Paperback: 978-1-78279-439-4 ebook: 978-1-78279-438-7

Goddess Calling

Inspirational Messages & Meditations of Sacred Feminine Liberation Thealogy

Rev. Dr. Karen Tate

A book of messages and meditations using Goddess archetypes and mythologies, aimed at educating and inspiring those with the desire to incorporate a feminine face of God into their spirituality.

Paperback: 978-1-78279-442-4 ebook: 978-1-78279-441-7

The Master Communicator's Handbook

Teresa Erickson, Tim Ward

Discover how to have the most communicative impact in this guide by professional communicators with over 30 years of experience advising leaders of global organizations.

Paperback: 978-1-78535-153-2 ebook: 978-1-78535-154-9

Meditation in the Wild

Buddhism's Origin in the Heart of Nature

Charles S. Fisher Ph.D.

A history of Raw Nature as the Buddha's first teacher, inspiring some followers to retreat there in search of truth.

Paperback: 978-1-78099-692-9 ebook: 978-1-78099-691-2

Ripening Time

Inside Stories for Aging with Grace

Sherry Ruth Anderson

Ripening Time gives us an indispensable guidebook for growing into the deep places of wisdom as we age.

Paperback: 978-1-78099-963-0 ebook: 978-1-78099-962-3

Striking at the Roots

A Practical Guide to Animal Activism

Mark Hawthorne

A manual for successful animal activism from an author with first-hand experience speaking out on behalf of animals.

Paperback: 978-1-84694-091-0 ebook: 978-1-84694-653-0

Readers of ebooks can buy or view any of these bestsellers by clicking on the live link in the title. Most titles are published in paperback and as an ebook. Paperbacks are available in traditional bookshops. Both print and ebook formats are available online.

Find more titles and sign up to our readers' newsletter at
http://www.johnhuntpublishing.com/transformation
Follow us on Facebook at
https://www.facebook.com/Changemakersbooks